Commercial Design
Using Autodesk® Revit® 2021

Daniel John Stine

SDC
PUBLICATIONS

SDC Publications
P.O. Box 1334
Mission, KS 66222
913-262-2664
www.SDCpublications.com
Publisher: Stephen Schroff

ISBN-13: 978-1-63057-351-5
ISBN-10: 1-63057-351-5

Printed and bound in the United States of America.

FOREWORD

To the Student:

This book has been written with the assumption that the reader has no prior experience using Autodesk® Revit®. The intent of this book is to provide the student with a well-rounded knowledge of architectural tools and techniques for use in both school and industry.

The book consists of a series of tutorials which primarily focus on the development of a single project. When you finish the book you will have learned how to document and model all of the major architectural aspects of a commercial project. This includes floor plans, interior and exterior elevations, wall and building sections, door and room finish schedules and organizing drawings on sheets for printing.

There are several videos that supplement the book. Studying these videos along with the book will help the reader to better understand Revit's architectural features. The book and videos will also help prepare for the Autodesk Revit Architecture Certification Exam; see Appendix A for more information.

Be sure to check out the bonus material available online.

> **Errata:**
> Please check the publisher's website from time to time for any errors or typos found in this book after it went to the printer. Simply browse to www.SDCpublications.com, and then navigate to the page for this book. Click the **View/Submit errata** link in the upper right corner of the page. If you find an error, please submit it so we can correct it in the next edition.
>
> You may contact the publisher with comments or suggestions at **service@SDCpublications.com**.
> *Please do not email with Revit questions unless they relate to a problem with this book.*

To the Instructor:

This book was designed for the architectural student using Revit 2021. *Note:* Students can successfully work through this book using Revit Architecture 2021 as well. Throughout the book, the student develops a two story commercial building. The drawings start with the floor plans and develop all the way to photo-realistic renderings similar to the one on the cover of this book.

Throughout the book many Revit tools and techniques are covered while creating the model. Also, in a way that is applicable to the current exercise, industry standards and conventions are covered. Access to the internet is required for some exercises.

Each chapter concludes with a self-exam and review questions. The answers to the self-exam questions are provided, but review question answers are not (they can only be found in the Instructor's Guide available from the publisher).

This text is updated every year for the latest version of Revit. The printed text has always been available for the fall term.

An Instructor's Resource Guide is available for this book. It contains:
- Answers to the questions at the end of each chapter
- Example images of each exercise to be turned in (can be used to grade students' work)
- Outline of tools and topics to be covered in each lesson's lecture
- Suggestions for additional student work (for each lesson)
- Author's direct contact information

About the Author:

Daniel John Stine AIA, CSI, CDT, is a registered architect with over twenty years of experience in the field of architecture. He is the BIM Administrator at LHB, a 250-person full service design firm. In addition to providing training and support for four offices, Dan implemented BIM-based lighting analysis using ElumTools, early energy modeling using Autodesk Insight, virtual reality (VR) using the HTC Vive/Oculus Rift along with Fuzor & Enscape, Augmented Reality (AR) using the Microsoft Hololens, and the Electrical Productivity Pack for Revit (sold by CTC Express Tools). Dell, the world-renowned computer company, created a video highlighting his implementation of VR at LHB.

Dan has presented internationally on BIM in the USA, Canada, Ireland, Denmark, Australia and Singapore. He was ranked multiple times as a top-ten speaker by attendees at Autodesk University, RTC/BILT, Midwest University, AUGI CAD Camp, NVIDIA GPU Technology Conference, Lightfair, and AIA-MN Convention. By invitation, he spent a week at Autodesk's largest R&D facility in Shanghai, China, to beta test and brainstorm new Revit features in 2016.

Committed to furthering the design profession, Dan teaches graduate architecture students at North Dakota State University (NDSU) and has lectured for interior design programs at NDSU, Northern Iowa State, and University of Minnesota, as well as Dunwoody's new School of Architecture in Minneapolis. As an adjunct instructor, Dan previously taught AutoCAD and Revit for twelve years at Lake Superior College. Dan is a member of the American Institute of Architects (AIA), Construction Specifications Institute (CSI), and Autodesk Developer Network (ADN), and is a Construction Document Technician (issued by CSI). He has presented live webinars for ElumTools, ArchVision, Revizto and NVIDIA. Dan writes about design on his blog, BIM Chapters, and in his textbooks published by SDC Publications:

- *Autodesk Revit 2021 Architectural Command Reference*
- *Residential Design Using Autodesk Revit 2021*
- *Commercial Design Using Autodesk Revit 2021*
- *Design Integration Using Autodesk Revit 2021 (Architecture, Structure and MEP)*
- *Interior Design Using Autodesk Revit 2021 (with co-author Aaron Hansen)*
- *Residential Design Using AutoCAD 2021*
- *Commercial Design Using AutoCAD 2013*
- *Chapters in Architectural Drawing (with co-author Steven H. McNeill, AIA, LEED AP)*
- *Interior Design Using Hand Sketching, SketchUp and Photoshop (also with Steven H. McNeill)*
- *Google SketchUp 8 for Interior Designers; Just the Basics*

You may contact the publisher with comments or suggestions at **service@SDCpublications.com**

Social Media:

Students can use social media, such as Twitter and LinkedIn, to start developing professional contacts and knowledge. Follow the author on social media for new articles, tips and errata updates. Consider following the design firms and associations (AIA, CSI, etc.) in your area; this could give you an edge in an interview!

Author's Blog: http://bimchapters.blogspot.com/

 Twitter
@DanStine_MN

 LinkedIn
https://www.linkedin.com/in/danstinemn

Certification Study Guides

The author of this book has also prepared the following certification study guides students can use to achieve successful results when taking certification exams which can help make the difference when trying to secure that dream job!

- *Microsoft Office Specialist Excel Associate 365/2019 Exam Preparation*
- *Microsoft Office Specialist Word Associate 365/2019 Exam Preparation*
- *Autodesk Revit for Architecture Certified User Exam Preparation (Revit 2021 Edition)*

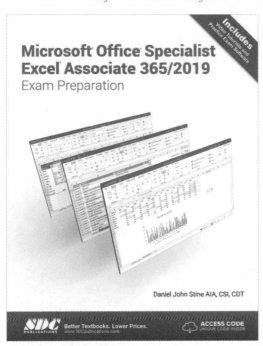

Thanks:

I could not have written this book without support from my family: Cheri, Kayla & Carter. Many thanks go out to Stephen Schroff and *SDC Publications* for making this book possible! Thanks to Tim Waldock, http://revitcat.blogspot.com/, for reviewing my vertical circulation chapter and providing valuable feedback.

Notes:

Table of Contents

Exclusive Online Content

See the inside-front cover of this book for download instructions and your unique access code.

Bonus Chapters

Bonus Videos

This book comes with several short videos, approximately 3-5 minutes long, which can be watched in order while working through this book. The full index of video titles can be found on the next page.

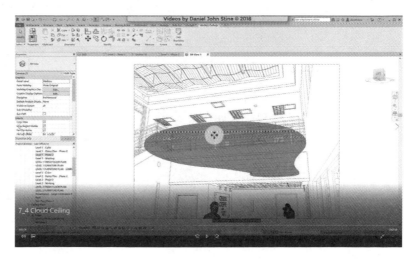

These videos are supplementary and not required to complete the exercises in this book. Rather, the videos are intended to cover many of the fundamental Revit topics in a different context to help the reader better understand this important material. A few videos cover intermediate topics, such as the sloped "cloud" celling shown in the image to the right.

Bonus Videos Index:

Lesson 1
Getting Started with Autodesk Revit 2021:

This chapter will introduce you to Autodesk® Revit® 2021. You will study the User Interface and learn how to open and exit a project and adjust the view of the drawing on the screen. It is recommended that you spend an ample amount of time learning this material, as it will greatly enhance your ability to progress smoothly through subsequent chapters.

Exercise 1-1:
What is Revit 2021?

What is Autodesk Revit used for?

Autodesk Revit (Architecture, Structure and MEP) is the world's first fully parametric building design software. This revolutionary software, for the first time, truly takes architectural computer aided design beyond simply being a high-tech pencil. Revit is a product of Autodesk, makers of AutoCAD, Civil 3D, Inventor, 3DS Max, Maya and many other popular design programs.

Revit can be thought of as the foundation of a larger process called **Building Information Modeling** (BIM). The BIM process revolves around a virtual, information rich 3D model. In this model all the major building elements are represented and contain information such as manufacturer, model, cost, phase and much more. Once a model has been developed in Revit, third-party add-ins and applications can be used to further leverage the data. Some examples are Facilities Management, Analysis (Energy, Structural, Lighting), Construction Sequencing, Cost Estimating, Code Compliance and much more!

Revit can be an invaluable tool to designers when leveraged to its full potential. The iterative design process can be accomplished using special Revit features such as *Phasing* and *Design Options*. Material selections can be developed and attached to various elements in the model, where one simple change adjusts the wood from oak to maple throughout the project. The power of schedules may be used to determine quantities and document various parameters contained within content (this is the "I" in BIM, which stands for Information). Finally, the three-dimensional nature of a Revit-based model allows the designer to present compelling still images and animations. These graphics help to more clearly communicate the design intent to clients and other interested parties. This book will cover many of these tools and techniques to **assist** in the creative process.

What is a parametric building modeler?

Revit is a program designed from the ground up using state-of-the-art technology. The term parametric describes a process by which an element is modified and an adjacent element(s) is automatically modified to maintain a previously established relationship. For example, if a wall is moved, perpendicular walls will grow, or shrink, in length to remain attached to the related wall. Additionally, elements attached to the wall will move, such as wall cabinets, doors, windows, air grilles, etc.

Revit stands for **Rev**ise **I**nstantly; a change made in one view is automatically updated in all other views and schedules. For example, if you move a door in an interior elevation view, the floor plan will automatically update. Or, if you delete a door, it will be deleted from all other views and schedules.

A major goal of Revit is to eliminate much of the repetitive and mundane tasks traditionally associated with CAD programs allowing more time for design, coordination and visualization. For example, all sheet numbers, elevation tags and reference bubbles are updated automatically when changed anywhere in the project. Therefore, it is difficult to find a mis-referenced detail tag.

The best way to understand how a parametric model works is to describe the Revit project file. A single Revit file contains your entire building project. Even though you mostly draw in 2D views, you are actually drawing in 3D. In fact, the entire building project is a 3D model. From this 3D model you can generate 2D elevations, 2D sections and perspective views. Therefore, when you delete a door in an elevation view you are actually deleting the door from the 3D model from which all 2D views are generated and automatically updated.

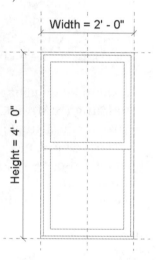

Window model size controlled by parameters *Width* and *Height*

Another way in which Revit is a parametric building modeler is that **parameters** can be used to control the size and shape of geometry. For example, a window model can have two *parameters* set up which control the size of the window. Thus, from a window's properties it is possible to control the size of the window without using any of the drawing modify tools such as *Scale* or *Move*. Furthermore, the *parameter* settings (i.e., width and height in this example) can be saved within the window model (called a *Family*). You could have the 2′ x 4′ settings saved as "Type A" and the 2′ x 6′ as "Type B." Each saved list of values is called a *Type* within the *Family*. Thus, this one double-hung window *Family* could represent an unlimited number of window sizes! You will learn more about this later in the book.

What Disciplines does Revit Support?

Revit supports **Architecture, Interior Design, Structure** and **MEP** (which stands for Mechanical, Electrical and Plumbing). There used to be three discipline specific versions of Revit and an all-in-one version—now there is just the all-in-one version. You can download the free 30-day trial from autodesk.com. Students may download a free 3-year version of Revit at www.autodesk.com/education.

Revit is not meant to support professional Civil design.

Now is as good a time as any to make sure the reader understands that Revit is not, nor has it ever been, backwards compatible. This means there is no *Save-As* back to a previous version of Revit. Also, an older version of Revit cannot open a file, project or content saved in a newer format. So, make sure you consider what version your school or employer is currently standardized on before upgrading any projects or content. Revit will automatically upgrade an older

Revit model of an existing building, with architecture, structural, mechanical, plumbing and electrical all modeled.

Image courteous of LHB, Inc.   www.LHBcorp.com

format when opened in a newer version of the software. This is a onetime process which can take several minutes.

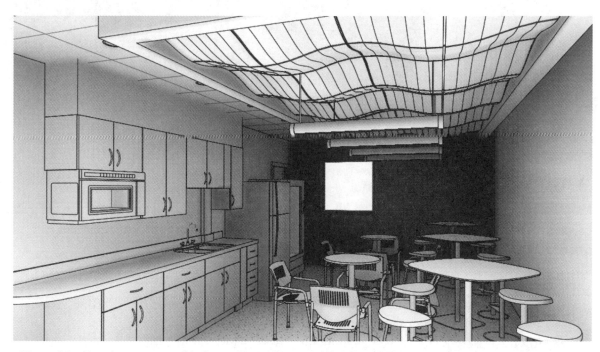

3D model of lunchroom created in *Interior Design Using Autodesk Revit 2021*

Download Autodesk Content

Be sure to download the additional Autodesk content pack found via the Get Autodesk Content command on the Insert tab. Starting with Revit 2021, some content is no longer installed with the software. Some of the additional content is required to successfully complete the exercises in the book. **Important Step!**

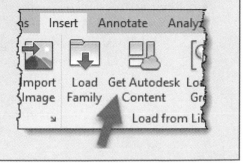

Lobby rendering from *Interior Design Using Autodesk Revit 2021*

Why use Revit?

Many people ask the question, why use Revit versus other programs? The answer can certainly vary depending on the situation and particular needs of an individual or organization.

Generally speaking, this is why most companies use Revit:

- Many designers and drafters are using Revit to streamline repetitive drafting tasks and focus more on designing and detailing a project.
- Revit is a very progressive program and offers many features for designing buildings. Revit is constantly being developed and Autodesk provides incremental upgrades and patches on a regular basis; Revit 2021 was released about a year after the previous version.
- Revit was designed specifically for architectural design and includes features like:
  - *Autodesk Renderer* Photo Realistic Rendering
  - Phasing; model changes over time
  - Design Options; model changes during the same time period
  - Live schedules
  - Cloud Rendering via *Autodesk A360*
  - Conceptual Energy Analysis via *Autodesk A360*
  - Daylighting Analysis via *Autodesk A360*

A few basic Revit concepts

The following is meant to be a brief overview of the basic organization of Revit as a software application. You should not get too hung up on these concepts and terms as they will make more sense as you work through the tutorials in this book. This discussion is simply laying the groundwork so you have a general frame of reference on how Revit works.

The Revit platform has three fundamental types of elements:
- Model Elements
- Datum Elements
- View-Specific Elements

Model Elements

Think of *Model Elements* as things you can put your hands on once the building has been constructed. They are typically 3D but can sometimes be 2D. There are two types of *Model Elements*:

- **Host Elements**: walls, floors, slabs, roofs, ceilings.

- **Model Components**: Stairs, Doors, Furniture, Casework, Beams, Columns, Pipes, Ducts, Light Fixtures, Model Lines, etc.

  o Some *Model Components* require a host before they can be placed within a project. For example, a window can only be placed in a host, which could be a wall, roof or floor depending on how the element was created. If the host is deleted, all hosted or dependent elements are automatically deleted.

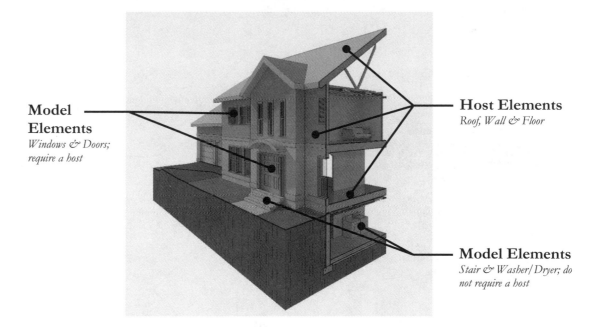

Model Elements
Windows & Doors; require a host

Host Elements
Roof, Wall & Floor

Model Elements
Stair & Washer/Dryer; do not require a host

Datum Elements

Datum Elements are reference planes within the building that graphically and parametrically define the location of various elements within the model. These are the three types of *Datum Elements,* all of which define 3D planes:

- **Grids**
  - Typically laid out in a plan view to locate structural elements such as columns and beams, as well as walls. Grids show up in plan, elevation and section views. Moving a grid in one view moves it in all other views as it is the same element. (See the next page for an example of a grid in plan view.)

- **Levels**
  - Used to define vertical relationships, mainly surfaces that you walk on. They only show up in elevation, section and 3D views. Most elements are placed relative to a *Level*; when the *Level* is moved those elements move with it (e.g., doors, windows, casework, ceilings). ***WARNING:*** *If a Level is deleted, those same "dependent" elements will also be deleted from the project!*

- **Reference Planes**
  - These are similar to grids in that they show up in plan and elevation or sections. They do not have reference bubbles at the end like grids. Revit breaks many tasks down into simple 2D tasks which result in 3D geometry. *Reference Planes* are used to define 2D surfaces on which to work within the 3D model. They can be placed in any view, horizontally, vertically or at an angle.

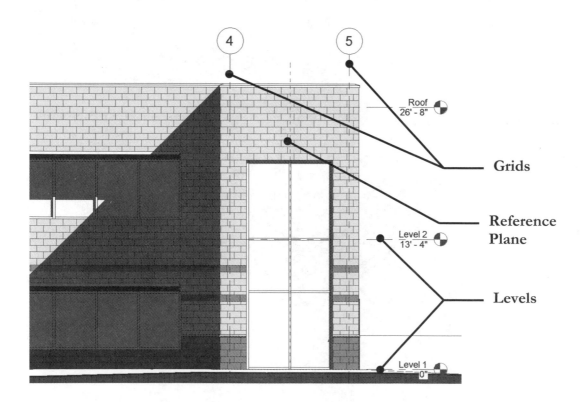

View-Specific Elements

As the name implies, the items about to be discussed only show up in the specific view in which they are created. For example, notes and dimensions added in the architectural floor plans will not show up in the structural floor plans. These elements are all 2D and are mainly communication tools used to accurately document the building for construction or presentations.

- **Annotation elements** (text, tags, symbols, dimensions)
    - o Size automatically set and changed based on selected view scale

- **Details** (detail lines, filled regions, 2D detail components)

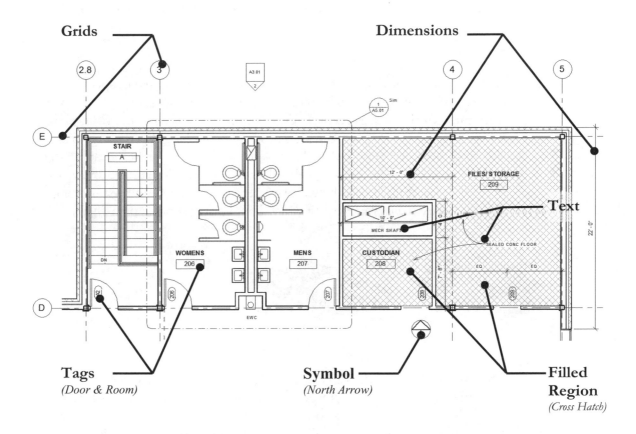

Grids

Dimensions

Text

Tags
(Door & Room)

Symbol
(North Arrow)

Filled Region
(Cross Hatch)

File Types (and extensions):

Revit has four primary types of files that you will work with as a Revit user. Each file type, as with any Microsoft Windows based program, has a specific three letter file name extension; that is, after the name of the file on your hard drive you will see a period and three letters:

.RVT	Revit project files; the file most used (also for backup files)
.RFA	Revit family file; loadable content for your project
.RTE	Revit template; a project starter file with office standards preset
.RFT	Revit family template; a family starter file with parameters

The Revit platform has three fundamental ways in which to work with the elements (for display and manipulation):

- Views
- Schedules
- Sheets

The following is a cursory overview of the main ideas you need to know. This is not an exhaustive study on views, schedules and sheets.

Views

Views, accessible from the *Project Browser*, is where most of the work is done while using Revit. Think of views as slices through the building, both horizontal (plans) and vertical (elevations and sections).

- **Plans**
  - A *Plan View* is a horizontal slice through the building. You can specify the location of the **cut plane** which determines if certain windows show up or how much of the stair is seen. A few examples are architectural floor plan, reflected ceiling plan, site plan, structural framing plan, HVAC floor plan, electrical floor plan, lighting [ceiling] plan, etc. The images below show this concept; the image on the left is the 3D BIM. The middle image shows the portion of building above the cut plane removed. Finally, the last image on the right shows the plan view in which you work and place on a sheet.

- **Elevations**
  - An "exterior" elevation is a vertical slice, but where the slice lies outside the floor plan as in the middle image below. Each elevation created is listed in the *Project Browser*. The image on the right is an example of a South exterior elevation view, which is a "live" view of the 3D model. If you select a window here and delete it, the floor plans will update instantly.

  - Similarly, an "interior" elevation is a vertical slice, but inside the building. This view is cropped at the perimeter of a given room.

- **Sections**
    - o Similar to elevations, sections are also vertical slices. However, these slices cut through the building. A section view can be cropped down to become a wall section or even look just like an elevation. The images below show the slice, the portion of building in the foreground removed, and then the actual view created by the slice. A setting exists, for each section view, to control how far into that view you can see. The example on the right is "seeing" deep enough to show the doors on the interior walls.

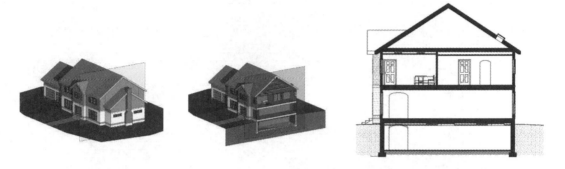

- **3D and Camera**
    - o In addition to the traditional "flattened" 2D views that you will typically work in, you are able to see your designs more naturally via 3D and Camera views. A 3D view is simply an orthogonal view of the project viewed from the exterior. A Camera view is a perspective view; cameras can be created both in and outside of the building. Like the 2D views, these 3D/Camera views can be placed on a sheet to be printed. Revit provides a number of tools to help explore the 3D view, such as Section Box, Steering Wheel, Temporary Hide and Isolate, and Render.

        The image on the left is a 3D view set to "shade mode" and has shadows turned on. The image on the right is a camera view set up inside the building; the view is set to "hidden line" rather than shaded, and the camera is at eyelevel.

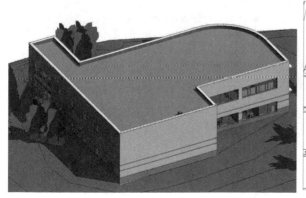

Schedules

Schedules are lists of information generated based on content that has been placed, or modeled, within the project. A schedule can be created, such as the door schedule example shown below, that lists any of the data associated with each door that exists in the project. Revit allows you to work directly in the schedule views. Any change within a schedule view is a change directly to the element being scheduled. Again, if a door were to be deleted from this schedule, that door would be instantly deleted from the project.

DOOR AND FRAME SCHEDULE												
DOOR NUMBER	DOOR				FRAME		DETAIL				FIRE RATING	HDWR GROUP	
	WIDTH	HEIGHT	MATL	TYPE	MATL	TYPE	HEAD	JAMB	SILL	GLAZING			
1000A	3' - 8"	7' - 2"	WD		HM		11/A8.01	11/A8.01					
1046	3' - 0"	7' - 2"	WD	D10	HM	F10	11/A8.01	11/A8.01 SIM				34	
1047A	6' - 0"	7' - 10"	ALUM	D15	ALUM	SF4	6/A8.01	6/A8.01	1/A8.01 SIM	1" INSUL		2	CARD READER N. LEAF
1047B	8' - 0"	7' - 2"	WD	D10	HM	F13	12/A8.01	11/A8.01 SIM			60 MIN	85	MAG HOLD OPENS
1050	3' - 0"	7' - 2"	WD	D10	HM	F21	8/A8.01	11/A8.01		1/4" TEMP		33	
1051	3' - 0"	7' - 2"	WD	D10	HM	F21	8/A8.01	11/A8.01		1/4" TEMP		33	
1052	3' - 0"	7' - 2"	WD	D10	HM	F21	8/A8.01	11/A8.01		1/4" TEMP		33	
1053	3' - 0"	7' - 2"	WD	D10	HM	F21	8/A8.01	11/A8.01		1/4" TEMP		33	
1054A	3' - 0"	7' - 2"	WD	D10	HM	F10	8/A8.01	11/A8.01		1/4" TEMP	-	34	
1054B	3' - 0"	7' - 2"	WD	D10	HM	F21	8/A8.01	11/A8.01		1/4" TEMP		33	
1055	3' - 0"	7' - 2"	WD	D10	HM	F21	8/A8.01	11/A8.01		1/4" TEMP	-	33	
1056A	3' - 0"	7' - 2"	WD	D10	HM	F10	9/A8.01	9/A8.01			20 MIN	33	
1056B	3' - 0"	7' - 2"	WD	D10	HM	F10	11/A8.01	11/A8.01			20 MIN	34	
1056C	3' - 0"	7' - 2"	WD	D10	HM	F10	20/A8.01	20/A8.01			20 MIN	33	
1057A	3' - 0"	7' - 2"	WD	D10	HM	F10	8/A8.01	11/A8.01			20 MIN	34	
1057B	3' - 0"	7' - 2"	WD	D10	HM	F30	9/A8.01	9/A8.01		1/4" TEMP	20 MIN	33	
1058A	3' - 0"	7' - 2"	WD	D10	HM	F10	9/A8.01	9/A8.01			-	33	

Sheets

You can think of sheets as the pieces of paper on which your views and schedules will be printed. Views and schedules are placed on sheets and then arranged. Once a view has been placed on a sheet, its reference bubble is automatically filled out and that view cannot be placed on any other sheet. The setting for each view, called "view scale," controls the size of the drawing on each sheet; view scale also controls the size of the text and dimensions.

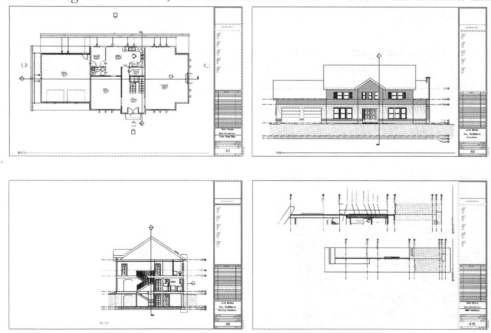

Exercise 1-2:
Overview of the Revit User Interface

Revit is a powerful and sophisticated program. Because of its powerful feature set, it has a measurable learning curve, though its intuitive design makes it easier to learn than other CAD or BIM based programs. However, like anything, when broken down into smaller pieces, we can easily learn to harness the power of Revit. That is the goal of this book.

This section will walk through the different aspects of the User Interface (UI). As with any program, understanding the user interface is the key to using the program's features.

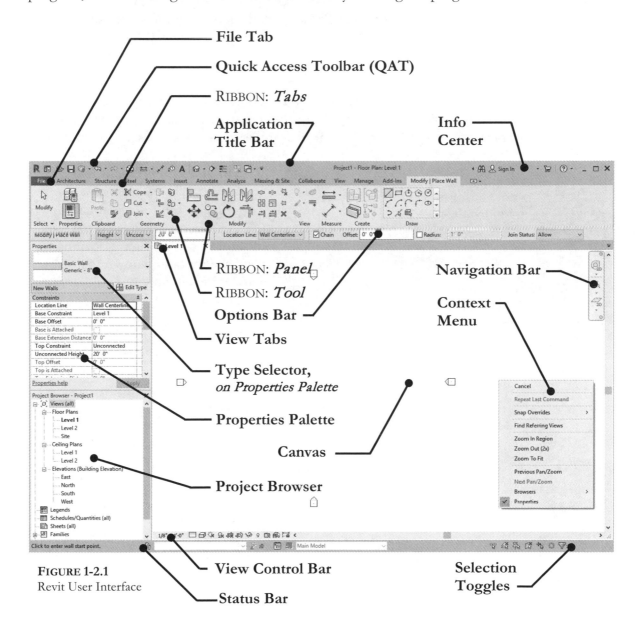

File Tab

Quick Access Toolbar (QAT)

RIBBON: *Tabs*

Application Title Bar

Info Center

RIBBON: *Panel*

RIBBON: *Tool*

Options Bar

View Tabs

Type Selector, *on Properties Palette*

Properties Palette

Navigation Bar

Context Menu

Canvas

Project Browser

View Control Bar

Selection Toggles

Status Bar

FIGURE 1-2.1
Revit User Interface

The Revit User Interface

TIP: See the online videos on the User Interface.

Application Title Bar

In addition to the *Quick Access Toolbar* and *Info Center*, which are all covered in the next few sections, you are also presented with the product name, version and the current **file-view** in the center. As previously noted already, the version is important as you do not want to upgrade unless you have coordinated with other staff and/or consultants; everyone must be using the same version of Revit. **Note:** the "R" icon on the left is not the same as in previous versions—the *Application Menu* has been replaced with the File Tab covered next.

File Tab

Access to *File* tools such as *Save*, *Plot*, *Export* and *Print* (both hardcopy and electronic printing). You also have access to tools which control the Revit application as a whole, not just the current project, such as *Options* (see the end of this section for more on *Options*).

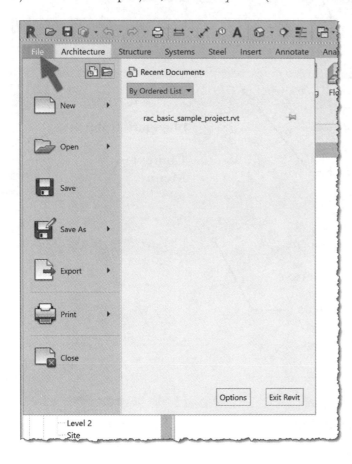

Recent and Open Documents:

 These two icons (from the *File tab*) toggle the entire area on the right to show either the recent documents you have been in (icon on the left) or a list of the documents you currently have open.

In the *Recent Documents* list you click a listed document to open it. This saves time as you do not have to click *Open* → *Project* and browse for the document (*Document* and *Project* mean the same thing here). Finally, clicking the "Pin" keeps that project from getting bumped off the list as additional projects are opened.

In the *Open Documents* list the "active" project you are working in is listed first; clicking another project switches you to that open project.

The list on the left, in the *File Tab* shown above, represents three different types of buttons: *button*, *drop-down button* and *split button*. Save and Close are simply **buttons**. Save-As and Export are **drop-down buttons**, which means to reveal a group of related tools. If you click or hover your cursor over one of these buttons, you will get a list of tools on the right. Finally, **split buttons** have two actions depending on what part of the button you click on; hovering over the button reveals the two parts (see bottom image to the right). The main area is the most used tool; the arrow reveals additional related options.

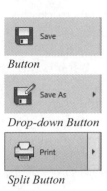

Quick Access Toolbar

Referred to as *QAT* in this book, this single toolbar provides access to often used tools (Home, *Open*, *Save*, *Undo*, *Redo*, *Measure*, *Tag*, etc.). It is always visible regardless of what part of the *Ribbon* is active.

The *QAT* can be positioned above or below the *Ribbon* and any command from the *Ribbon* can be placed on it; simply right-click on any tool on the *Ribbon* and select *Add to Quick Access Toolbar*. Moving the *QAT* below the *Ribbon* gives you a lot more room for your favorite commands to be added from the *Ribbon*. Clicking the larger down-arrow to the far right reveals a list of common tools which can be toggled on and off.

The first icon toggles back to the Home screen seen when Revit first opens. Some of the icons on the *QAT* have a down-arrow on the right. Clicking this arrow reveals a list of related tools. In the case of *Undo* and *Redo*, you can undo (or redo) several actions at once.

Ribbon – Architecture Tab

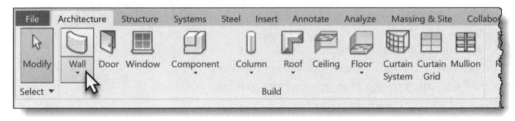

The *Architecture* tab on the *Ribbon* contains most of the tools the architect needs to model a building, essentially the things you can put your hands on when the building is done. The specific discipline versions of Revit omit some of the other discipline tabs.

Each tab starts with the *Modify* tool, i.e., the first button on the left. This tool puts you into "selection mode" so you can select elements to modify. Clicking this tool cancels the current tool and unselects elements. With the *Modify* tool selected you may select elements to view their properties or edit them. Note that the *Modify* tool, which is a button, is different than the *Modify* tab on the *Ribbon*.

The *Ribbon* has three types of buttons: *button, drop-down button* and *split*, as covered on the previous page. In the image above you can see the *Wall* tool is a **split button**. Most of the time you would simply click the top part of the button to draw a wall. Clicking the down-arrow part of the button, for the *Wall* tool example, gives you the option to draw a *Wall, Structural Wall, Wall by Face, Wall Sweep,* and a *Reveal*.

TIP: *The Model Text tool is only for placing 3D text in your model, not for adding notes!*

Ribbon – Annotate Tab

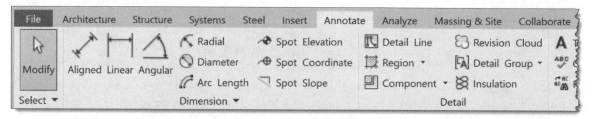

To view this tab, simply click the label "Annotate" near the top of the *Ribbon*. This tab presents a series of tools which allow you to add notes, dimensions and 2D "embellishments" to your model in a specific view, such as a floor plan, elevation, or section. All of these tools are **view specific**, meaning a note added in the first floor plan will not show up anywhere else, not even another first floor plan: for instance, a first floor electrical plan.

Notice, in the image above, that the *Dimension* panel label has a down-arrow next to it. Clicking the down-arrow will reveal an **extended panel** with additional related tools.

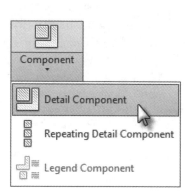

Finally, notice the *Component* tool in the image above; it is a **split button** rather than a *drop-down button*. Clicking the top part of this button will initiate the *Detail Component* tool. Clicking the bottom part of the button opens the fly-out menu revealing related tools. Once you select an option from the fly-out, that tool becomes the default for the top part of the split button for the current session of Revit (see image to right).

Ribbon – Modify Tab

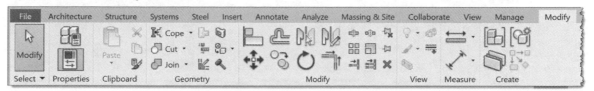

Several tools which manipulate and derive information from the current model are available on the *Modify* tab. Additional *Modify* tools are automatically appended to this tab when elements are selected in the model (see *Modify Contextual Tab* on the next page).

> ***TIP:*** *Do not confuse the Modify <u>tab</u> with the Modify <u>tool</u> when following instructions in this book.*

Ribbon – View Tab

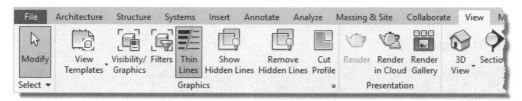

The tools on the *View* tab allow you to create new views of your 3D model; this includes views that look 2D (e.g., floor plans, elevations and sections) as well as 3D views (e.g., isometric and perspective views).

The *View* tab also gives you tools to control how views look, everything from what types of elements are seen (e.g., Plumbing Fixture, Furniture or Section Marks) as well as line weights.

> ***NOTE:*** *Line weights are controlled at a project wide level but may be overridden.*

Finally, notice the little arrow in the lower-right corner of the *Graphics* panel. When you see an arrow like this you can click on it to open a dialog box with settings that relate to the panel's tool set (*Graphics* in this example). Hovering over the arrow reveals a tooltip which will tell you what dialog box will be opened.

Ribbon – Modify Contextual Tab

The *Modify* tab is appended when certain tools are active or elements are selected in the model; this is referred to as a *contextual tab*. The first image below shows the *Place Wall* tab which presents various options while adding walls. The next example shows the *Modify Walls* contextual tab which is accessible when one or more walls are selected.

Place Wall contextual tab – visible when the Wall tool is active.

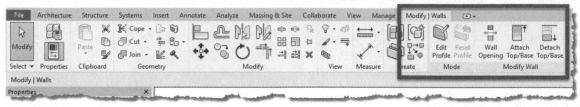

Modify Walls contextual tab – visible when a wall is selected.

Ribbon – Customization

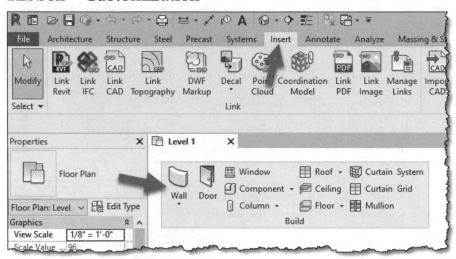

There is not too much customization that can be done to the *Ribbon*. One of the only things you can do is pull a panel off the *Ribbon* by clicking and holding down the left mouse button on the titles listed, e.g. "Link" or "Build," across the bottom. This panel can be placed within the *drawing window* or on another screen if you have a dual monitor setup.

The image above shows the *Build* panel, from the *Architecture* tab, detached from the *Ribbon* and floating within the drawing window. Notice that the *Insert* tab is active. Thus, you have constant access to the *Build* tools while accessing other tools. Note that the *Build* panel is not available on the *Architecture* tab as it is literally moved, not just copied.

When you need to move a detached panel back to the *Ribbon* you do the following: hover over the detached panel until the sidebars show up and then click the "Return panels to ribbon" icon in the upper right (identified in the image above).

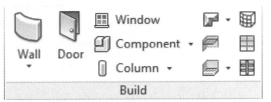

FYI: Whenever the resolution of your monitor is too low or you don't have the Revit application maximized on the screen, the buttons may be modified to take up less room on the Ribbon; typically the words are removed. Compare the image to the right with the Build panel above.

If you install an **add-in** for Revit on your computer, you will likely see a new tab appear on the Ribbon. Some add-ins are free while others require a fee. If an add-in only has one tool, it will likely be added to the catch-all tab called Add-Ins (shown in the image below).

Tab Visibility

This is not really customizing the User Interface, but in the Options dialog there are several adjustments one can make – such as turning off tabs and tools not used.

TIP: When Revit is first opened, you are prompted to indicate your role, or what type of work you do. This will turn off tabs here.

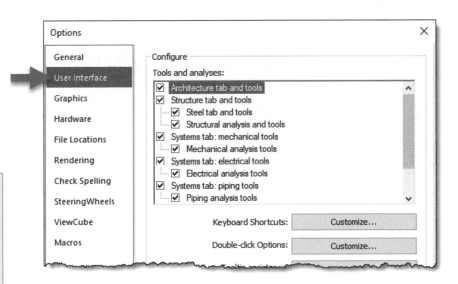

Ribbon – States

The *Ribbon* can be displayed in one of four states:

- Full Ribbon (default)
- Minimize to Tabs
- Minimize to Panel Tiles
- Minimize to Panel Buttons

The intent of this feature is to increase the size of the available drawing window. It is recommended, however, that you leave the *Ribbon* fully expanded while learning to use the program. The images in this book show the fully expanded state. The images below show the other three options. When using one of the minimized options you simply hover (or click) your cursor over the Tab or Panel to temporarily reveal the tools.

FYI: Double-clicking on a Ribbon tab will also toggle the states.

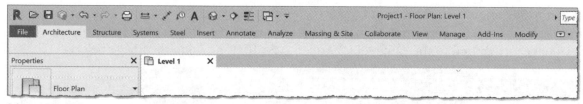

Minimize to Tabs

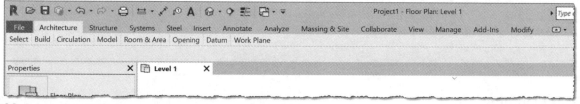

Minimize to Panel Tiles

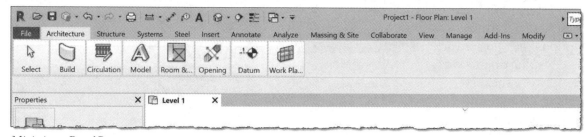

Minimize to Panel Buttons

Ribbon – References in this Book

When the exercises make reference to the tools in the *Ribbon* a consistent method is used to eliminate excessive wording and save space. Take a moment to understand this so that things go smoothly for you later.

Throughout the textbook you will see an instruction similar to the following:

> 23. Select **Architecture → Build → Wall**

This means click the ***Architecture*** tab, and within the ***Build*** panel, select the ***Wall*** tool. Note that the *Wall* tool is actually a split button, but a subsequent tool was not listed so you are to click on the primary part of the button. Compare the above example to the one below:

> 23. Select **Architecture → Build → Wall → Structural Wall**

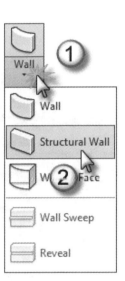

The above example indicates that you should click the down-arrow part of the *Wall* tool in order to select the *Structural Wall* option.

Thus the general pattern is this:

Tab → Panel → Tool → drop-down list item

#1 *Tab:* This will get you to the correct area within the *Ribbon*.

#2 *Panel:* This will narrow down your search for the desired tool.

#3 *Tool:* Specific tool to select and use.

Drop-down list item: This will only be specified for drop-down buttons and sometimes for split buttons.

The image below shows the order in which the instructions are given to select a tool; note that you do not actually click the panel title.

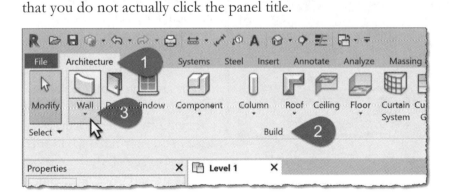

Options Bar

This area dynamically changes to show options that complement the current operation. The *Options Bar* is located directly below the *Ribbon*. When you are learning to use Revit you should keep your eye on this area and watch for features and options appearing at specific times. The image below shows the *Options Bar* example with the *Wall* tool active.

Properties Palette – Element Type Selector

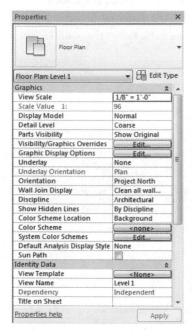

Properties Palette; nothing selected

The *Properties Palette* provides instant access to settings related to the element selected or about to be created. When nothing is selected, it shows information about the current view. When multiple elements are selected, the common parameters are displayed.

The *Element Type Selector* is an important part of the *Properties Palette*. Whenever you are adding elements or have them selected, you can select from this list to determine how a wall to be drawn will look, or how a wall previously drawn should look (see lower left image). If a wall type needs to change, you never delete it and redraw it; you simply select it and pick a new type from the *Type Selector*.

The **Selection Filter** drop-down list below the *Type Selector* lets you know the type and quantity of the elements currently selected. When multiple elements are selected you can narrow down the properties for just one element type, such as *wall*. Notice the image below shows four walls are in the current

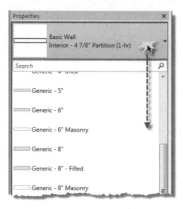

Type Selector; Wall tool active or a Wall is selected

Selection Filter; multiple elements selected

selection set. Selecting **Walls (4)** will cause the *Palette* to only show *Wall* properties even though many other elements are selected (and remain selected).

The width of the *Properties Palette* and the center column position can be adjusted by dragging the cursor over that area. You may need to do this at times to see all the information. However, making the *Palette* too wide will reduce the useable drawing area.

The *Properties Palette* should be left open; if you accidentally close it you can reopen it by **View → Window → User Interface → Properties** or by typing **PP** on the keyboard.

Project Browser

The *Project Browser* is the "Grand Central Station" of the Revit project database. All the views, schedules, sheets and content are accessible through this hierarchical list. The first image to the left shows the seven major categories; any item with a "plus" next to it contains sub-categories or items.

Double-clicking on a View, Legend, Schedule or Sheet will open it for editing; the current item open for editing is bold (**Level 1** in the example to the left). Right-clicking will display a pop-up menu with a few options such as *Delete* and *Copy*.

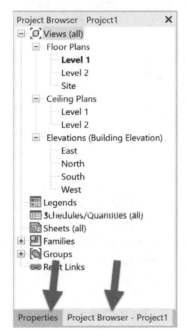

Right-click on *Views (all)*, at the top of the *Project Browser*, and you will find a **Search** option in the pop-up menu. This can be used to search for a *View, Family*, etc., a very useful tool when working on a large project with 100s of items to sift through.

Like the *Properties Palette*, the width of the *Project Browser* can be adjusted. When the two are stacked above each other, they both move together. You can also stack the two directly on top of each other; in this case you will see a tab for each at the bottom as shown in the second image to the left.

The *Project Browser* should be left open; if you accidentally close it by clicking the "X" in the upper right, you can reopen it by **View → Window → User Interface → Project Browser**.

The *Project Browser* and *Properties Palette* can be repositioned on a second monitor, if you have one, when you want more room to work in the drawing window.

Status Bar

This area will display information, on the far left, about the current command or list information about a selected element. The right-hand side of the *Status Bar* shows the number of elements selected. The small funnel icon to the left of the selection number can be clicked to open the *Filter* dialog box, which allows you to reduce your current selection to a specific category; for example, you could select the entire floor plan, and then filter it down to just the doors. This is different than the *Selection Filter* in the *Properties Palette* which keeps everything selected.

On the *Status Bar*, the first five icons on the right, shown in the image below, control how elements are **selected**. From left to right these are

- Select Links
- Select Underlay Elements
- Select Pinned Elements
- Select Elements by Face
- Drag Elements on Selection

Hover your cursor over an icon for the name and for a brief description of what it does. These are toggles that are on or off; **the red 'X' in the upper right of each icon means you cannot select that type of element within the model**. These controls help prevent accidentally moving or deleting things. Keep these toggles in mind if you are having trouble selecting something; you may have accidentally toggled one of these on.

Finally, the two drop-down lists towards the center of the *Status Bar* control **Design Options** and **Worksets** (see image on previous page). Design Options facilitate multiple design solutions in the same area, and *Worksets* relate to the ability for more than one designer to be in the model at a time.

View Control Bar

This is a feature which gives you convenient access to tools which control each view's display settings (i.e., scale, shadows, detail level, graphics style, etc.). The options vary slightly between view types: 2D View, 3D view, Sheet and Schedule. The important thing to know is that these settings only affect the current view, the one listed on the *Application Title Bar*. Most of these settings are available in the *Properties Palette*, but this toolbar cannot be turned off (i.e. hidden) like the *Properties Palette* can.

Context Menu

The *context menu* appears near the cursor whenever you right-click on the mouse (see image at right). The options on that menu will vary depending on what tool is active or what element is selected.

Canvas (aka Drawing Window)

This is where you manipulate the Building Information Model (BIM). Here you will see the current view (plan, elevation or section), schedule or sheet. Any changes made are instantly propagated to the entire database.

Context menu example with a wall selected

Elevation Marker

 This item is not really part of the Revit UI but is visible in the drawing window by default via the various templates you can start with, so it is worth mentioning at this point. The four elevation markers point at each side of your project and ultimately indicate the drawing sheet on which you would find an elevation drawing of each side of the building. All you need to know right now is that you should draw your floor plan generally in the middle of the four elevation markers that you will see in each plan view; DO NOT delete them as this will remove the related view from the *Project Browser*.

Revit Options

There are several settings, related to the *User Interface*, which are not tied to the current model. That is, these settings apply to the installation of Revit on your computer, rather than applying to just one model or file on your computer. A few of these settings will be briefly discussed here. It is recommended that you don't make any changes here right now.

Click on **Options** in the *File* tab. The options under File Locations tell Revit where to find templates and content. In the Hardware section you can verify your display driver information and compare with the "Supported Hardware" link. If Revit is crashing or exhibits strange behavior, it often relates to your graphics driver. You can bypass this problem by turning off *Use Hardware Acceleration*, but this will make Revit run slower when panning and zooming.

This concludes your brief overview of the Revit user interface. Many of these tools and operations will be covered in more detail throughout the book.

Efficient Practices

The *Ribbon* and menus are really helpful when learning a program like Revit; however, most experienced users rarely use them! The process of moving the mouse to the edge of the screen to select a command and then back to where you were is very inefficient, especially for those who do this all day long, five days a week. Here are a few ways experienced BIM operators work:

- Use the wheel on the mouse to Zoom (spin the wheel), Pan (press and hold the wheel button while moving the mouse) and Zoom Extents (double-clicking the wheel button). All this can be done while in another command; so, if you are in the middle of drawing walls and need to zoom in to see which point you are about to Snap to, you can do it without canceling the Line command and without losing focus on the area you are designing by having to click an icon near the edge of the screen.

- Revit conforms to many of the Microsoft Windows operating system standards. Most programs, including Revit, have several standard commands that can be accessed via keyboard shortcuts. Here are a few examples (press both keys at the same time):

  - Ctrl + S   Save   *(saves the current model)*
  - Ctrl + A   Select All   *(selects everything in text editor)*
  - Ctrl + Z   Undo   *(undoes the previous action)*
  - Ctrl + X   Cut   *(Cut to Windows clipboard)*
  - Ctrl + C   Copy   *(does not replace Revit's Copy tool)*
  - Ctrl + V   Paste   *(used to copy between models/views/levels)*
  - Ctrl + Tab   Change View   *(toggles between open views)*
  - Ctrl + P   Print   *(opens print dialog)*
  - Ctrl + N   New   *(create new project file)*
  - F7   Spelling   *(launch spell check feature)*
  - ENTER   Previous Command   *(repeat previous command)*

- If you recall, the *Open Documents* area, on the *File Tab,* lists all the views that are currently open on your computer. By clicking one of the names in the list you "switch" to that view. A shortcut is to press **Ctrl + Tab** to quickly cycle through the open drawings.

- Many Revit commands also have keyboard shortcuts. So, with your right hand on the mouse (and not moving from the "design" area), your left hand can type **WA** when you want to draw a Wall, for example. You can see all the

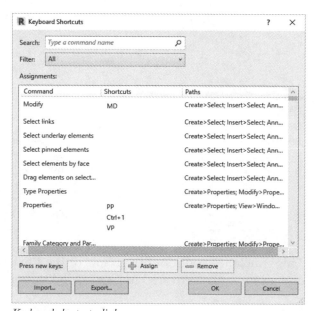

Keyboard shortcuts dialog

preloaded shortcuts and add new ones by clicking *View (tab)* → *User Interface (drop-down)* → *Keyboard Shortcuts*.

It should be noted that any customized keyboard shortcuts are specific to the computer you are working on, not the project. You can use the *Export* button (see image to right) to save the entire keyboard shortcuts list to a file, and then *Import* it into another computer's copy of Revit.

View Tabs

Revit displays a tab for each open view. Clicking a tab is a quick way to switch between open views. Click the "X" in each view tab to close that view. Drag a tab to change its position. Tabs can also be pulled outside of the main Revit application window – even to a second monitor. In the second image below, the schedule could be filling a second computer screen while reviewing the same information in the floor plan.

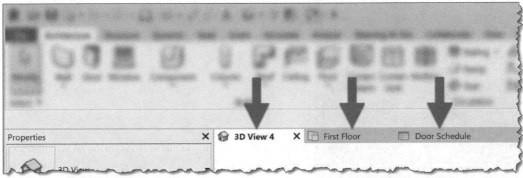

View Tabs; one for each open view

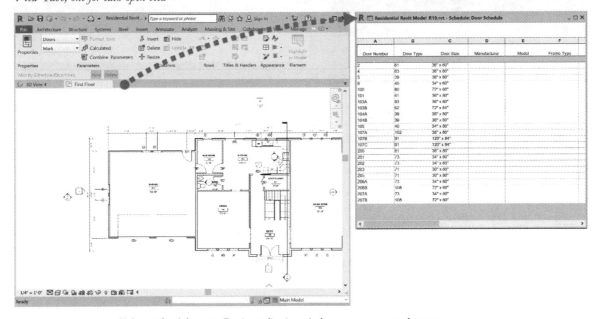

View Tabs can be pulled outside of the main Revit application window, even to a second screen

This concludes your brief overview of the Revit user interface. Many of these tools and operations will be covered in more detail throughout the book.

Exercise 1-3:
Open, Save and Close a Revit Project

To *Open* **Revit 2021**:

Start → Autodesk → Revit 2021

Or double-click the Revit icon from your desktop.

Revit 2021

This may vary slightly on your computer; see your instructor or system administrator if you need help.

How to Open a Revit Project

By default, Revit will open in the *Home* screen, which will display thumbnails of recent projects you have worked on. Clicking on the preview will open the project.

1. Click the **Open** button as shown to the right.

FYI: If the preview is for a *Central* file (aka Worksharing) it will automatically open a *Local* file. Worksharing allows multiple people to work in the same file.

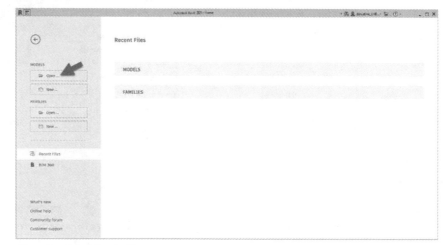

FIGURE 1-3.1 Revit Home screen

Next you will open an existing Revit project file. You will select a sample file provided online at SDCpublications.com; see the inside front cover for more information.

2. Click the drop-down box at the top of the *Open* dialog (Figure 1-3.2). Browse to your downloaded files (provided files need to be downloaded from www.SDCpublications.com; see backside cover of this book for more info).

    *TIP: If you cannot access the online sample files, you can substitute any Revit file you can find. Some sample files may be found here on your computer's hard drive: C:\Program Files\Autodesk\Revit 2021\Samples, or via **File** (tab) → **Open** (fly-out) → **Sample Files**.*

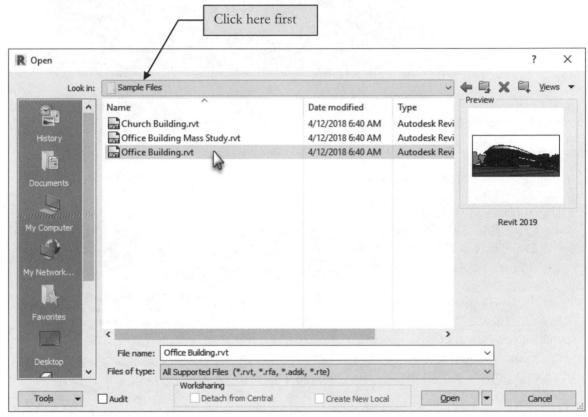

FIGURE 1-3.2 Open dialog; downloaded content from SDCpublications.com

3. Select the file named **Office Building.rvt** and click **Open**.

FYI: Notice the preview of the selected file. This will help you select the correct file before taking the time to open it.

The *Office Building.rvt* file is now open and the last saved view is displayed in the drawing window (Figure 1-3.3).

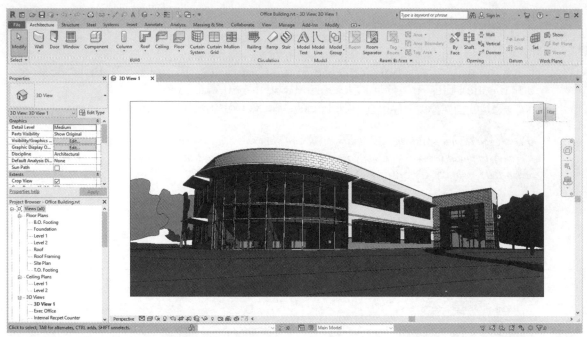

FIGURE 1-3.3 Sample file "Office Building.rvt"

The *File Tab* lists the projects and views currently open on your computer (Figure 1-3.4).

4. Click **File Tab → Open Documents (icon)** (Figure 1-3.4).

Notice that the *Office Building.rvt* project file is listed. Next to the project name is the name of a view open on your computer (e.g., floor plan, elevation).

FIGURE 1-3.4 File Tab: Open Documents view

Additional views will be added to the list as you open them. Each view has the project name as a prefix. The current view, the view you are working in, is always at the top of the list. You can quickly toggle between opened views from this menu by clicking on them.

You can also use the *Switch Windows* tool on the *View* tab; both do essentially the same thing.

Opening Another Revit Project

Revit allows you to have more than one project open at a time.

5. Click **Open** from the *Quick Access Toolbar.*

6. Per the instructions previously covered, browse the downloaded online files.

7. Select the file named **Church Building.rvt** and click **Open** (Figure 1-3.5).

> *TIP: If you cannot locate the online files, substitute one of the sample files.*

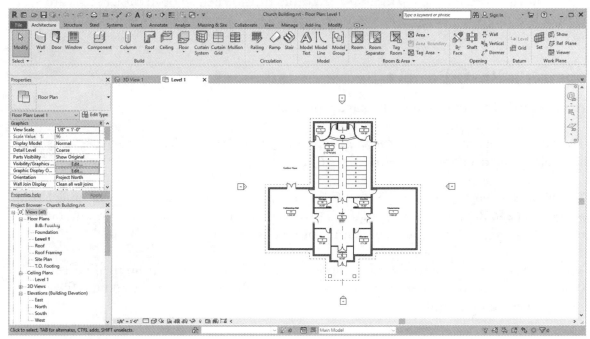

FIGURE 1-3.5 Sample file "Church Building.rvt"

8. Click **Open Documents** from the *File Tab* (Figure 1-3.6).

Notice that the *Church Building.rvt* project is now listed along with a view: Floor Plan: Level 1.

Try toggling between projects by clicking on *Office Building.rvt – 3D View: 3D View 1.*

FIGURE 1-3.6 Open Documents

Close a Revit Project

9. Select **File Tab → Close**; click **No** if prompted to save.

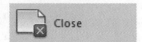

This will close the current project/view. If more than one view is open for a project, only the current view will close. The project and the other opened views will not close until you get to the last open view.

10. Repeat the previous step to close the other project file.

If you made changes and have not saved your project yet, you will be prompted to do so before Revit closes the view. **Do not save at this time**. When all open project files are closed, you find yourself back in the *Recent Files* screen – which is where you started.

Saving a Revit Project

At this time we will not actually save a project.

To save a project file, simply select *Save* from the *Quick Access Toolbar*. You can also select *Save* from the *File Tab* or press *Ctrl + S* on the keyboard.

You should get in the habit of saving often to avoid losing work due to a power outage or program crash. Revit offers a reminder to save; this notification is based on the amount of time since your last save. The **Save Time Interval** is set in *Options*.

You can also save a copy of the current project by selecting **Save As** from the File Tab. Once you have used the *Save As* command you are in the new project file and the file you started in is then closed.

Closing the Revit Program

Finally, from the *File Tab* select **Exit Revit**. This will close any open projects/views and shut down Revit. Again you will be prompted to save, if needed, before Revit closes the view. **Do not save at this time**.

You can also click the "X" in the upper right corner of the *Revit Application* window.

Exercise 1-4:
Creating a New Project

Open **Autodesk Revit**.

Creating a New Project File

The steps required to set up a new Revit model project file are very simple. As mentioned earlier, simply opening the Revit program starts you in the *Recent Files* window.

To manually create a new project (maybe you just finished working on a previous assignment and want to start the next one):

1. Select **File tab → New → Project**.

   FYI: You can also select the New *button on the Home screen.*

After clicking *New → Project* you will get the ***New Project*** dialog box (Figure 1-4.1).

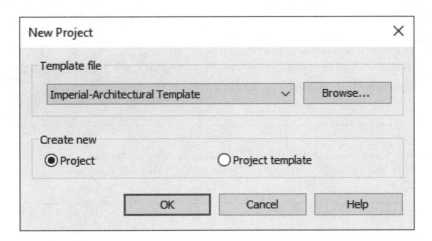

FIGURE 1-4.1 New Project dialog box

The *New Project* dialog box lets you specify the template file you want to use, or not use a template at all via the <None> option (which is not recommended). You can also specify whether you want to create a new project or template file.

2. Select the **Imperial-Architectural Template** option in the drop-down list. Leave *Create new* set to **Project** (Figure 1-4.1).

3. Click **OK**. You now have a new unnamed/unsaved project file.

To name an unnamed project file you simply *Save*. The first time an unnamed project file is saved you will be prompted to specify the name and location for the project file.

4. Select **File Tab → Save** from the *Menu Bar*.

5. Specify a **name** and **location** for your new project file.
   Your instructor may specify a location or folder for your files if in a classroom setting.

What is a Template File?

A template file allows you to start your project with specific content and certain settings preset the way you like or need them.

For example, you can have the units set to *Imperial* or *Metric*. You can have the door, window and wall families you use most loaded and eliminate other less often used content. Also, you can have your company's title block preloaded and even have all the sheets for a project set up.

A custom template is a must for design firms and contractors using Revit and will prove useful to the student as he or she becomes more proficient with the program.

> **Be Aware:**
> It will be assumed from this point forward that the reader understands how to create, open and save project files. Please refer back to this section as needed. If you still need further assistance, ask your instructor for help.

Exercise 1-5:
Using Zoom and Pan to View Your Drawings

Learning to *Pan* and *Zoom* in and out of a drawing is essential for accurate and efficient drafting and visualization. We will review these commands now so you are ready to use them with the first design exercise.

Open **Revit**.

You will select a sample file from the provided online files.

1. Select **Open** from the *Quick Access Toolbar*.

2. Browse to the **downloaded online files**.

3. Select the file named **Church Building.rvt** and click **Open** (Figure 1-5.1).

   TIP: If you cannot locate the online files, substitute any of the training files that come with Revit, found at C:\Program Files\Autodesk\Revit 2021\Samples.

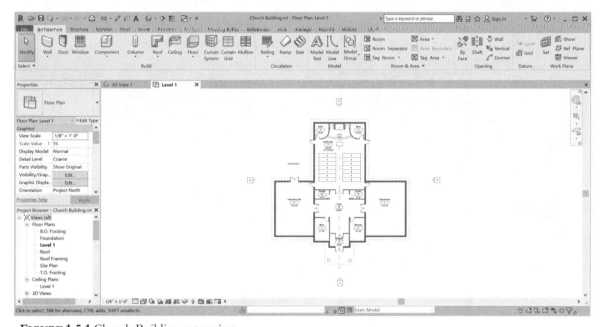

FIGURE 1-5.1 Church Building.rvt project

If the default view that is loaded is not **Floor Plan: Level 1**, double-click on **Level 1** under **Views\Floor Plans** in the *Project Browser*. Level 1 will be bold when it is the active or current view in the drawing window.

Using Zoom and Pan Tools

You can access the zoom tools from the *Navigation Bar*, or the *scroll wheel* on your mouse.

The *Zoom* icon contains several *Zoom* related tools:
- The default (i.e., visible) icon is *Zoom in Region*, which allows you to window an area to zoom into.
- The *Zoom* icon is a **split button**.
- Clicking the down-arrow part of the button reveals a list of related *Zoom* tools.
- You will see the drop-down list on the next page.

Zoom In

4. Select the top portion of the *Zoom* icon (see image to right).

5. Drag a window over your plan view (Figure 1-5.2).

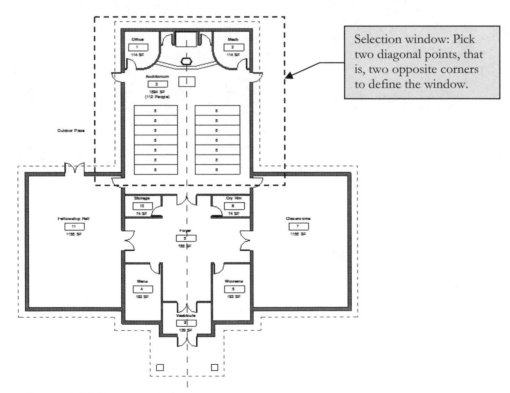

Selection window: Pick two diagonal points, that is, two opposite corners to define the window.

FIGURE 1-5.2 Zoom In window

You should now be zoomed in to the specified area (Figure 1-5.3).

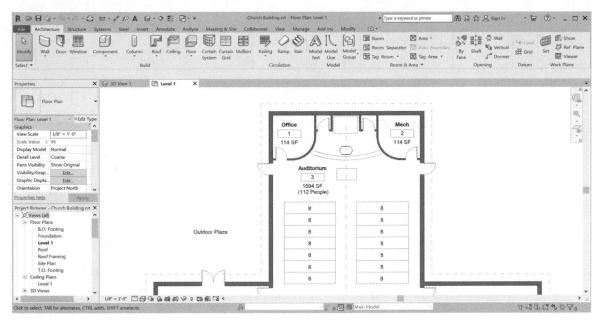

FIGURE 1-5.3 Zoom In results

Zoom Out

6. Click the down-arrow next to the zoom icon (Figure 1-5.4). Select **Previous Pan/Zoom**.

You should now be back where you started. Each time you select this icon you are resorting to a previous view state. Sometimes you have to select this option multiple times if you did some panning and multiple zooms.

7. **Zoom** into a smaller area, and then **Pan**, i.e., adjusting the portion of the view seen, by holding down the scroll wheel button.

The Pan tool just changes the portion of the view you see on the screen; it does not actually move the model.

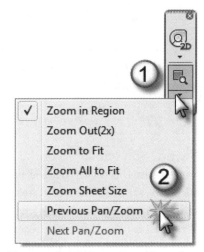

FIGURE 1-5.4 Zoom Icon drop-down

Take a minute and try the other *Zoom* tools to see how they work. When finished, click **Zoom to Fit** before moving on.

TIP: You can double-click the wheel button on your mouse to Zoom Extents *in the current view.*

Default 3D View

Clicking on the *Default 3D View* icon, on the *QAT*, loads a 3D View. This allows you to quickly switch to a 3D view.

8. Click on the **Default 3D View** icon.

9. Go to the **Open Documents** listing in the *File Tab* and notice the *3D View* and the *Floor Plan* view are both listed at the bottom.

   REMEMBER: You can toggle between views here.

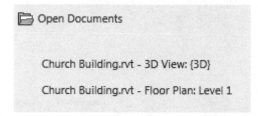

10. Click the **Esc** key to close the *File Tab*.

ViewCube

The *ViewCube* gives you convenient view control over the 3D view. This technology has been implemented in many of Autodesk's programs to make the process seamless for the user.

11. You should notice the ***ViewCube*** in the upper right corner of the drawing window (Figure 1-5.5). If not, you can turn it on by clicking *View → Windows → User Interface → ViewCube*.

   TIP: The ViewCube only shows up in 3D views.

Hovering your cursor over the *ViewCube* activates it. As you move about the *Cube* you see various areas highlight. If you click, you will be taken to that highlighted area in the drawing window. You can also click and drag your cursor on the *Cube* to "roll" the model in an unconstrained fashion. Clicking and dragging the mouse on the disk below the *Cube* allows you to spin the model without rolling. Finally, you have a few options in a right-click menu, and the **Home** icon, just above the *Cube*, gets you back to where you started if things get disoriented!

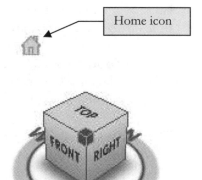

FIGURE 1-5.5 ViewCube

12. Give the *ViewCube* a try, then click the **Home** icon when you are done.

   REMEMBER: The Home icon only shows up when your cursor is over the ViewCube.

Navigation Wheel

Similar to the *ViewCube*, the *Navigation Wheel* aids in navigating your model. With the *Navigation Wheel* you can walk through your model, going down hallways and turning into rooms. Revit has not advanced to the point where the doors will open for you; thus, you walk through closed doors or walls as if you were a ghost!

The *Navigation Wheel* is activated by clicking the upper icon on the *Navigation Bar*.

Unfortunately, it is way too early in your Revit endeavors to learn to use the *Navigation Wheel*. You can try this in the chapter on creating photorealistic renderings and camera views. You would typically use this tool in a camera view.

FIGURE 1-5 6 Navigation Wheel

13. **Close** the *Church Building* project **without** saving.

Using the Scroll Wheel on the Mouse

The scroll wheel on the mouse is a must for those using BIM software. In Revit you can *Pan* and *Zoom* without even clicking a zoom icon. You simply **scroll the wheel to zoom** and **hold the wheel button down to pan**. This can be done while in another command (e.g., while drawing walls). Another nice feature is that the drawing zooms into the area near your cursor, rather than zooming only at the center of the drawing window. Give this a try before moving on. Once you get the hang of it, you will not want to use the icons. Also, double-clicking the wheel button does a *Zoom to Fit* so everything is visible on the screen.

TIP: Avoid a mouse with a wheel that tilts left and right as this makes using the wheel-button harder to use, making it not ideal for CAD/BIM users.

Exercise 1-6:
Using Revit's Help System

This section of your introductory chapter will provide a quick overview of Revit's *Help System*. This will allow you to study topics in more detail if you want to know how something works beyond the introductory scope of this textbook.

1. Click the **round question mark** icon in the upper-right corner of the screen.

You are now in Revit's Help site (Figure 1-6.1). This is a website which opened in your web browser. The window can be positioned side by side with Revit, which is especially nice if you have a dual-screen computer system. This interface requires a connection to the internet. As a website, Autodesk has the ability to add and revise information at any time, unlike files stored on your hard drive. This also means that the site can change quite a bit, potentially making the following overview out of date. If the site has changed, just follow along as best you can for the next three pages.

FIGURE 1-6.1 Autodesk Help site

In the upper left you can click the magnifying glass and search the *Help System* for a word or phrase. You may also click any one of the links to learn more about the topic listed. The next few steps will show you how to access the *Help System's* content on the Revit user interface, a topic you have just studied.

2. On the left, click the **plus symbol** next to **Get Started**.

3. Now, click the **plus symbol** next to **User Interface** (Figure 1-6.2).

Notice the tree structure on the left. You can use this to quickly navigate the help site.

4. Finally, click directly on **Ribbon**.

You now see information about the *Ribbon* as shown in the image below. Notice additional links are provided below on the current topic. You can use the browser's *Back* and *Forward* buttons to move around in the *Help System*.

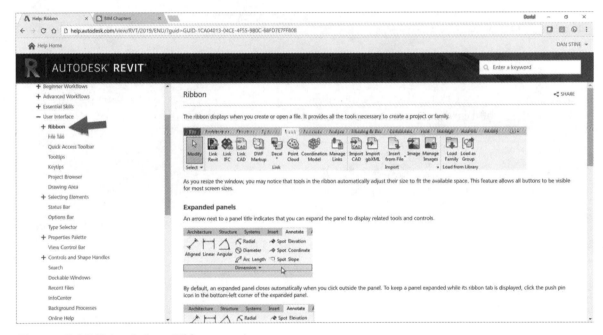

FIGURE 1-6.2 Help window; Ribbon overview

Next, you will try searching the Revit *Help System* for a specific Revit feature. This is a quick way to find information if you have an idea of what it is you are looking for.

5. In the upper right corner of the current *Help System* web page, click in the search text box and then type **gutter**.

6. Press **Enter** on the keyboard.

The search results are shown in Figure 1-6.3. Each item in the list is a link which will take you to information on that topic.

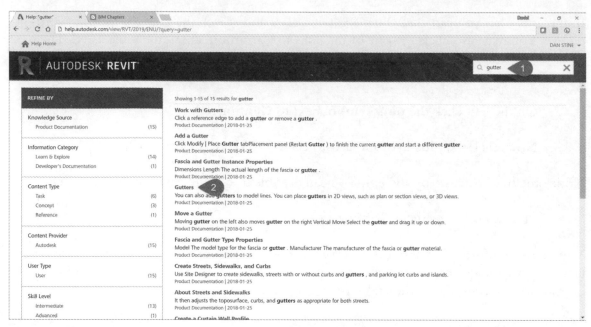

FIGURE 1-6.3 Help search results

The *Help System* can be used to complement this textbook. It is a reference resource, not a tutorial. As you are working through the tutorials in this book, you may want to use the *Help System* to fill in the blanks if you are having trouble or want more information on a topic.

Exercise 1-7:
Introduction to Autodesk 360

We will finish this chapter with a look at Autodesk A360, which is "ground zero" for all of Autodesk's **Cloud Services**. It is important that the student read this information in order to follow along in the book when specific steps related to using these cloud services are covered. The reader does not necessarily need to use Autodesk's cloud services to successfully complete this book.

The main features employed in the book are:
- Saving your work to *Autodesk A360* so you can access the data anywhere and know that the files are in a secure, backed up location. This feature is free to anyone, with some limitations to be discussed later.
- Sending your photorealistic rendering project to the *Cloud* to dramatically reduce the overall processing time. This feature is free to students and a free trial is available to everyone else.

Here is how Autodesk describes *Autodesk A360* on their website:

> The Autodesk® A360 cloud-based framework provides tools and services to extend design beyond the desktop. Streamline your workflows, effectively collaborate, and quickly access and share your work anytime, from anywhere. With virtually infinite computing power in the cloud, Autodesk 360 scales up or down to meet business needs without the infrastructure or upfront investment required for traditional desktop software.

Before we discuss *Autodesk A360* with more specificity, let's define what the *Cloud* is. **The Cloud is a service, or collection of services, which exists partially or completely online.** This is different from the *Internet*, which mostly involves downloading static information, in that you are creating and manipulating data. Most of this happens outside of your laptop or desktop computer. This gives the average user access to massive amounts of storage space, computing power and software they could not otherwise afford if they had to actually own, install and maintain these resources in their office, school or home. In one sense, this is similar to a *Tool Rental Center*, in that the average person could not afford, nor would it be cost-effective to own, a jackhammer. However, for a fraction of the cost of ownership and maintenance, we can rent it for a few hours. In this case, everyone wins!

Creating a Free Autodesk A360 Account

The first thing an individual needs to do in order to gain access to *Autodesk A360* is create a free account at https://a360.autodesk.com/ (students: see tip below); the specific steps will be covered later in this section, so you don't need to do this now. This account is for an individual person, not a computer, not an installation of Revit or AutoCAD, nor does it come from your employer or school. Each person who wishes to access *Autodesk A360* services must create an account, which will give them a unique username and password.

> *TIP:* Students should first create an account at
> http://www.autodesk.com/education/home. This is the same place you go
> to download free Autodesk software. Be sure to use your school email
> address as this is what identifies you as a qualifying student. Once you create
> an account there, you can use this same user name and password to access
> *Autodesk A360.* Following these steps will give you access to more storage
> space and unlimited cloud rendering!

Generally speaking, there are three ways you can access *Autodesk A360* cloud services:
- Autodesk A360 website
- Within Revit or AutoCAD; local computer
- Mobile device, smart phone or tablet

Autodesk A360 Website

When you have documents stored in the *Cloud* you can access them via your web browser. Here you can manage your files, view them without the full application (some file formats not supported) and share them. These features use some advanced browser technology, so you need to make sure your browser is up to date; Chrome works well.

Using the website, you can upload files from your computer to store in the *Cloud.* To do this, you create or open a project (Fig, 1-7.1) and click the **Upload** option (Fig. 1-7.2).

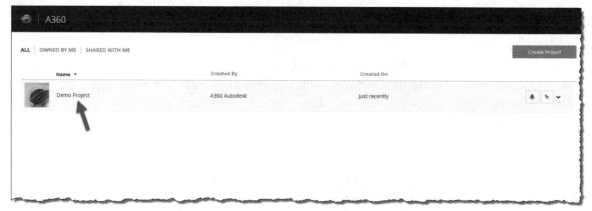

FIGURE 1-7.1 Viewing files stored in the cloud

Tip: If using Firefox or Chrome, you can drag and drop documents into the *Upload Documents* window. This is a great way to create a secure backup of your documents.

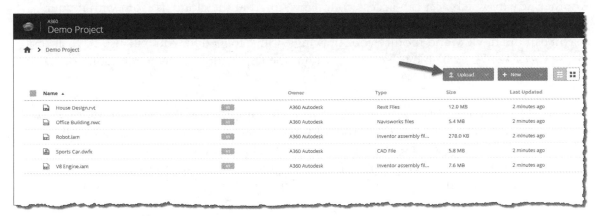

FIGURE 1-7.2 Upload files to the cloud

You can share files stored in the *Cloud* with others. <u>Private sharing</u> with others who have an *Autodesk A360* account is very easy. Another option is <u>public sharing</u>, which allows you to send someone a link and they can access the file, even if they don't have an A360 account. Simply hover over a file within *Autodesk A360* and click the **Share** icon to see the Share dialog (Figure 1-7.3).

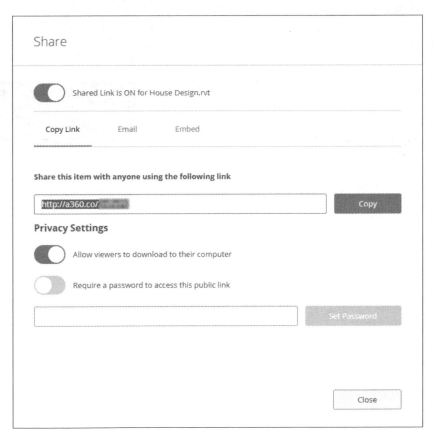

FIGURE 1-7.3 Sharing files stored in the cloud

Autodesk A360 within Revit or AutoCAD; local computer

Another way in which you can access your data, stored in the cloud, is from within your Autodesk application; for example, Revit or AutoCAD. This is typically the most convenient as you can open, view and modify your drawings. Once logged in, you will also have access to any *Cloud Services* available to you from within the application, such as rendering or *Green Building Studio.*

To sign in to *Autodesk A360* within your application, simply click the **Sign In** option in the notification area in the upper right corner of the window. You will need to enter your student email address and password (or personal email if you are not a student) as discussed in the previous section. When properly logged in, you will see your username or email address listed as shown in Figure 1-7.4 below. You will try this later in this section.

FIGURE 1-7.4 Example of user logged into Autodesk A360

Autodesk A360 Mobile

Once you have your files stored in the *Cloud* via Autodesk A360, you will also be able to access them on your tablet or smart phone if you have one. Autodesk has a free app called **A360 – View & Markup CAD files** for both the Apple or Android phones and tablets (Figure 1-7.5).

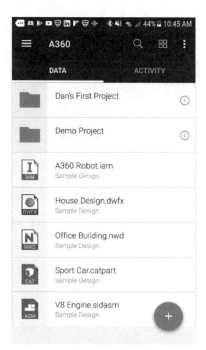

Some of the mobile features include:
- Open and view files stored in your Autodesk 360 account
- 2D and 3D DWG™ and DWF™ files
- Revit® and Navisworks® files
- Use multi-touch to zoom, pan, and rotate drawings
- View meta data and other details about elements within your drawing
- Find tools that help you communicate changes with your collaborators

The **Android** app is installed via *Google Play* and the **iPhone** app comes from the *Apple App Store.*

FIGURE 1-7.5
Viewing files on a smart phone

Setting up your Autodesk A360 account

The next few steps will walk you through the process of setting up your free online account at *Autodesk A360*. These steps are not absolutely critical to completing this book, so if you have any reservations about creating an *Autodesk A360* account – don't do it.

1. To create a free Autodesk 360 account, do one of the following:

    a. If you are a **student**, create an account at
       https://www.autodesk.com/education/home

    b. If you work for a company who has their Autodesk software on
       subscription, ask your *Contract Administrator* (this is a person in your office)
       to create an account for you and send you an invitation via
       https://manage.autodesk.com.

    c. **Everyone else**, create an account at https://a360.autodesk.com.

    FYI: The "s" at the end of "http" in the *Autodesk A360* URL means this is a secure
    website.

2. Open your application: AutoCAD or Revit.

3. Click the **Sign In** option in the upper-right corner of the application window (Figure
   1-7.6).

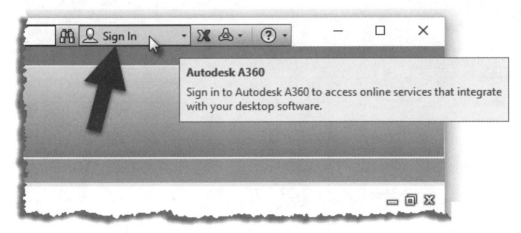

FIGURE 1-7.6 Signing in to Autodesk A360

It is recommended, as you work though this book, that you save all of your work in the *Cloud*, via Autodesk A360, so you will have a safe and secure location for your files. These files can then be accessed from several locations via the two methods discussed here. It is still important to maintain a separate copy of your files on a flash drive, portable hard drive or in another *Cloud*-type location such as *Dropbox*. This will be important if your main files ever become corrupt. You should manually back up your files to your backup location so a corrupt file does not automatically corrupt your backup files.

TIP: If you have a file that will not open try one of the following:

- **In Revit:** Open Revit and then, from the *File tab*, select Open, browse to your file, and select it. Click the **Audit** check box, and then click Open. Revit will attempt to repair any problems with the project database. Some elements may need to be deleted, but this is better than losing the entire file.

- **In AutoCAD:** Open AutoCAD and then, from the *Application Menu*, select Drawing Utilities → Recover → **Recover**. Then browse to your file and open it. AutoCAD will try and recover the drawing file. This may require some things to be deleted but is better than losing the entire file.

Be sure to check out the Autodesk website to learn more about Autodesk A360 Mobile and the growing number of cloud services Autodesk is offering.

Autodesk BIM 360 Design

With the proper Autodesk entitlements, not currently available to students, a Revit project can be worked on, via the cloud, by multiple people at the same time using **BIM 360 Design**. Thus, multiple firms can work together with their files linked together.

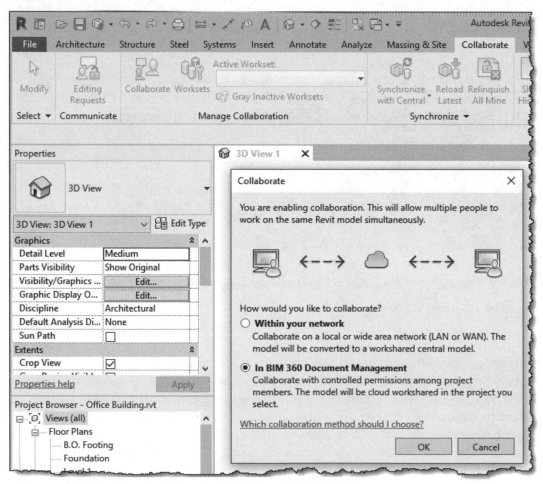

FIGURE 1-7.7 Enabling cloud-based collaboration in Revit

Self-Exam:

The following questions can be used as a way to check your knowledge of this lesson. The answers can be found at the bottom of this page.

1. The *View* tab allows you to save your project file. (T/F)

2. You can zoom in and out using the wheel on a wheel mouse. (T/F)

3. Revit is a parametric architectural design program. (T/F)

4. A _____ file allows you to start your project with specific content and certain settings preset the way you like or need them.

5. Autodesk 360 allows you to save your files safely in the _____ .

Review Questions:

The following questions may be assigned by your instructor as a way to assess your knowledge of this section. Your instructor has the answers to the review questions.

1. The *Options Bar* dynamically changes to show options that complement the current operation. (T/F)

2. Revit is strictly a 2D drafting program. (T/F)

3. The Projects/Views listed in the *Open Documents* list allow you to see which Projects/Views are currently open. (T/F)

4. When you use the *Pan* tool you are actually moving the drawing, not just changing what part of the drawing you can see on the screen. (T/F)

5. Revit was not originally created for architecture. (T/F)

6. The icon with the floppy disk picture ( ) allows you to _____ a project file.

7. Clicking on the _____ next to the *Zoom* icon will list additional zoom tools not currently shown in the *View* toolbar.

8. You do not see the *ViewCube* unless you are in a _____ view.

9. Creating an *Autodesk 360* account is free. (T/F)

10. The *Autodesk A360* App, for mobile devices, is expensive. (T/F)

Notes:

Autodesk User Certification Exam

Be sure to check out Appendix A for an overview of the official Autodesk User Certification Exam and the available study book and software, sold separately, from SDC Publications and this author.

Successfully passing this exam is a great addition to your resume and could help set you apart and help you land a great a job. People who pass the official exam receive a certificate signed by the CEO of Autodesk, the makers of Revit.

Lesson 2
Quick Start: Small Office:

In this lesson you will get a down and dirty overview of the functionality of Autodesk Revit. The very basics of creating the primary components of a floor plan will be covered: Walls, Doors, Windows, Roof, Annotation and Dimensioning. This lesson will show you the amazing "out-of-the-box" powerful, yet easy to use, features in Revit. It should get you very excited about learning this software program. Future lessons will cover these features in more detail while learning other editing tools and such along the way.

Exercise 2-1:
Walls, Grids and Dimensions

In this exercise you will draw the walls, starting with the exterior. Read the directions carefully; everything you need to do is clearly listed.

Exterior Walls

1. Start a new project named **Small Office** per the following instructions:

    a. **File Tab → New → Project**

    b. Click **Browse…** (Figure 2-1.1)

    c. Select the template file named **Commercial-Default.rte**. *(You should be brought to the correct folder automatically.)*

    d. Click **Open**.

    e. With the template file just selected and *Create new* "Project" selected, click **OK** (Figure 2-1.1).

    See Lesson 1 for more information on creating a new project.

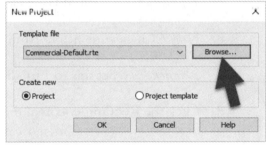

FIGURE 2-1.1 New Project

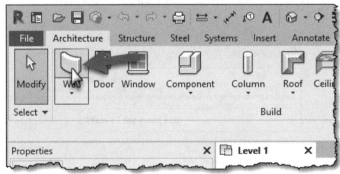

FIGURE 2-1.2 Wall tool

2. Select **Architecture → Build → Wall** on the *Ribbon*. (See Figure 2-1.2.)

Notice that the *Ribbon*, *Options Bar* and *Properties Palette* have changed to show settings related to walls. Next you will modify those settings.

FYI: By default, the bottoms of new walls will be at the current floor level and the tops of the walls are set via the Options Bar as shown in the next step.

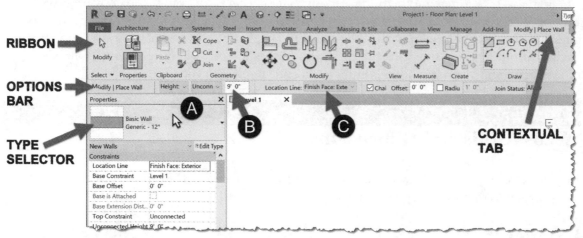

FIGURE 2-1.3 Ribbon and Options Bar

3. Modify the *Ribbon*, *Options Bar* and *Type Selector* to the following (Figure 2-1.3):

    a. *Type Selector:* Click in this area and select <u>*Basic Wall:*</u> **Generic – 12″**.

    b. *Height:* Change the height from 14′-0″ to **9′-0″**.

    c. *Location Line:* Set this to **Finish Face: Exterior**.

    d. Click the **Rectangle** icon in the *Draw* panel on the *Ribbon*. *(This allows you to draw four walls at once [i.e., a rectangle], rather than one wall at a time.)*

You are now ready to draw the exterior walls.

4. In the *Drawing Window*, click in the upper left corner.

    TIP: Make sure to draw within the four elevation markers (see image to right).

5. Start moving the mouse down and to the right. **Click** when the two temporary on-screen listening dimensions are approximately **100′** (East to West) and **60′** (front to back).

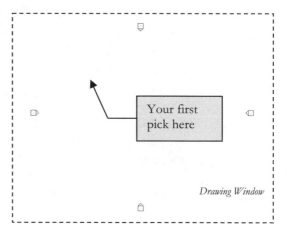

Getting the dimensions exact is not important as they will be revised later on.

Your drawing should look similar to Figure 2-1.4; similar in that the dimensions do not have to be exact right now and the building's location relative to the four elevation tags may vary slightly.

The *Temporary Dimensions* are displayed until the next action is invoked by the user. While the dimensions are displayed, you can click on the dimension text and adjust the wall dimensions. Also, by default the *Temporary Dimensions* reference the center of the wall – you can change this by simply clicking on the grips located on each *Witness Line*; each click toggles the witness line location between center, exterior face and interior face.

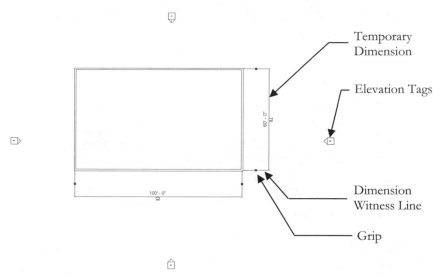

Temporary
Dimension

Elevation Tags

Dimension
Witness Line

Grip

FIGURE 2-1.4 Exterior walls

In the next few steps you will create grid lines and establish a relationship between the walls and the grids such that moving a grid causes the wall to move with it.

Grids:

Grids are used to position structural columns and beams in a building. Adding a grid involves selecting the *Grid* tool and then picking two points in the drawing window.

6. Click **Modify** on the *Ribbon* (to finish using the *Wall* tool) and then select the **Architecture → Datum → Grid**.

Grid

> *FYI: The same* Grid *tool is also found on the* Structure *tab.*

Next you will draw a vertical grid off to the left of your building. Once you have drawn all the grids you will use a special tool to align the grid with the walls and lock that relationship.

7. [*first pick*] **Click** down and to the left of your building as shown in Figure 2.1-5.

   FYI: 'Click' always means left-click, unless a right-click is specifically called for.

8. [*second pick*] Move the cursor straight up (i.e., vertically) making sure you see a dashed cyan line, which indicates you are snapped to the vertical plane, and the angle dimension reads 90 degrees. Just past the North edge of the building (as shown in Figure 2.1-5), click.

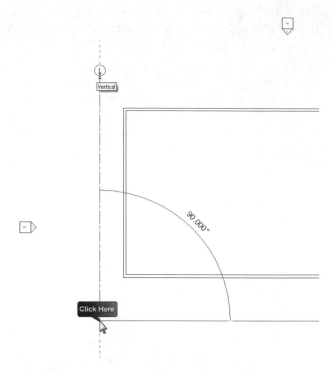

FIGURE 2-1.5 Drawing a grid

You have now drawn your first grid line. Next you will quickly draw four more grid lines, two horizontal and two vertical.

   NOTE: The Grid tool will remain active until you select another tool or select Modify (selecting Modify allows you to select other elements in the drawing window).

9. Draw another vertical grid approximately centered on your building. BEFORE YOU PICK THE FIRST POINT, make sure you see a dashed cyan reference line indicating the grid line will align with the previous grid line (you will see this before clicking the mouse at each end of the grid line), then go ahead and pick both points (Figure 2-1.6).

10. Draw the remaining grid lines shown in Figure 2-1.6. Again, do not worry about the exact location of the grid lines, just make sure the ends align with each other.

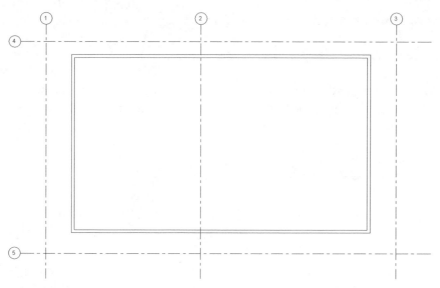

FIGURE 2-1.6 Grids added

Next you will change the two horizontal grid lines to have letters instead of numbers.

11. Zoom in on the grid bubble for the upper horizontal grid line.

12. Click *Modify* and then click on the grid line to select it.

13. With the grid line selected, click on the dark blue text within the bubble.

14. Type **A** and press *Enter* on the keyboard (Figure 2-1.7).

15. Click **Modify** again.

16. Change the other horizontal grid to **B**.

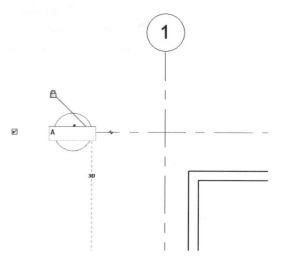

FIGURE 2-1.7 Grid edit

Align

Next you will use the *Align* tool to reposition the grid lines so they "align" with the exterior face of the adjacent walls. The steps are simple: select the *Align* tool from the *Ribbon*; pick the reference line (i.e., the exterior wall face); and then you select the item to move (i.e., the grid line). This tool works on many Revit objects!

17. Select **Modify** → **Modify** → **Align** from the *Ribbon*.

   REMINDER: *Ribbon instructions are* Tab → Panel → Tool

18. [*Align: first pick*] With the *Align* tool active notice the prompt on the *Status Bar*, select **Wall faces** on the *Options Bar*, next to Prefer, and then select the exterior face of the wall adjacent to grid line 1.

19. [*Align: second pick*] Select **grid line 1**. Be sure to see the next step before doing anything else!

The grid line should now be aligned with the exterior face of the wall. Immediately after using the *Align* tool you have the option of "locking" that relationship; you will do that next. The ability to lock this relationship is only available until the next tool is activated. After that you would need to use the *Align* tool again.

20. Click the un-locked **padlock** symbol to "lock" the relationship between the grid line and the wall (see Figure 2-1.8).

21. Use the steps just outlined to **Align** and **Lock** the remaining grid lines with their adjacent walls. Do not worry about the location of grid line 2 (i.e., the vertical grid in the center).

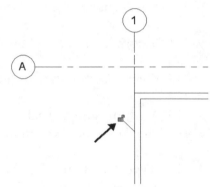

FIGURE 2-1.8 Align "lock"

Dimensions

Next you will add dimensions to the grid lines and use them to drive the location of the walls and grids and learn how to lock them.

22. Select **Modify** and then **right-click** anywhere within the *Drawing Window*; click **Zoom To Fit**.

23. Select **Annotate** → **Dimensions** → **Aligned** tool.

Aligned

At this point you are in the *Dimension* tool. Notice the various controls available on the *Ribbon* and *Options Bar*. You can set things like the dimension style (via the *Element Type Selector*) and the kind of dimension (linear, angle, radius, etc.) and which portion of the wall to *Prefer* (e.g., face, center, core face).

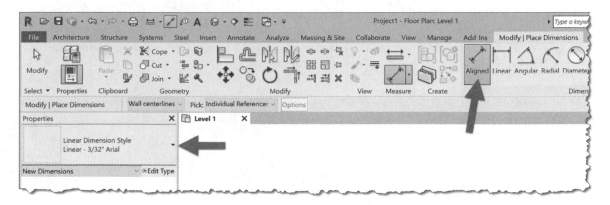

24. With the *Ribbon*, *Options Bar* and *Properties Palette* as shown above, which should be the default settings, **select grid line 1**.

    FYI: The grid line will pre-highlight before you select it, which helps make sure you select the correct item (e.g., the grid line versus the wall). Be careful not to select the wall.

25. Now **select grid line 2** and then **select grid line 3**.

Your last pick point is to decide where the dimension line should be.

26. Click in the location shown in Figure 2-1.9 to position the dimension line.

    TIP: Do not click near any other objects or Revit will continue the dimension string to that item.

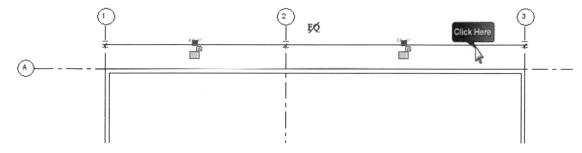

FIGURE 2-1.9 Adding dimensions

Notice that while the dimension string is selected, you see an EQ symbol with a slash through it. This symbol indicates that the individual components of the dimension string are not equal in length. The next step will show you how easy it is to make these dimensions equal!

27. With the dimension string selected, click the EQ symbol located near the middle of the dimension.

The grid lines are now equally spaced (Figure 2-1.10) and this relationship will be maintained until the EQ symbol is selected again to toggle the "dimension equality" feature off.

> *NOTE: When dimension equality is turned off or the dimension is deleted the grid line **will not** move back to its original location; Revit does not remember where the grid was.*

Typically, you would not want to click the padlock icons here because that would lock the current dimension and make it so the grid lines could not be moved at all.

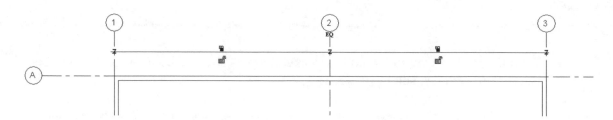

FIGURE 2-1.10 Toggling dimension equality

Next you will add an "overall building" dimension from grid line 1 to grid line 3. This dimension can be used to drive the overall size of your building (all the time keeping grid line 2 equally spaced).

28. Using the **Aligned** *Dimension* tool, add a dimension from grid line 1 to grid line 3 and then pick to position the dimension line (Figure 2-1.11).

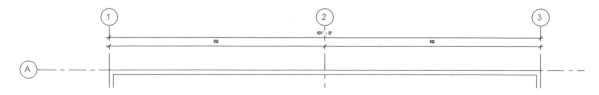

FIGURE 2-1.11 Overall building dimension added

When using a dimension to drive the location of geometry, you need to select the item you want to move and then select the dimension text to enter the new value. You cannot just select the dimension because Revit does not know whether you want the left, right or both grid lines to move. The only thing you can do, graphically, by selecting the dimension directly is "lock" that dimension by clicking on the padlock symbol and then click the blue dimension text to add a suffix if desired. Next you will adjust the overall building size.

29. Click **Modify** (or press the *Esc* key twice) to make sure you cancel or finish the *Dimension* tool and that nothing is selected.

30. **Select grid line 3**.

31. With grid line 3 selected, click the dimension text and type **101** and then press *Enter*.

    FYI: Notice that Revit assumes feet if you do not provide a foot or inch symbol.

32. Repeating the previous steps, add a dimension between grid lines A and B, and then adjust the model so the dimension reads **68'-0"**.

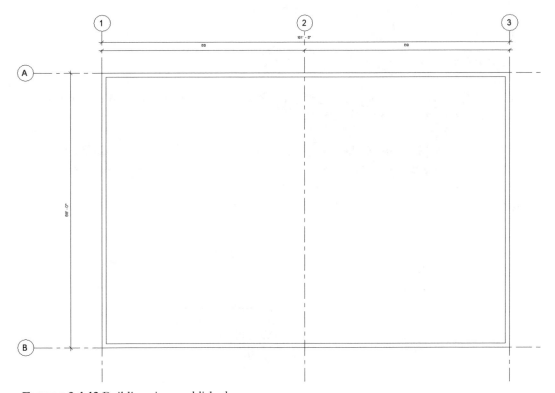

FIGURE 2-1.12 Building size established

Your project should now look similar to Figure 2-1.12. You should notice that dimensions must "touch" two or more items (the grid lines in this case). Also, because the walls were aligned and locked to the grids, moving the grids caused the walls to move.

The last thing you will do before moving on to the interior walls is to swap out the generic walls with a more specific wall. This would be a common situation in a design firm; a generic wall is added as a "place holder" until the design is refined to the point where the exterior wall system is selected.

The process for swapping a wall is very simple: select the wall and pick a different type from the *Type Selector*. The next steps will do this but will also show you how to quickly select all the exterior walls so you can change them all at once!

33. Click **Modify** and then hover your cursor over one of the exterior walls so it highlights. (Do not click yet.)

34. *With an exterior wall highlighted,* i.e., pre-selected, take your hand off the mouse and tap the **Tab** key until all four walls pre-highlight.

    FYI: The Tab *key cycles through the various items below your cursor. The current options should include one wall, a chain of walls, and a grid line.*

35. *With all four walls highlighted,* **click** to select them.

36. *With all four walls selected,* pick *Basic Wall:* **Exterior – Brick on CMU** from the *Type Selector* on the *Properties Palette.*

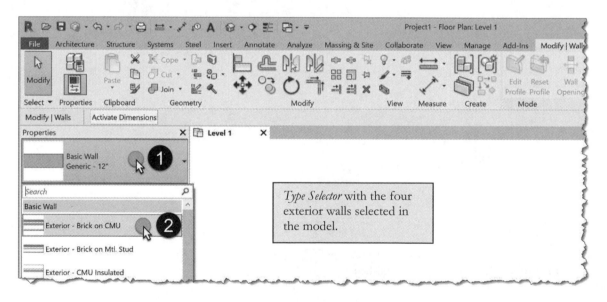

Type Selector with the four exterior walls selected in the model.

Detail Level

Revit allows you to control how much detail is shown in the walls.

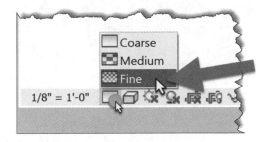

37. On the *View Control Bar*, located in the lower left corner of the *Drawing Window* on the *View Control Bar*, set the *Detail Level* to **Fine**.

As you can see in the two images below, *Coarse* simply shows the outline of a wall type and *Fine* shows the individual components of the wall (i.e., brick, insulation, concrete block, etc.).

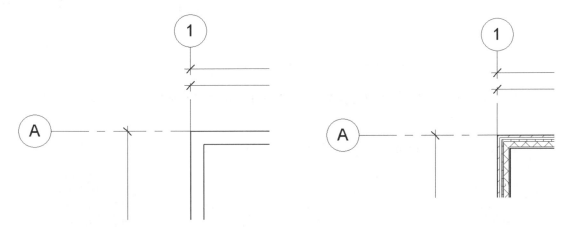

FIGURE 2-1.13 Detail level – coarse (left) vs. fine (right)

Now that you have the exterior walls established, the grid lines properly placed, and their relationships locked in the project, you can now proceed with the layout of the interior spaces.

Interior Walls

38. With the **Wall** tool selected, modify the *Ribbon*, *Options Bar* and *Properties Palette* to the following (also, see image on the following page):

   a. *Type Selector:* set to <u>*Basic Wall:*</u> **Interior – 4 7/8″ Partition (1-hr)**.

   b. *Height:* **Roof**

   c. *Location Line:* Set this to **Wall Centerline.**

   d. *Turn off Chain.*

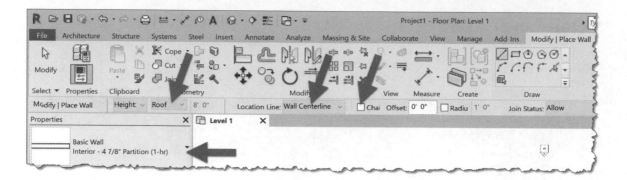

39. Draw a wall from the West wall (i.e., "vertical" wall on the left) to the East wall (on the right). See Figure 2-1.14.

   a. Make sure your cursor "snaps" to the wall before clicking.

   b. Before clicking the second point of the wall, make sure the dashed cyan line is visible so you know the wall will be truly horizontal.

   c. The exact position of the wall is not important at this point as you will adjust it in the next step.

   d. With the temporary dimensions still active, proceed to the next step.

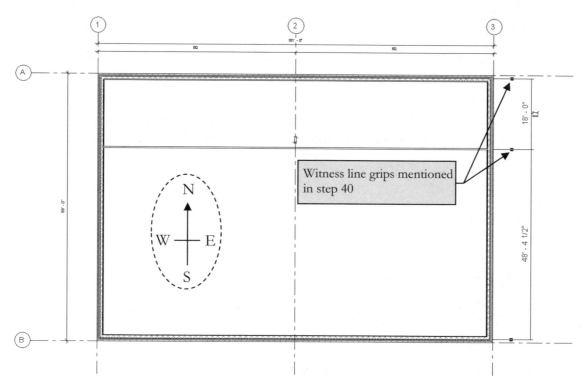

FIGURE 2-1.14 Adding interior walls – North indicator added for reference only

40. Click the **witness line grips** (see Figure 2-1.14) until the "clear" space of the room is listed (see Figure 2-1.15).

41. Now click the blue text of the temporary dimension, type **22**, and then press *Enter* (Figure 2-1.15).

42. Click **Modify** to finish the current task.

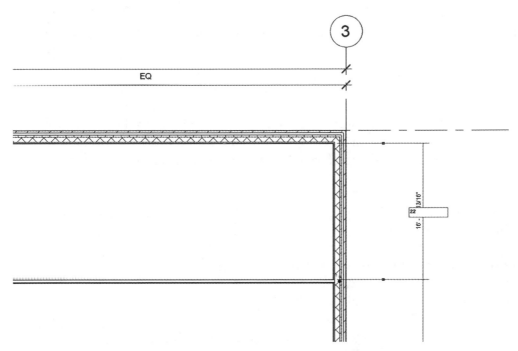

FIGURE 2-1.15 Repositioning interior wall via temporary dimensions

The clear space between the interior wall and the north wall is now 22'-0". Next you will add additional interior walls to create equally spaced rooms in this area.

43. Using the same settings as the interior wall just added, draw five (5) vertical walls as shown in Figure 2-1.16.

    a. Make sure they are orthogonal (i.e., the dashed cyan line is visible before picking the walls endpoint).

    b. Make sure you "snap" to the perpendicular walls (at the start and endpoint of the walls you are adding).

    c. Do not worry about the exact position of the walls.

    TIP: Uncheck Chain on the Options Bar.

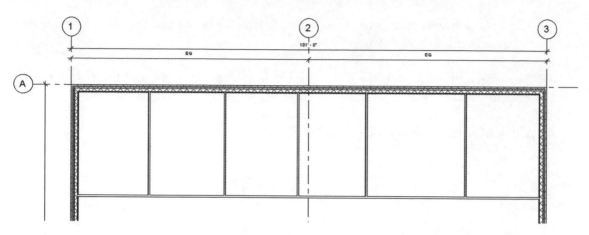

FIGURE 2-1.16 Adding additional interior walls

In the next step you will use a dimension string to reposition the walls so they are equally spaced. This process is similar to what you did to reposition grid line 2. However, you have to specify which part of the wall you want to dimension to (center, face, core center, core face).

FYI: *The "core" portion of a wall system typically consists of the structural element(s) such as the concrete block (in your exterior walls) or the metal studs (in your interior walls).*

44. Select **Annotate → Dimension → Aligned** tool on the *Ribbon.*

45. On the *Options Bar,* select **Wall Faces**.

This setting will force Revit to only look for the face of a wall system. You can select either face depending on which side of the wall you favor with your cursor. This feature lets you confidently pick specific references without needing to continually zoom in and out all over the floor plan.

46. Select the interior face of the West wall to start your dimension string.

The next several picks will need to reference the wall centerlines. Revit allows you to toggle the wall position option on the fly via the *Options Bar.*

47. Change the setting to **Wall Centerlines**.

48. The next five picks will be on the five "vertical" interior walls. Make sure you see the dashed reference line centered on the wall to let you know you are about to select the correct reference plane.

49. Change the wall location setting back to **Wall Faces** and select the interior face of the East wall (the wall at grid line 3).

50. Your last pick should be away from any elements to position the dimension string somewhere within the rooms (Figure 2-1.17).

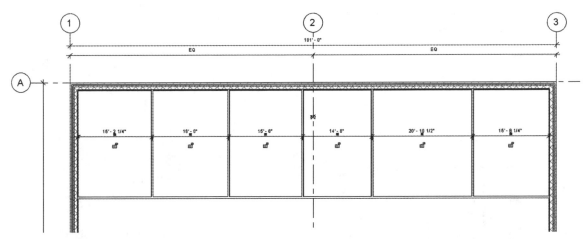

FIGURE 2-1.17 Adding a dimension string

51. Click the **EQ** symbol to reposition the interior walls.

52. Click **Modify**.

The interior walls are now equally spaced (Figure 2-1.18)!

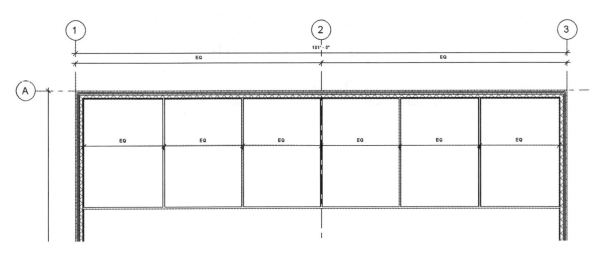

FIGURE 2-1.18 Enabling dimension equality

53. Add a vertical "clear" dimension to indicate the depth of the rooms. Set *Prefer* to **Wall Faces** for both ends of the dimension line. See Figure 2-1.19.

54. Click the **padlock** symbol (🔒) to tell Revit this dimension should not change (Figure 2-1.19).

55. Click **Modify**.

Next you will adjust the overall building dimensions and notice how the various parametric relationships you established cause the model to update!

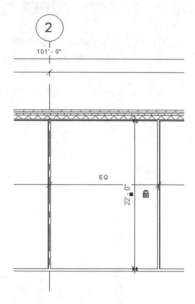

FIGURE 2-1.19
Locking dimensions

56. Click grid line 3 and change the overall dimension from 101 to **40**, by clicking on the dimension text and then pressing *Enter*.

TIP: When adjusting the building footprint via the dimensions, you need to select the grid line, not the east wall, because the dimension references the grid line.

Notice the interior walls have adjusted to remain equal, and grid line 2 is still centered between grids 1 and 3 (Figure 2-1.20).

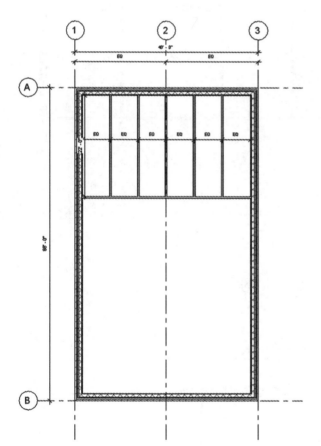

FIGURE 2-1.20 Adjusting dimensions

57. Change the 40'-0" dimension to **110'-0"**.

58. Select grid line A and change the 68'-0" dimension to **38'-0"**.

Your model should now look similar to Figure 2-1.21. Notice the interior wall maintained its 22'-0" clear dimension because the interior wall has a dimension which is locked to the exterior wall, and the exterior wall has an alignment which is locked to grid line A.

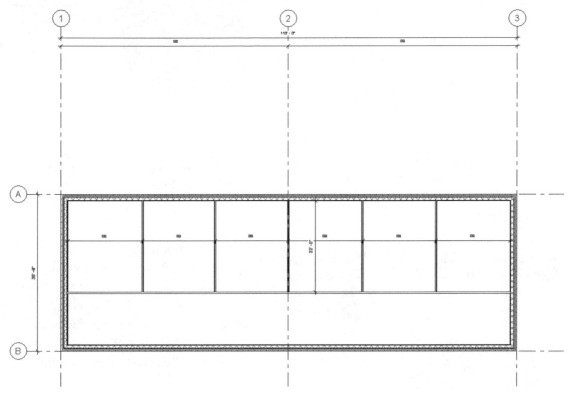

FIGURE 2-1.21 Adjusting dimensions

59. Click **Undo** icon on the *QAT* to restore the 68' dimension.

Your building should now be 110'-0" x 68'-0".

60. **Save** your project (*Small Office.rvt*).

> *TIP: You can use the Measure tool to list the distance between two points. This is helpful when you want to quickly verify the clear dimension between walls. Simply click the icon and snap to two points and Revit will temporarily display the distance. You can also click "chain" on the Options Bar and have Revit add up the total length of several picks.*

Exercise 2-2:
Doors

In this exercise you will add doors to your small office building.

1. Open **Small Office.rvt** created in Exercise 2-1.

Placing Doors

2. Select **Architecture → Build → Door** tool on the *Ribbon* (Figure 2-2.1).

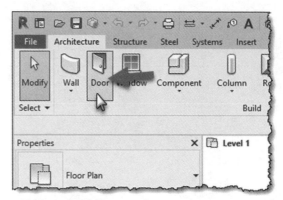

FIGURE 2-2.1 Door tool

Notice that the *Ribbon, Options Bar & Properties Palette* have changed to show options related to Doors. Next you will modify those settings.

The *Type Selector* indicates the door style, width and height. Clicking the down arrow to the right lists all the doors pre-loaded into the current project.

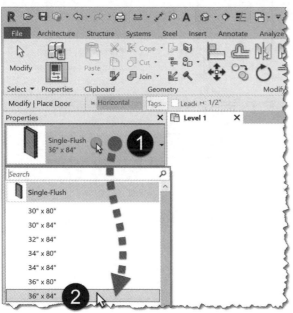

FIGURE 2-2.2 Type selector: Doors

The default template project that you started from has several sizes for a single flush door. Notice, in Figure 2-2.2, that there are two standard heights in the list. The 80″ (6′-8″) doors are the standard residential height and the 84″ (7′-0″) doors are the standard commercial door height.

3. Change the *Element Type Selector* to *Single-Flush:* **36″ x 84″**, and click **Tag on Placement** on the *Ribbon*.

4. Move your cursor over a wall and position the door as shown in **Figure 2-2.3**. (Do not click yet.)

Notice that the swing of the door changes depending on what side of the wall your cursor is favoring.

Also notice Revit displays listening dimensions to help you place the door in the correct location.

5. Click to place the door. Revit automatically trims the wall and adds a door tag.

   TIP: Press the spacebar before clicking to flip the door swing if needed.

6. While the door is still selected, click on the *change swing (control arrows)* symbol to make the door swing against the wall if it is not already (Figure 2-2.4).

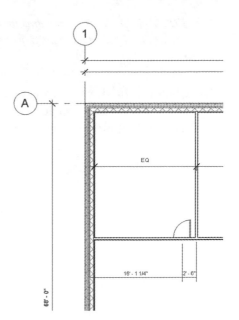

FIGURE 2-2.3 Adding door

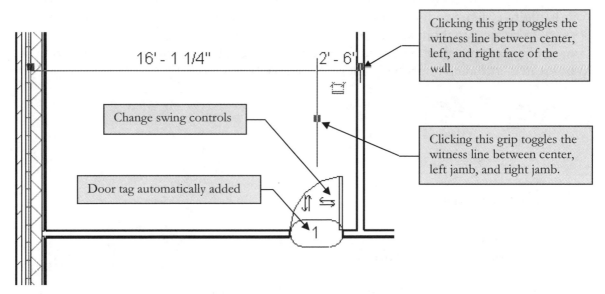

FIGURE 2-2.4 Door just placed

Next you will reposition the door relative to the adjacent wall.

7. Click **Modify** to finish the *Door* tool.

8. Click the door (not the door tag) you just placed to select it.

9. Click the **witness line grips** so the temporary dimension references the right door jamb and the wall face as shown in Figure 2-2.5.

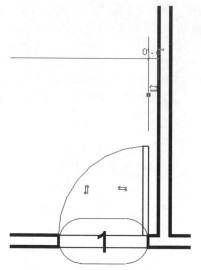

TIP: You can also click and drag the witness line grip to another wall or line if the default location was not what you are concerned with.

10. Click on the dimension text, type **4″** and press *Enter*. Make sure you add the inch symbol or you will get feet rather than inches (Figure 2-2.5).

FIGURE 2-2.5 Edit door location

Unfortunately, the door *Families* loaded in the commercial template do not have frames. So the 4″ dimension just entered provides for a 2″ frame and 2″ of wall. The library installed on your hard drive, along with the Revit "web library," do provide some doors with frames. It is possible to create just about any door and frame combination via the *Family Editor*. The *Family Editor* is a special mode within Autodesk Revit that allows custom parametric content to be created, including doors with sidelights, transoms and more!

Mirroring Doors

The *Mirror* command will now be used to quickly create another door opposite the adjacent perpendicular wall.

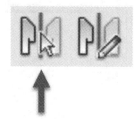

11. With the door selected, click the **Mirror → Pick Mirror Axis** on the *Contextual Tab*.

12. On the *Options Bar*, click **Copy**. If copy was not selected, then the door would be relocated rather than copied.

13. With the door selected and the *Mirror* command active, hover the cursor over the adjacent wall until the dashed reference line appears centered on the wall, keep moving the mouse until you see this, and then click. (Figure 2-2.6)

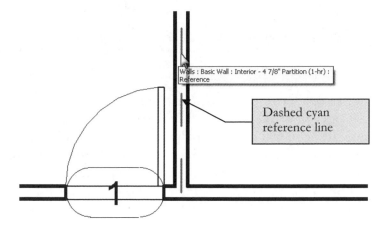

FIGURE 2-2.6 Mirroring a door

As you can see in Figure 2-2.7, the door has been mirrored into the correct location.

Revit does not automatically add door tags to mirrored or copied doors. These will be added later.

TIP: The size of the door tag is controlled by the view's scale.

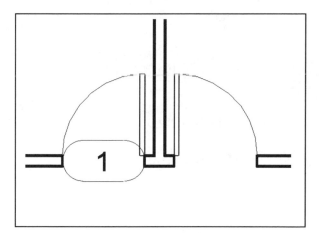

FIGURE 2-2.7 Door mirrored

Copying Doors

Now you will copy the two doors so the other rooms have access.

14. Click to select the first door (not the door tag) and then press and hold the **Ctrl** key. *While holding the Ctrl key,* click to add the second door to the selection set.

15. With the two doors selected, click the **Copy** tool.

16. On the *Options Bar*, select **Multiple**.

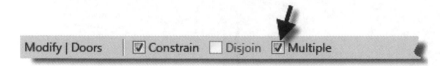

At this point you need to pick two points: a "copy from here" point and a "copy to there" point. The first point does not have to be directly on the elements(s) to be copied. The next step will demonstrate this; you will pick the midpoint of the wall adjacent to the two doors (first pick) and then you will pick the midpoint of the wall where you want a set of doors (second pick). With "multiple" checked, you can continue picking "second points" until you are finished making copies (pressing *Esc* or *Modify* to end the command).

17. Pick three points:

    a. *First pick:* midpoint/centerline of wall (see Figure 2-2.8);

    b. *Second pick:* midpoint/centerline of wall shown in Figure 2-2.9;

    c. *Third pick:* midpoint/centerline of wall shown in Figure 2-2.9.

18. Pick **Modify** to end *Copy*.

The doors are now copied.

FIGURE 2-2.8 Copy – first point with midpoint symbol visible

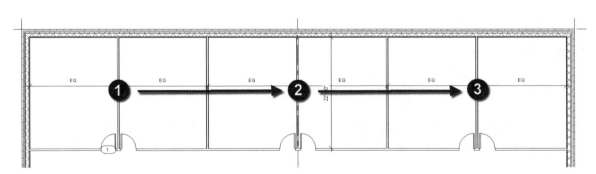

FIGURE 2-2.9 Numbers indicate pick-points listed in step #17

You will now add two exterior doors using the same door type.

19. Using the **Door** tool, add two exterior doors approximately located per Figure 2-2.10. Match the swing and hand shown.

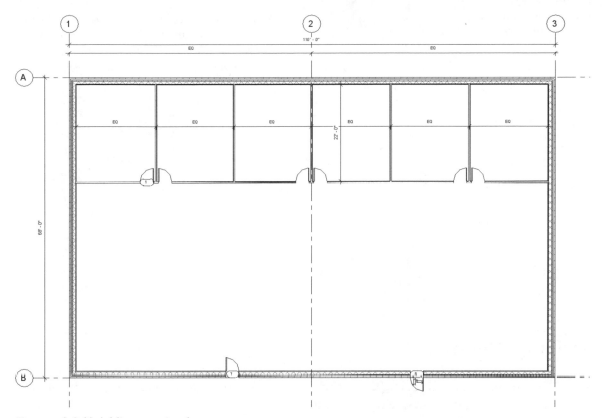

FIGURE 2-2.10 Adding exterior doors

Tag All (Not Tagged)

Revit provides a command to quickly add a tag (e.g., a door tag) to any door that does not currently have one in the current view. The tag might have to be moved or rotated once placed, but this still saves time and the possibility of missing a door tag.

20. Select **Annotate → Tag → Tag All**.

21. In the *Tag All Not Tagged* dialog box, check the box for **Door Tags** under *Category* and set *Tag Orientation* to **Vertical**. Click **OK** (see image on next page).

All the doors should now be tagged in your floor plan.

FYI: Door tags can be deleted at any time and added again later at any time. Tags simply display information in the element being tagged – thus, no information about the element is being deleted; the building information integrity remains intact.

Deleting Doors

Next you will learn how to delete a door when needed. This process will work for most elements (i.e., walls, windows, text, etc.) in Revit.

22. Click **Modify**.

23. Click on door number 7 (the door on the left, not the door tag) and press the **Delete key** on your keyboard.

As you can see, the door is deleted and the wall is automatically filled back in. Also, a door tag can only exist by being attached to a door; therefore, the door tag was also deleted.

One last thing to observe: Revit numbers the doors in the order in which they have been placed (regardless of level). Doors are not automatically renumbered when one is deleted. Also, doors can be renumbered to just about anything you want.

24. **Save** your project *(Small Office.rvt)*.

Exercise 2-3:
Windows

In this exercise you will add windows to your small office building.

1. Open **Small Office.rvt** created in Exercise 2-2.

Placing Windows

2. Select **Architecture → Build → Window**.

Notice that the *Ribbon*, *Options Bar* and *Properties Palette* have changed to show options related to windows. Next you will modify those settings.

The *Type Selector* indicates the window style, width and height. Clicking the down arrow to the right lists all the windows loaded in the current project.

3. With the *Window* tool active, do the following (Figure 2-3.1):

   a. Change the *Type Selector* to *Fixed:* **36″ x 48″**.

   b. Verify *Tag on Placement* is toggled off on the *Ribbon*.

   c. Note the **Sill Height** value in the *Properties Palette*.

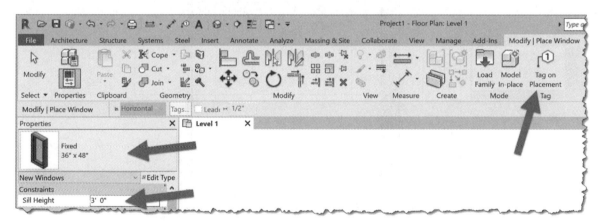

FIGURE 2-3.1 Ribbon and Options Bar: Window tool active

4. Move your cursor over a wall and place **two windows** as shown in **Figure 2-3.2**. *Notice that the position of the window changes depending on what side of the wall your cursor favors.*

FYI: *The window sill height is controlled by Properties Palette which you will study later in this book. For now, the default dimension was used.*

5. Adjust the **temporary dimensions** per the following:

    a. Dimensions per Figure 2-3.2.

    b. Use the witness line grip to adjust the witness line position.

    REMEMBER: *The selected item moves when temporary dimensions are adjusted. Pick the left window to set the 6'-0" dimension and the right for the 8'-0" dimension.*

6. Using the **Copy** command, in a way similar to copying the doors in the previous exercise, copy the two windows into each office as shown in Figure 2-3.3. (Do not worry about exact dimensions.)

7. **Save** your project *(Small Office.rvt).*

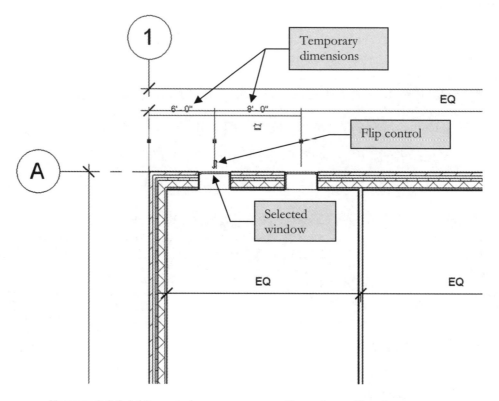

FIGURE 2-3.2 Adding windows – temporary dimensions still active

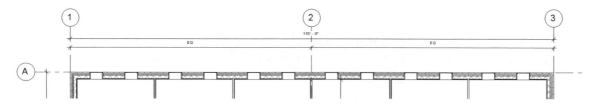

FIGURE 2-3.3 Windows added to north wall

The windows will not adjust with the grid lines and interior walls; it is possible to add dimensions and get this to work. If you tried to change the overall dimension from 110′ to 40′ again, Revit would let you know it needed to delete some windows before the change could be made. Like doors, windows need their host to exist.

The windows can all be adjusted via the temporary dimensions. The window selected is the largest width available in the project (based on the template file from which the project was started), but it is not a masonry dimension. However, additional window sizes can be added on the fly at any time. Additionally, you can create your own template file that has the doors, windows, walls, etc. that you typically need for the kind of design work you do.

In addition to the preloaded windows, several window styles are available via the family library loaded on your hard drive and the *Autodesk Web Library* (e.g., dbl-hung, casement, etc.). It is also possible to create just about any window design in the *Family Editor*.

Object Snap Symbols:

By now you should be well aware of the snaps that Revit suggests as you move your cursor about the drawing window.

If you hold your cursor still for a moment while a snap symbol is displayed, a tooltip will appear on the screen. However, when you become familiar with the snap symbols you can pick sooner (Figure 2-3.4).

The TAB key cycles through the available snaps near your cursor.

The keyboard shortcut turns off the other snaps for one pick. For example, if you type SE on the keyboard while in the Wall command, Revit will only look for an endpoint for the next pick.

Finally, typing SO (snaps off) turns all snaps off for one pick.

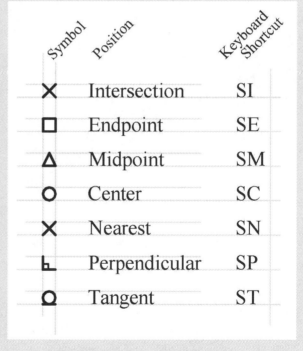

Symbol	Position	Keyboard Shortcut
✕	Intersection	SI
▢	Endpoint	SE
△	Midpoint	SM
○	Center	SC
✕	Nearest	SN
∟	Perpendicular	SP
Ω	Tangent	ST

FIGURE 2-3.4 Snap Reference Chart

Exercise 2-4:
Roof

You will now add a simple roof to your building.

    1.   Open **Small Office.rvt** created in Exercise 2-3.

The first thing you will do is take a quick look at a 3D view of your building and notice an adjustment that needs to be made to the exterior walls.

    2.   Click the **Default 3D View** icon on the *QAT*.

The 3D icon switches you to the default 3D view in the current project. Your view should look similar to Figure 2-4.1. Notice the exterior walls are not high enough, which is due to a previous decision to set the wall height to 9′-0″. Next you will change this, which can be done in the plan view or the current 3D view.

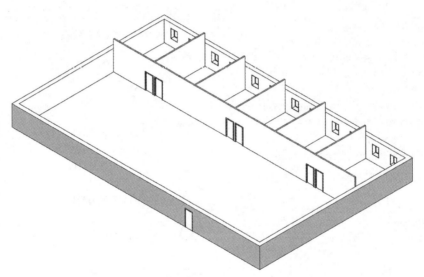

FIGURE 2-4.1 3D view of small office

    3.   In the 3D view, hover your cursor over one of the exterior walls to pre-highlight it, then (before clicking) press the **Tab** key to pre-highlight a "chain of walls" (i.e., all the exterior walls), and then click to select them.

Next you will access the properties of the selected walls so you can adjust the wall height. In Revit, most any design decisions that are made can be adjusted at any time.

4. Change the following in the *Properties Palette*:

    a. *Top Constraint:* **Up to level: Roof**

    b. *Top Offset:* **2'-0"**

    c. Click **Apply** (Figure 2-4.2B).

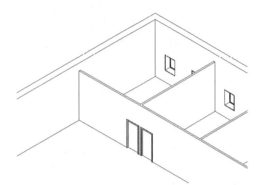

FIGURE 2-4.2A Exterior wall heights adjusted

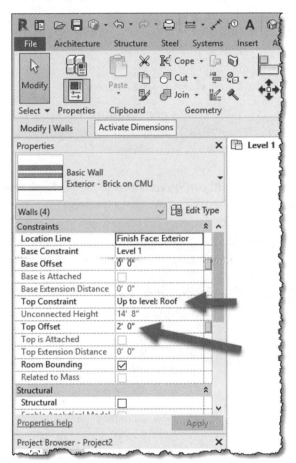

Setting the top of wall to be associated with a *Level* establishes a parametric relationship that causes the wall height to automatically adjust if the level datum is adjusted (e.g., from 12'-0" to 14'-0").

Plus, the *Top Offset* at 2'-0" creates a 2'-0" parapet, which will always be 2'-0" high no matter what the roof elevation is set to. There are instances when you would want the height to be fixed.

FIGURE 2-4.2B Selected wall properties

All of the settings related to the selected wall show up here; these are called instance parameters.

Sketching a Roof

Now that the exterior walls are the correct height, you will now add the roof. This building will have a flat roof located at the roof level.

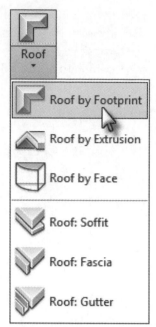

5. Double-click **Level 1** in the *Project Browser* to switch back to that view.

6. Click **Architecture → Build → Roof** (Figure 2-4.3).

The fly-out prompts you to choose the method you want to use to create the roof.

7. Click **Roof by Footprint**.

FYI: If you get a prompt to move the completed roof to the "Roof" level that means you started this exercise with the wrong template.

FIGURE 2-4.3 Roof tool

At this point you have entered *Sketch Mode* where the Revit model is grayed out so the perimeter you are about to sketch stands out.

Also notice the *Ribbon, Options Bar* and *Properties Palette* have temporarily been replaced with Sketch options relative to the roof (Figure 2-4.4).

8. Click **Extend to Core** on the *Options Bar*, and make sure **Defines Slope** is not checked (Figure 2-4.4).

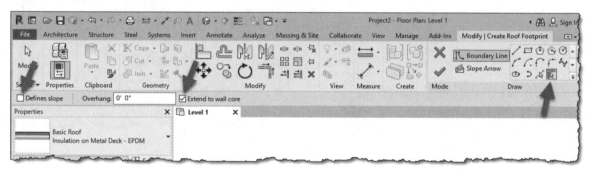

FIGURE 2-4.4 Roof sketch tools

9. Select the exterior walls:

    a. Hover your cursor over one of the exterior walls to pre-highlight the wall.

    b. Press **Tab** to select a "Chain of Walls" (i.e., all the exterior walls).

    c. **Click** to select the exterior walls, which will add the sketch lines.

At this point you should have four magenta sketch lines, one on each wall, which represent the perimeter of the roof you are creating. When sketching a roof footprint, you need to make sure that lines do not overlap and corners are cleaned up with the *Trim* command if required. Your sketch lines require no additional edits because of the way you added them (i.e., Pick walls and Tab select).

Notice for the roof sketch you need to adjust the level on which the roof will be created. By default, the top surface of the roof element will be parametrically aligned with the current level (i.e., Level 1 in this case). You will change this to the roof level.

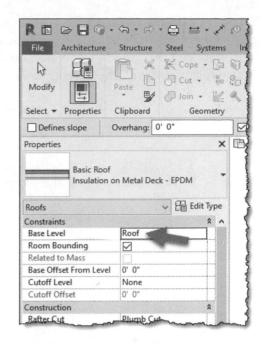

FIGURE 2-4.5 Roof instance properties

10. Set *Base Level* to **Roof** (Figure 2-4.5).

Now you are ready to finish the roof and exit sketch mode.

11. Click the **green check mark** on the *Ribbon*.

12. Click **Yes** to the join geometry prompt (Figure 2-4.6).

The join geometry option will make the line work look correct in sections. If you clicked "No," the wall and floor lines would just overlap each other and look messy.

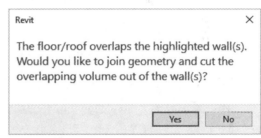

FIGURE 2-4.6 Join geometry prompt

The roof is now created and, in section, will extend through the finishes to the concrete block because "extend to core" was selected when the sketch lines were added. Also, because you used the "Pick Walls" option on the *Ribbon* (which was the default), the roof edge will move with the exterior walls.

13. To see the roof, click the **Default 3D View** icon.

14. To adjust the 3D view, press and hold the **Shift** key while pressing the **wheel button** and dragging the mouse around.

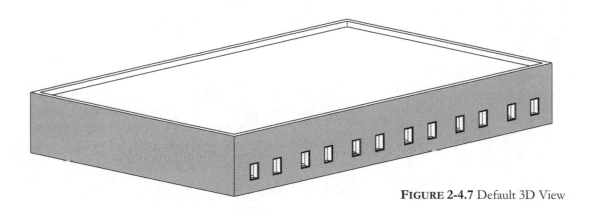

FIGURE 2-4.7 Default 3D View

15. Click the **X** on the {3D} drawing tab to close that view. This will close the 3D view but not the project or the Level 1 view.

> **REMEMBER:** *Clicking the **X** in the upper right of the application title bar will close Revit, but it will prompt you to save first if needed.*

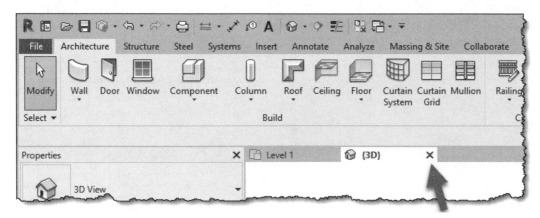

16. **Save** your project.

Exercise 2-5:
Annotation, Room Tags & Schedules

Adding text is very simple in Revit. In this exercise you will add a title below the floor plan. You will also place room tags.

Placing Text

1. Open **Small Office.rvt** created in Exercise 2-4.

2. Make sure your current view is **Level 1**. The word "Level 1" will be bold under the *Floor Plans* heading in your *Project Browser*. If Level 1 is not current, simply double-click on the Level 1 text in the *Project Browser*.

3. Select **Annotate → Text → Text** tool on the *Ribbon*.

Once again, notice the *Ribbon* has changed to display some options related to the active tool (Figure 2-5.1).

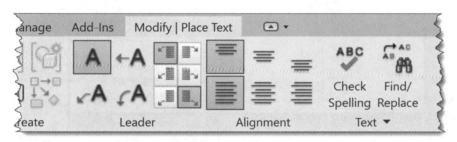

FIGURE 2-5.1 Ribbon with Text tool active

The *Type Selector* indicates the text style (which determines the font style, height and more); users can create additional text styles. From this *Contextual Tab*, on the Ribbon, your alignment (i.e., Left justified, Centered or Right justified) can also be set.

4. Set the *Ribbon* settings to match those shown above, and **Click** below the floor plan to place the text (Figure 2-5.2).

5. Type **OFFICE BUILDING – Option A**, then click somewhere in the plan view to finish the text (do not press *Enter*).

The text height, in the *Type Selector*, refers to the size of the text on a printed piece of paper. For example, if you print your plan you should be able to place a ruler on the text and read ¼" when the text is set to ¼" in the *Type Selector*.

Text size can be a complicated process in CAD programs; Revit makes it very simple. All you need to do is change the **view scale** for **Level 1** and Revit automatically adjusts the text and annotation to match that scale – so it always prints ¼" tall on the paper.

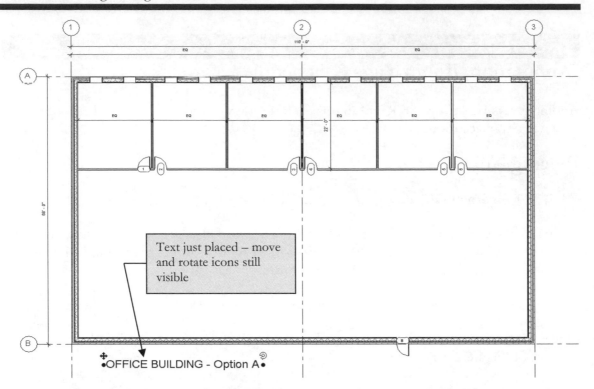

FIGURE 2-5.2 Placing text

You will not change the scale now, but it can be done via the *View Control Bar* (Figure 2-5.3). If you want to try changing it, just make sure it is set back to ⅛″ = 1′-0″ when done.

You should now notice that your text and even your door and window symbols are half the size they used to be when changing from ⅛″ to ¼″.

You should understand that this scale adjustment will only affect the current view (i.e., Level 1). If you switched to Level 2 (if you had one) you would notice it is still set to ⅛″=1′-0″. This is nice because you may, on occasion, want one plan at a larger scale to show more detail.

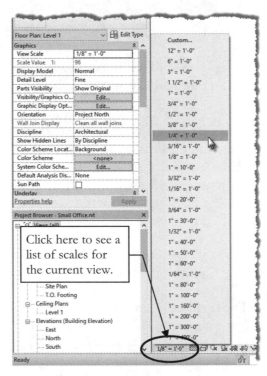

FIGURE 2-5.3 Set View Scale

Placing Room Tags

Placing *Room Tags* must be preceded by placing a *Room*. A *Room* element is used to define a space and hold information about a space (e.g., floor finish, area, department, etc.).

The *Room* feature searches for areas enclosed by walls; a valid area is pre-highlighted before you click to create it.

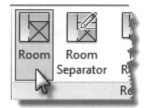

By default, Revit will automatically place a *Room Tag* at the cursor location when you click to add the *Room* element.

6. Select **Architecture → Room & Area → Room**.

7. Set the *Type Selector* to **Room Tag: Room Tag With Area** and make sure **Tag on placement** is selected on the *Ribbon*.

8. Click within each room in the order shown in Figure 2-5.4; watch for the dashed reference line to align the tags.

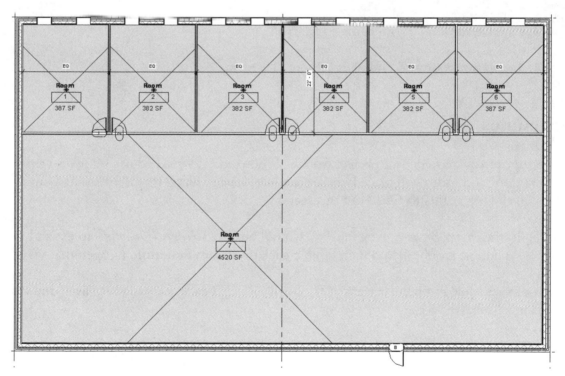

FIGURE 2-5.4 Placing rooms and room tags

While the *Room* tool is active, the placed rooms in the model are shaded light blue so you can see which spaces already have rooms placed. The large "X" is also part of the room. When the *Room* tool is not active, you can hover the cursor over the approximate location of the "X" until it pre-highlights, then you can click to select the room. With the *Room* object selected you can add information or delete it via the *Properties Palette*.

9. Click **Modify** to end the current tool.

Notice the rooms are not visible and the "X" is gone. Also notice the *Room Tag* selected shows the following information stored within the *Room* object: Name, Number and Area.

> *FYI: The area updates automatically when the walls move. Next you will change the room names.*

10. Click on the *Room Tag* for **room number 1** to select it.

When a *Room Tag* is selected the "dark blue" text is editable and the "lighter blue" text is not. An example of text that cannot be edited would be the actual text "Sheet Number" next to the sheet number on a sheet border.

11. Click on the room name text, type **OFFICE**, and then press **Enter** on the keyboard.

12. Change rooms 2-4 to also be named **OFFICE**.

13. Change the large room name to **LOBBY**.

14. Leave two rooms (5 and 6) as "Room" for now.

Schedules

The template you started your project from had room and door schedules set up. So from the first door and room you placed, these schedules started filling themselves out! You will take a quick look at this to finish out this section.

15. In the *Project Browser*, click the "**+**" symbol next to *Schedules/Quantities* to expand that section (if required) and then double-click on **Room Schedule** to open that view.

The room schedule is a tabular view of the Revit model. This information is "live" and can be changed (Figure 2-5.5).

Level 1 | Room Schedule

<Room Schedule>

A	B	C	D	E	F	G	H
Room Number	Room Name	Floor Finish	Base Finish	Wall Finish	Ceiling Finish	Ceiling Height	Comments
1	OFFICE						
2	OFFICE						
3	OFFICE						
4	OFFICE						
5	Room						
6	Room						
7	LOBBY						

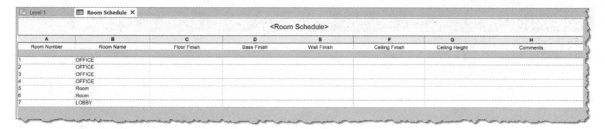

FIGURE 2-5.5 Room Schedule

Next you will change the two rooms named "room," and see that the floor plan is automatically updated!

16. Click in the *Room Name* column for room number 5 and change the text to read **MEN'S TOILET RM.**

17. Click in the *Room Name* column for room number 6 and change the text to read **WOMEN'S TOILET RM.**

Room Number	Room Name
1	OFFICE
2	OFFICE
3	OFFICE
4	OFFICE
5	MEN'S TOILET RM.
6	WOMEN'S TOILET RM
7	LOBBY

18. Click the lower "X" in the upper right of the drawing window to close the room schedule view.

19. Switch to *Level 1* (if required) and **zoom in** on rooms 5 and 6 (Figure 2-5.6).

Notice that the room names have been updated because the two views (floor plan and schedule) are listing information from the same "parameter value" in the project database.

FIGURE 2-5.6 Room names updated

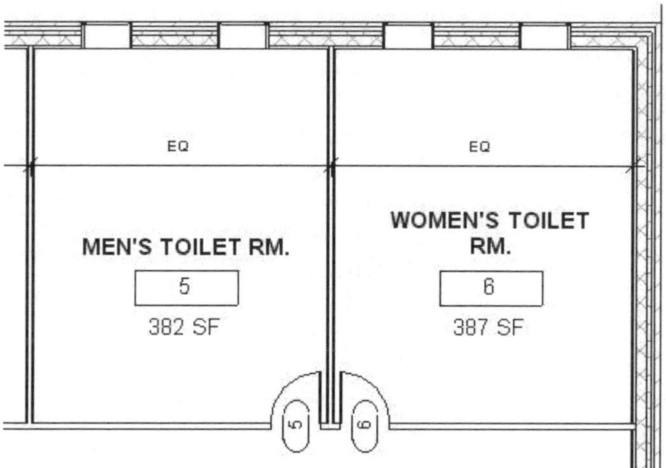

TIP: You can zoom in and out in a schedule by holding down the **CTRL** key and **scrolling the wheel** on your mouse. Use **CTRL + 0** to reset the view.

20. Open and Close the door schedule to view its current status.

21. **Save** your project.

Exercise 2-6:
Printing

The last thing you need to know to round off your basic knowledge of Revit is how to print the current view.

Printing the current view:

1. In *Level 1* view, right-click anywhere and select **Zoom to Fit**.

2. Select **File Tab → Print**.

3. Adjust your settings to match those shown in **Figure 2-6.1**.

   - Select a printer from the list that you have access to.
   - Set *Print Range* to **Visible portion of current window**.

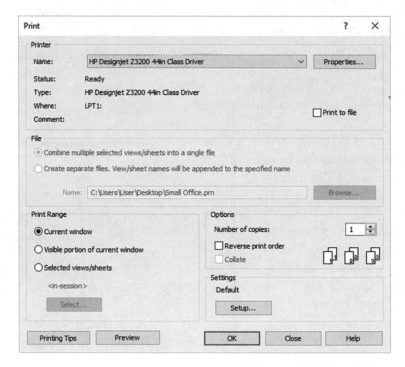

FIGURE 2-6.1 Print dialog

4. Click on the **Setup** button to adjust additional print settings.

5. Adjust your settings to match those shown in **Figure 2-6.2**.

   - *Set Zoom to:* **Fit to page**

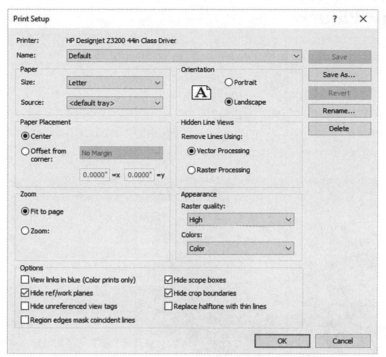

FIGURE 2-6.2 Print Setup dialog

6. Click **OK** to close the *Print Setup* dialog and return to *Print*.

7. Click the **Preview** button in the lower left corner. This will save paper and time by verifying the drawing will be correctly positioned on the page (Figure 2-6.3).

8. Click the **Print** button at the top of the preview window.

9. Click **OK** to print to the selected printer.

FYI: Notice you do not have the option to set the scale (i.e., ⅛″ = 1′-0″). If you recall from our previous exercise, the scale is set in the properties for each view. If you want a quick half-scale print you can change the zoom factor to 50%. You could also select "Fit to page" to get the largest image possible but not to scale.

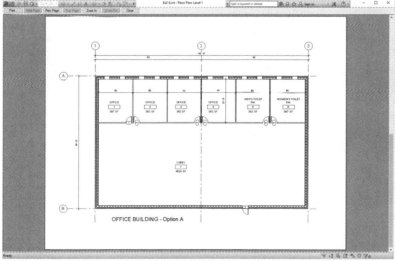

FIGURE 2-6.3 Print Preview

Printer versus Plotter?

Revit can print to any printer or plotter installed on your computer.

A <u>Printer</u> is an output device that uses smaller paper (e.g., 8½″x11″ or 11″x17″). A <u>Plotter</u> is an output device that uses larger paper; plotters typically have one or more rolls of paper ranging in size from 18″ wide to 36″ wide. A roll feed plotter has a built-in cutter that can – for example – cut paper from a 36″ wide roll to make a 24″x36″ sheet.

Plotter with three paper rolls

Color **printer** / copier

Self-Exam:

The following questions can be used as a way to check your knowledge of this lesson. The answers can be found at the bottom of this page.

1. The *Measure* tool is used to dimension drawings. (T/F)

2. Revit will automatically trim the wall lines when you place a door. (T/F)

3. Snap will help you to draw faster and more accurately. (T/F)

4. A 6′-8″ door is a standard door height in _____ construction.

5. While using the wall tool, the height can be quickly adjusted on the

   _____ *Bar*.

Review Questions:

The following questions may be assigned by your instructor as a way to assess your knowledge of this section. Your instructor has the answers to the review questions.

1. The *View Scale* for a view is set by clicking the scale listed on the *View Control Bar*. (T/F)

2. Dimensions are placed with only two clicks of the mouse. (T/F)

3. The relative size of text in a drawing is controlled by the *View Scale*. (T/F)

4. You can quickly switch to a different view by double-clicking on a View tab. (T/F)

5. You cannot select which side of the wall a window is offset to. (T/F)

6. The _____ key cycles through the available snaps near your cursor.

7. The _____ tool can be used to list the distance between two walls without drawing a dimension.

8. While in the *Door* tool you can change the door style and size via the

   _____ within the *Properties Palette*.

Notes:

Download Autodesk Content

Be sure to download the additional Autodesk content pack found via the Get Autodesk Content command on the Insert tab. Starting with Revit 2021, some content is no longer installed with the software. Some of the additional content is required to successfully complete the exercises in the book. **Important Step!**

Lesson 3
FLOOR PLAN (First Floor):

In this lesson you will draw the first floor plan of an office building. The office building will be further developed in subsequent chapters. It is recommended that you spend adequate time on this lesson as later lessons build on this one.

Exercise 3-1:
Project Overview

A program statement is created in the pre-design phase of a project. Working with the client (or user group), the architect gathers as much information as possible about the project before starting to design.

The information gathered includes:
- Rooms: What rooms are required?
- Size: How big do the rooms need to be? (E.g., toilets for a convention center are much bigger than for a dentist's office.)
- Adjacencies: This room needs to be next to that room. (E.g., the public toilets need to be accessible from the public lobby.)

With the project statement in hand, the architect can begin the design process. Although modifications may (and will) need to be made to the program statement, it is used as a goal to meet the client's needs.

You will not have a program statement, per se, with this project. However, the same information will be provided via step-by-step instructions in this book.

Project Overview

You will model a three-story office building located in a rural setting. Just to the North of the building site is a medium-sized lake. For the sake of simplicity, the property is virtually flat.

The main entry and parking is from the south side of the building. You enter the building into a three-story atrium. Levels 2 and 3 have guard railings that look down into Level 1 in the atrium. The atrium is enclosed on three sides by full height curtain walls (glass walls). See the image on the front cover.

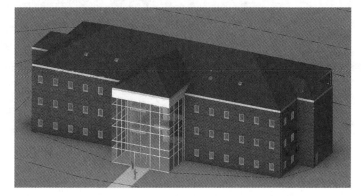

This building is not meant to meet any particular building code. It is strictly a tool to learn how to use Revit. Having said that, however, there are several general comments as to how codes may impact a particular part of the design.

The floor plans are mostly open office areas with a few smaller rooms for toilets, private offices, work and break rooms, etc. These areas have several "punched" window openings on the exterior walls (punched as opposed to ribbon windows).

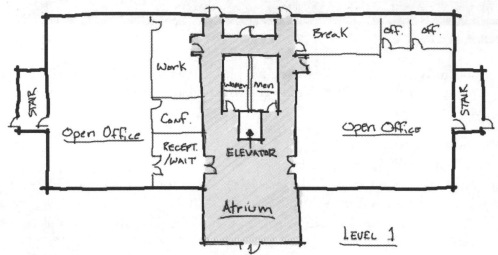

FIGURE 3-1.1 Level 1 floor plan sketch

FIGURE 3-1.2 South elevation sketch

Exercise 3-2:
Exterior Walls

You will begin the first floor plan by drawing the exterior walls. Like many projects, early on you might not be certain what the exterior walls are going to be. So, we will start out using the generic wall styles. Then we will change them to a custom wall style (that you will create) once we have decided what the wall construction is.

Adjust Wall Settings

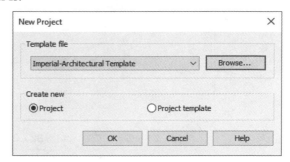

1. Start a new project using the **Imperial-Architectural Template** (aka default.rte). *Warning:* Be sure to select the correct template as this is the basis for the remainder of the book.

The previous chapter started with a more complete template. This chapter starts from the default template, so you have the opportunity to learn how to create things such as the room finish schedule, so you better understand how Revit works.

2. Select the **Wall** tool from the *Ribbon* and then make the following changes to the wall options (Figure 3-2.1):
   * Wall style: *Basic Wall:* **Generic – 12″**
   * *Height:* **Unconnected**
   * *Height:* **36′ 0″**
   * *Location Line:* **Finish Face: Exterior**
   * *Chain:* **Checked**

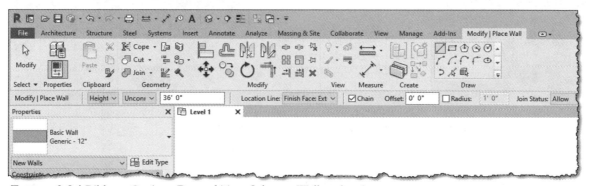

FIGURE 3-2.1 Ribbon, Options Bar and Type Selector: Wall tool active

Draw the Exterior Walls

3. Draw the walls shown in Figure 3-2.2. Make sure your dimensions are correct. Use the *Measure* tool to verify your dimensions. Do not add the dimensions.

   NOTE: If you draw in a clockwise fashion, your walls will have the exterior side of the wall correctly positioned. You can also use the spacebar to toggle which side the exterior face is on.

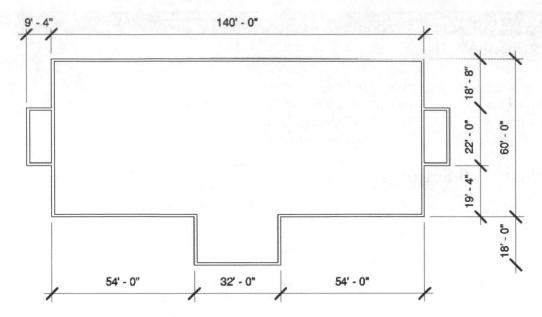

FIGURE 3-2.2 Exterior walls

Create a Custom Wall Style

Revit provides several predefined wall styles, from metal studs with gypsum board to concrete block and brick cavity walls. However, you will occasionally need a wall style that has not yet been predefined by Revit. You will study this feature next.

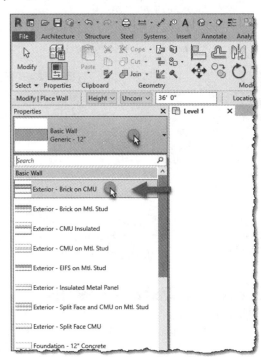

First, you will take a quick look at a more complex wall type that Revit provides so you can see how they are set up.

4. With the *Wall* tool selected, pick the wall type *Basic Wall:* **Exterior – Brick on CMU** from the *Type Selector* drop-down list. (See image at right.)

5. Click the **Edit Type** button on the Properties Palette (Fig. 3-2.3).

6. You should be in the *Type Properties* dialog box. Click the **Edit** button next to the *Structure* parameter (Figure 3-2.4).

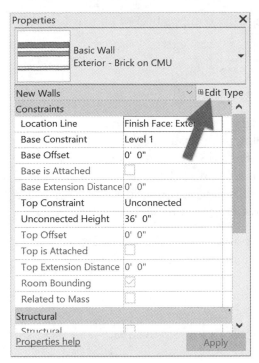

FIGURE 3-2.3 Properties Palette

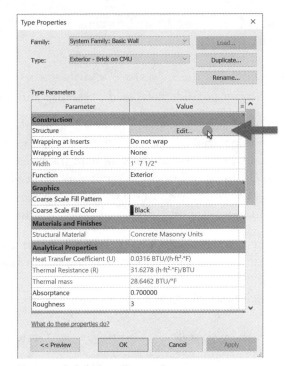

FIGURE 3-2.4 Type Properties

7. Finally, you are in the *Edit Assembly* dialog box. This is where you can modify existing wall types or create new ones. Click **<<Preview** to display a preview of the selected wall type (Figure 3-2.5).

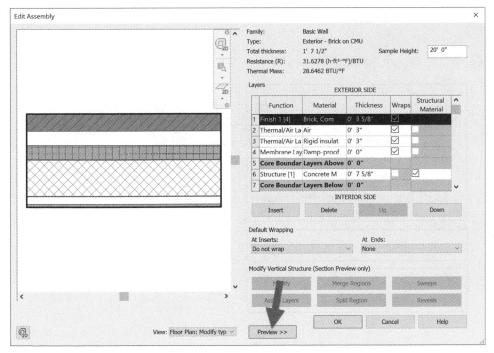

FIGURE 3-2.5 Edit Assembly

Here, the *Edit Assembly* dialog box allows you to change the composition of an existing wall or to create a new wall.

Things to notice in the *Edit Assembly* dialog box (Figure 3-2.5):
- The exterior side is labeled at the top and interior side at the bottom.

- You will see horizontal rows (i.e., Core Boundary) identifying the core material. The core material can be used to place walls and dimension walls. For example, the *Wall* tool will let you draw a wall with the interior or exterior core face as the reference line. On an interior wall you would typically dimension to the face of CMU rather than to the finished face of gypsum board. This is to work out coursing and give the contractor the information needed for the part of the wall he will build first.

- Each row is called a layer. By clicking on a layer and picking the **Up** or **Down** buttons, you can reposition materials, or layer, within the wall assembly.

8. Click **Cancel** in each open dialog box to close them.

9. Set the wall style back to *Basic Wall:* **Generic – 12″** in the *Type Selector*.

10. Click the **Edit Type** button again on the *Properties Palette*.

11. Click **Duplicate**.

12. Enter **Brick & CMU cavity wall** for the new wall type name, and then click **OK** (Figure 3-2.6).

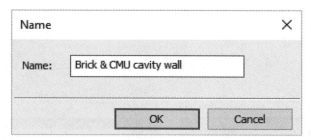

FIGURE 3-2.6 New wall type name

13. Click the **Edit** button next to the *Structure* parameter.

Using the **Insert** button and the **Up** and **Down** buttons, add the *layers* to your new wall style as shown below in **Figure 3-2.7**.

Function	Material	Thickness
Finish 1 [4]	Brick, Common	4″
Thermal/Air Layer [3]	Air	2″
Thermal/Air Layer [3]	Rigid insulation	2″
Membrane Layer	Damp-proofing	0″
Core Boundary	*Layers above wrap*	*0″*
Structure [1]	Concrete Masonry Units	8″
Core Boundary	*Layers below wrap*	*0″*
Substrate [2]	Metal Furring	2 1/2″
Finish 2 [5]	Gypsum Wall Board	5/8″

FIGURE 3-2.7 New wall layers

Masonry is typically drawn nominally in plans and smaller scaled details. This helps to figure out coursing for both drawing and dimensioning. For example, 8″ concrete block is actually 7⅝″.

Also, notice that the CMU, Rigid Insulation, Air Space and Brick add up to 16″ in thickness. This portion of the wall would sit on a 16″ concrete block (CMU) foundation wall directly below.

14. Your dialog should look like **Figure 3-2.8**. Click **OK** to close all dialog boxes.

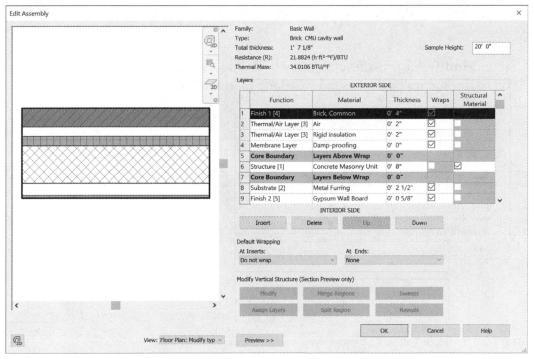

FIGURE 3-2.8 Edit Assembly for new wall type

The next step is to change the wall type for the walls previously drawn.

15. Select the **Modify** button on the *Ribbon*; this allows you to select elements in your drawing.

16. **Zoom out** so you can see the entire plan. Dragging your mouse from one corner to the other, make a window over the plan to select all the walls.

17. With the walls selected, pick *Basic Wall:* **Brick & CMU cavity wall** from the *Type Selector* drop down.

TIP: If, after selecting all the walls, the Type Selector is not active and does not show any wall types, you probably have some other elements selected such as text or dimensions. Try to find those elements and delete them (except the elevation tags). You can also click on the Filter button (located on the Ribbon when objects are selected) and uncheck the types of elements to exclude from the current selection.

You should notice the wall thickness change, but the wall cavity lines and hatch are not showing yet. This is controlled by the *Detail Level* option for each view.

18. Click on **Detail Level** icon in the lower-left corner of the *Drawing Window*, on the *View Control Bar*.

Detail Level; Set to Medium

19. Select **Medium**.

You should now see the brick and CMU thicknesses with hatching. If you did not pay attention when drawing the walls originally, some of your walls may show the brick to the inside of the building.

20. Select **Modify** (or press **Esc**); select a wall. You will see a symbol appear that allows you to flip the wall orientation by clicking on that symbol (Figure 3-2.9).

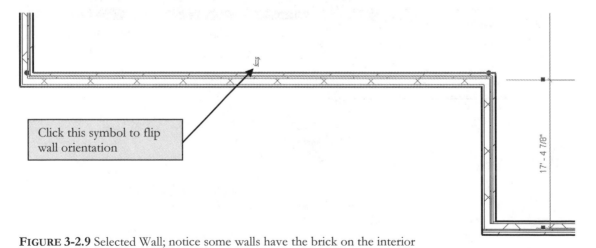

Click this symbol to flip wall orientation

FIGURE 3-2.9 Selected Wall; notice some walls have the brick on the interior

21. Whether you need to adjust walls or not, click on the flip symbol to experiment with its operation.

    TIP: The Flip symbol is always on the exterior side (or what Revit thinks is the exterior side) of the wall.

22. If some walls do need to be adjusted so the brick is to the exterior, do it now. You will probably have to select the wall(s) and use the *Move* tool to reposition the walls to match the required dimensions.

    TIP: If you set the "location line," via the Properties Palette, to "Wall Centerline," the wall will not move and mess up the overall dimensions when flipping it. You will want to set the "location line" back to Finish Face: Exterior when done.

23. **Save** your project as ex3-2.rvt.

TIP: You can use the MOVE tool Move on the Ribbon (when a wall is selected) to accurately move walls.

Follow these steps to move an object:
- *Select the wall*
- *Click the Move tool*
- *Pick any point on the wall*
- *Start your cursor in the desired direction (don't click)*
- *Start typing the distance you want to move the wall and press Enter.*

Finally, you will change the three walls at the atrium to be curtain walls (full glass). This will let lots of light into the atrium and better identify the main entry of the building.

24. Drag a selection window (from left to right) to select the three walls around the atrium, or hold the Ctrl key on the keyboard and select them.

25. With the walls selected, select *Curtain Wall:* **Curtain Wall 1** from the *Type Selector* drop-down.

Your atrium is now surrounded by curtain walls (Figure 3-2.10). In a later lesson we will add horizontal and vertical mullions to the curtain wall.

You can see your progress nicely with a 3D view. Click the **3D View** button on the *QAT.* Notice that Revit shows the curtain wall as transparent because it knows the curtain wall is glass. The other walls are shaded on the exterior side due to the brick pattern that is applied.

26. Save your project as **ex3-2.rvt**.

Revit automatically sets the hatch intensity and line weights.

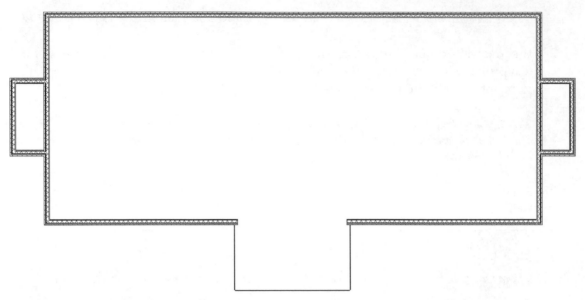

FIGURE 3-2.10 Completed exercise

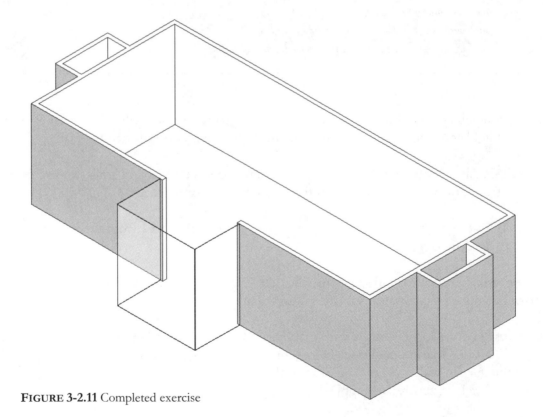

FIGURE 3-2.11 Completed exercise

Exercise 3-3:
Interior Walls

In this lesson you will draw the interior walls for the first floor.

Adjust Wall Settings

1. Select **Architecture → Build → Wall** from the *Ribbon*.

2. Make the following changes to the wall options on the *Ribbon, Options Bar* and *Type Selector* (Figure 3-2.1):
    *   Wall style: *Basic Wall*: **Interior – 4 7/8″ partition (1-hr)**
    *   *Height:*   **Level 2**
    *   *Location Line:*   **Wall Centerline**

Draw the Interior Walls

3. Draw a "vertical" wall approximately as shown in Figure 3-3.1. You will adjust its exact position in step #4.

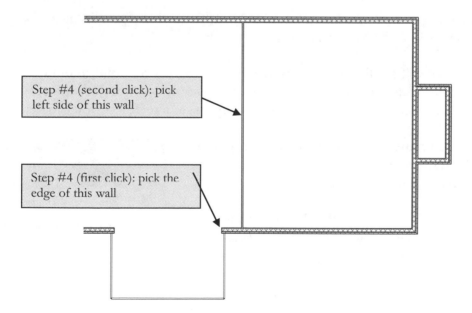

Step #4 (second click): pick left side of this wall

Step #4 (first click): pick the edge of this wall

FIGURE 3-3.1 First interior wall

4. Select **Modify → Modify → Align** tool to align it with the edge of the exterior wall in the atrium (Figure 3-3.1). When you are done, the wall should look like Figure 3-3.2.

5. Create the same wall for the West side of the atrium repeating the above steps.

Modify an Existing Wall Type

Next you will add some additional interior walls. You will be drawing 8″ CMU walls. Revit does have an 8″ Masonry wall type available in the default template file that you started your project from. However, the thickness for this wall type is 7⅝″, which is the actual size of a block. Floor plans are usually drawn nominally (i.e., 8″) not actual (7⅝″). This is done so you can figure out coursing so minimal cutting is required. Therefore, rather than creating a new wall type you can simply modify the existing wall type.

Note: some may disagree with this approach. However, these changes should still be made to this project so you learn more about how walls work and so all the tasks in this book work as expected or intended.

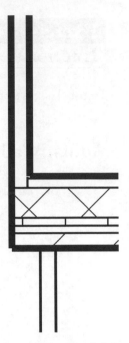

FIGURE 3-3.2
First interior wall

6. In the Wall command, select the wall type *Basic Wall:* **Generic – 8″ Masonry**.

7. Select the **Edit Type** button on the *Properties Palette* to view the wall's *Type Parameters*, and then click **Edit** next to the *Structure* parameter.

8. Change the masonry thickness from 7⅝″ to **8″** in the *Edit Assembly* dialog box, and then select **OK** to close each dialog.

Occasionally Revit will not list dimensions, relative to the walls you want to draw new walls from, while in the create wall mode. One way to deal with this is to draw temporary *Detail Lines* to use as a reference. After using the temporary line as a reference you can delete it.

9. Select **Annotate → Detail → Detail Line** from the *Ribbon*; the line type does not really matter, but select a continuous one via the *Type Selector*.

10. Draw the "vertical" line, **35'-0"** long, as shown in Figure 3-3.3; be sure to snap to the *Midpoint* of the atrium wall as your first point.

Next you will draw an elevator shaft, centered on the atrium and 35'-0″ back (thus the temp. line).

The <u>inside</u> dimensions of the elevator are **7'-4″ x 6'-10″**. Because you know the inside dimension you will want to adjust the location line to match the known info.

11. Use the *Wall* tool to draw *Basic Wall:* **Generic – 8″ Masonry**.

12. Set the *Location Line* to **Finish Face: Interior**.

13. Draw the elevator shaft. Make sure the location line is to the <u>inside</u> so your shaft is the correct dimension. Draw the shaft anywhere in the *Drawing Window*; you will adjust the exact position next.

   FYI: *The inside dimensions are listed above.*

14. Select the 4 walls that represent the elevator shaft, and then pick the **Move** tool.

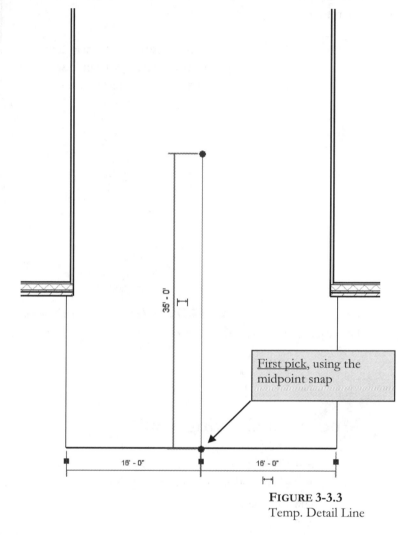

FIGURE 3-3.3
Temp. Detail Line

TIP: *Concrete blocks come in various widths, and most are 16″ long and 8″ high. When drawing plans there is a simple rule to keep in mind to make sure you are designing walls to coursing. This applies to wall lengths and openings within CMU walls.*

Dimension rules for CMU coursing in floor plans:
- *e′-0″ or e′8″*   *where e is any even number (e.g., 6′-0″ or 24′-8″)*
- *o′4″*   *where o is any odd number (e.g., 5′-4″)*

15. Snap to the *Midpoint* of the shaft as your first point, and then snap to the *Endpoint* of your temporary detail line (Figure 3-3.4). You should zoom in to verify your snaps. Do not draw the dimensions; they are for reference only.

The elevator shaft is now perfectly centered in the atrium and exactly 35′-0″ back from the South curtain wall.

16. At this point you can **delete** the temporary line. Select the line and then right-click and select delete, or press the Delete key on the keyboard.

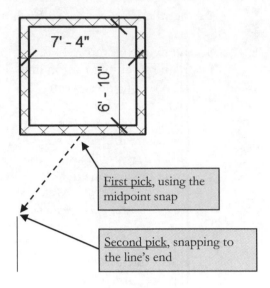

First pick, using the midpoint snap

Second pick, snapping to the line's end

FIGURE 3-3.4 Move elevator shaft into place

FYI: *When a wall is selected, you can see that wall's properties via the* Properties Palette *(Type "PP" to open the Palette if it is not visible). Click one of the elevator shaft walls and verify that it is 36′-0″ tall.*

Modify an Existing Wall

Next we want to change the portion of wall between the building and the East and West stair shafts. To do this you will need to split the current wall, trim the corners and then draw an 8″ masonry wall.

17. **Zoom** in on the West stair shaft and select the **Split Element** tool (*Modify* tab on the *Ribbon*).

18. Pick somewhere in the middle of the wall (Figure 3-3.5).

19. Select **Modify → Modify → Trim** to trim the corners so the exterior wall only occurs at exterior conditions (Figure 3-3.6).

TIP: Select the portion of wall you wish to retain.

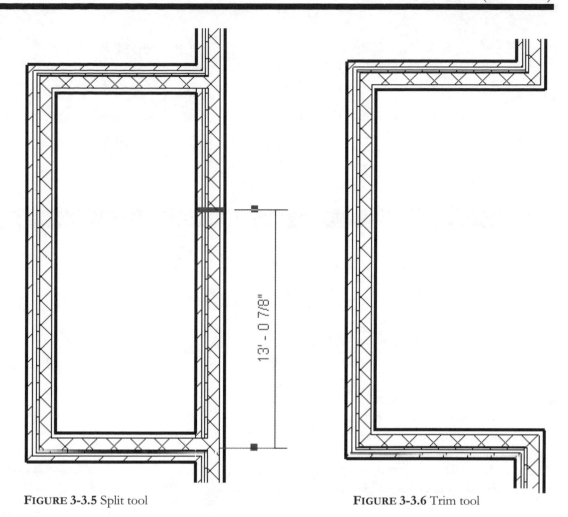

FIGURE 3-3.5 Split tool   **FIGURE 3-3.6** Trim tool

Additional Custom Wall Types

We decide that the stair shafts are mostly utilitarian and do not require gypsum board on the walls. In the next steps you will create a new exterior wall type just like the one previously created less the gypsum board and metal studs. Also, you will create a custom wall type to close the open side we created in the previous steps. This wall type will have gypsum board and metal studs on one side.

20. Using wall type *Basic Wall: Brick & CMU cavity wall* as a starting point, create a new wall type named **Brick & CMU cavity wall (no GWB)**. Remove the gypsum board and metal studs and save the new wall type. *(Remember to click Duplicate.)*

21. Change the **three exterior walls** around the west stair shaft to the new wall type created in the previous step.

22. Using wall type _Basic Wall: Brick & CMU cavity wall_ as a starting point, create a new wall type named **8″ Masonry with GWB 1S**. Remove the brick, air space, rigid insulation and membrane layer and save the new wall type (Figure 3-3.7).

> **FYI:** _It will be useful to come up with a standard naming system for your custom wall types. If the names get too long they are hard to read. The example above has:_
> - _GWB = Gypsum Wall Board (and would imply studs)_
> - _1S = finish only occurs on one side of the wall._

Function	Material	Thickness
_n/a_	_n/a_	_n/a_
Core Boundary	_Layers above wrap_	_0″_
Structure [1]	Concrete Masonry Units	8″
Core Boundary	_Layers below wrap_	_0″_
Substrate [2]	Metal Furring	2½″
Finish 2 [5]	Gypsum Wall Board	⅝″

FIGURE 3-3.7 New wall layers

23. Draw a wall so the gypsum finish continues on the office side, using the _Align_ tool if necessary (Figure 3-3.8).
    Use the Measure tool to make sure the stair shaft is the correct size; don't draw the dimensions.

Next you will use the _Mirror_ tool to update the east stair, but first you will draw a _Reference Plane_ to use as the _Axis of Reflection_ (more on this later while using the Mirror tool).

24. Select **Architecture → Work Plane → Ref Plane** from the _Ribbon_.

25. Draw a Reference Plane snapped to the vertical, centered on the South atrium wall (See Figure 3-3.9).

26. Erase the four walls of the East stair shaft; this will include the main east wall of the office building (Figure 3-3.9).

27. Select the six walls at the West stair (Figure 3-3.9).

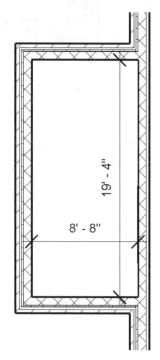

FIGURE 3-3.8 Revised west stair

> **TIP:** _Make sure the count is correct on the Status Bar._

28. Select the **Mirror - Draw Axis** tool (on the *Ribbon*) and then select the *Reference Plane* (Figure 3-3.9).

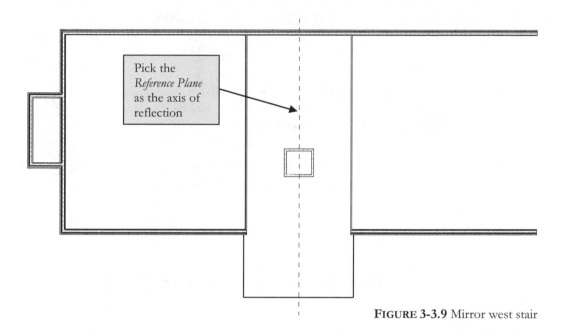

Pick the
Reference Plane
as the axis of
reflection

FIGURE 3-3.9 Mirror west stair

29. Use the **Measure** tool to verify the overall length of the building is 140'-0". Adjust as necessary (see Fig. 3-2.2).

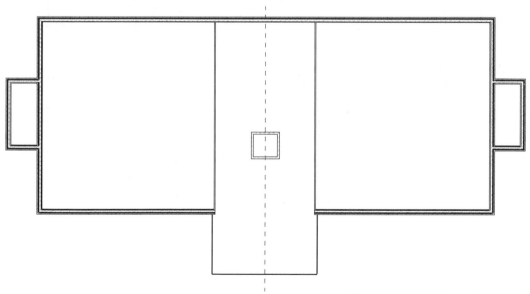

FIGURE 3-3.10 Mirrored west stair

Finally, you will draw a few more interior walls to complete the first floor plan. **Do not delete the *Reference Plane*** (note: the reference line may not show up in every image in this text).

30. Using the *Wall* tool, set the wall type to <u>*Basic Wall:*</u> **Interior – 4 7/8″ partition (1-hr)**.

31. Draw the additional walls shown in Figure 3-3.11. Make sure to position the walls per the dimensions shown. Use the *Measure* tool to verify accuracy. Also, modify the *Location Line* as required.

DRAWING TIPS: *Copy the existing atrium wall 6′-4⅞″ over (6′-0″ plus one wall thickness), draw a wall from the midpoint of the elevator shaft with centerline reference (Location Line), and use the Trim and Mirror tools. Do not draw the dimensions.* ***SAVE YOUR PROJECT as ex3-3.rvt.***

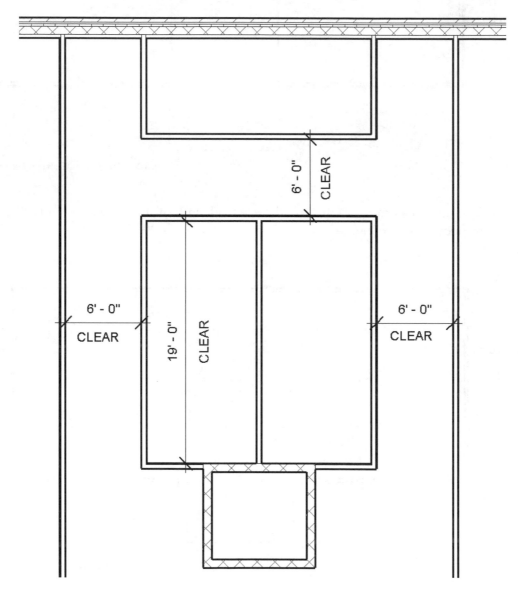

FIGURE 3-3.11 Remaining interior walls

Exercise 3-4:
Elevator

This lesson will show you how to insert an elevator into your elevator shaft.

Insert Elevator

Revit provides many *Families*, which are predefined elements ready to insert into your project. However, many elements are not readily available, like elevators for example. You will get an elevator family online in this exercise. The online library is where you will acquire an elevator family for use in your project.

1. Open project ex3-3.rvt and *Save As* **ex3-4.rvt**.

You will have to download the elevator from the web. *Of course you will need to be connected to the Internet.*

2. Click on the **Insert** tab on the *Ribbon*.

3. Select **Load from Library**
    → **Load Family** from the *Ribbon*.

Load Family

4. Load **Elevator-Electric.rfa** from the downloaded 'Required Families' folder; see the inside front cover for information about downloading the provided electronic content for this textbook.

5. In the *Project Browser*, click the plus next to **Families** to expand the list (Figure 3-4.1).

6. Expand the **Specialty Equipment** list, and then **Elevator-Electric**.

As you can see, four elevator types were loaded into your project. Similar to wall types, you can add one of these types as-is, or you can modify or create a new type. Next, you will add information in the *Type Properties* dialog to better document the elevator specified.

7. Right-click on the elevator type **2000 lbs**, and then select **Type Properties** from the pop-up menu.

You will now see a listing of the type properties for the selected elevator type.

8. Click the **Preview** button (if necessary) to see the

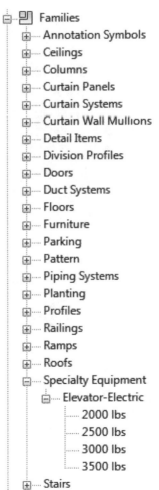

FIGURE **3-4.1** Elevator family:
Project Browser

graphical review of the elevator type. Set the *View* to 3D View: View 1.

9. Add the following information (Figure 3-4.2):

    *   *Model:*   **MadeUp 8864**
    *   *Manufacturer:*   **ThyssenKrupp Elevator**
    *   *URL:*   **www.thyssenelevator.com**

This data just entered represents the **I** in **BIM** (Building <u>Information</u> Modeling).

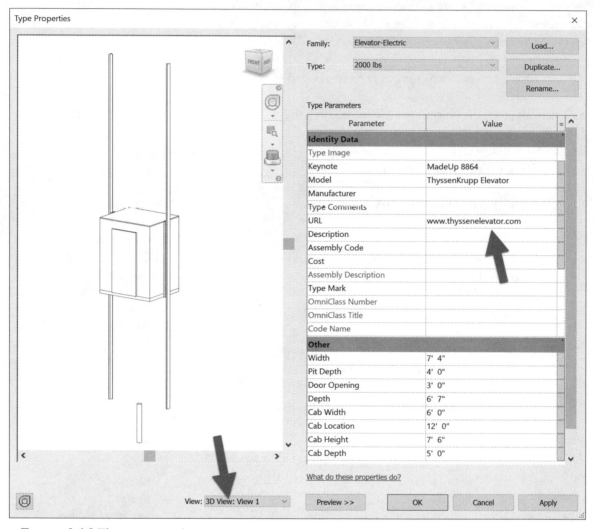

FIGURE 3-4.2 Elevator properties

10. Click **OK** to close the open dialog box.

11. Drag the **2000 lbs** elevator type from the *Project Browser* into the first floor plan.

    FYI: You can also use the Component tool from the Architecture tab to place the elevator.

The elevator type will be attached to your cursor, ready for insertion.

12. Move your cursor within the elevator shaft and adjust it until the elevator "snaps" in place, then click (Figure 3-4.10).

13. Press **Esc** twice to tell Revit you are finished placing elevators.

Now you have to add an elevator door in the shaft walls at each level; this is similar to a regular door in a wall.

14. Similar to the steps previously covered, load the **Elevator Door – Center** family from the provided online content.

15. Drag the **36" x 84"** elevator door type from the project browser into the first floor plan (Figure 3-4.3).

16. Place the elevator door at the center of the wall, aligned with the elevator door on the cab (Figure 3-4.4).

    TIP: If the door is inserted on the wrong side of the wall, select the door and click the Control Arrows to flip it within the wall.

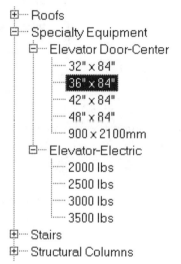

FIGURE 3-4.3 Elevator doors in Project Browser

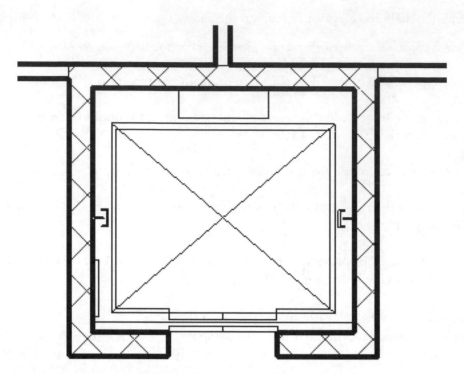

FIGURE 3-4.4 Elevator door added to plan

Notice when the elevator is selected, the flip icon (control arrow) is displayed. Similar to the doors and walls, you can click this icon to flip the orientation of the elevator within the shaft.

17. Save your project as **ex3-4.rvt**.

Exercise 3-5:
Doors and Windows

This lesson will take a closer look at inserting doors and windows.

Insert Doors

Revit has done an excellent job providing several different door families. This makes sense seeing as doors are an important part of an architectural project. Some of the provided families include bi-fold, double, pocket, sectional (garage), and vertical rolling, to name a few. In addition to the families found on your local hard drive, many more are available via other internet sites (some free, some not).

The default template you started with only provides the **Sgl Flush** (Single Flush) family in the *Doors* category. If you want to insert other styles you will need to load them from the library. The reason for this step is that, when you load a family, Revit actually copies the data into your project file. If every possible family was loaded into your project at the beginning, not only would it be hard to find what you want in a large list of doors, but also the files would be several megabytes in size before you even drew the first wall.

You will begin this section by loading a few additional families into your project.

1. Open project ex3-4.rvt and **Save-As ex3-5.rvt**.

2. From the Insert tab, select the **Load Family** button on the *Ribbon* (Figure 3-5.1).

3. Browse through the **English-Imperial** library folder for a moment.

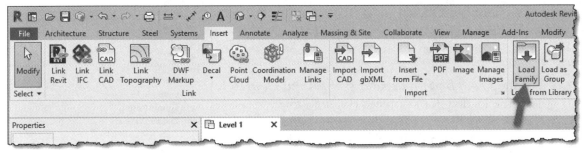

FIGURE 3-5.1 Load Family

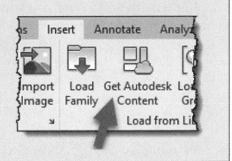

Download Autodesk Content
Be sure to download the additional Autodesk content pack found via the Get Autodesk Content command on the Insert tab. Starting with Revit 2021, some content is no longer installed with the software. Some of the additional content is required to successfully complete the exercises in the book. **Important Step!**

Each file represents a *Family*; next you will load four door *Families* into your project.

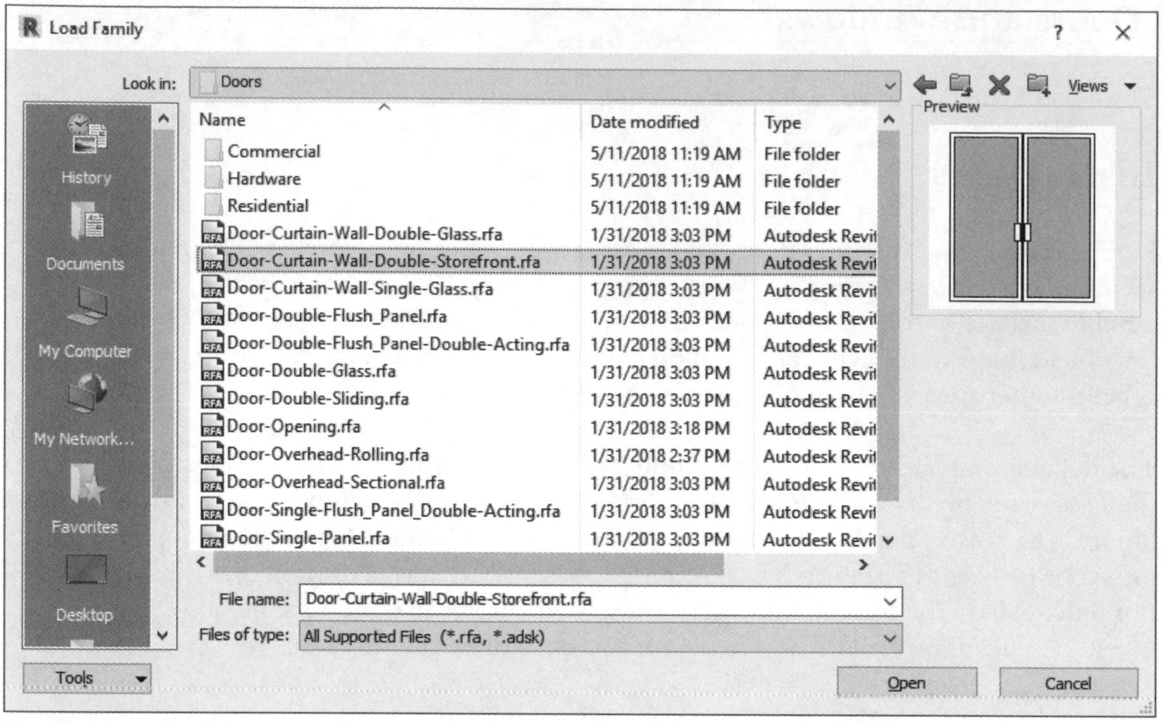

FIGURE 3-5.2 Door families on hard drive

4. Open the **Doors** folder and then select **Door-Curtain-Wall-Double-Storefront.rfa**, and then click **Open** (Figure 3-5.2).

5. Repeat steps 2 – 4 to load the following provided (online) door families:

    a. **Double-Glass 1**
    b. **Sidelights 1**
    c. **Single-Glass 1**

6. In the *Project Browser*, expand *Families* and *Doors* to see the loaded door families (Figure 3-5.3).

If you expand a door family itself in the *Project Browser* you see the predefined door sizes associated with that family. Right-clicking on a door size allows you to rename, delete or duplicate it. To add a door size you duplicate and then modify properties for the new item.

Next you will insert the doors into the stair shafts.

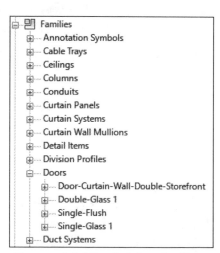

FIGURE 3-5.3 Loaded door families

7. Select the *Door* tool, from the *Architecture* tab, and then pick **Sgl Flush: 36″ x 84″** from the *Type Selector* on the *Properties Palette*.

8. Insert two doors in the West stair shaft as shown in Figure 3-5.4. Remember you are inserting a door into a masonry wall so your door position and size need to work with coursing, thus the 8″ dimension (however, you would also need to include the door frame into the equation).

9. Repeat the previous step to insert doors into the East stair shaft.

10. Finish inserting doors for the first floor (Figure 3-5.5).
Use the following guidelines:

    a. All doors should be 36″ wide and 7′-0″ tall.
    b. You will not insert doors into the curtain wall for now. You will do that in a later lesson when you design the curtain wall.
    c. Use the style and approximate location shown in Figure 3-5.5.
    d. Doors across from each other in the two atrium walls should align with each other.

    TIP: *While inserting the second set of doors, watch/wait for the reference line to show up, indicating alignment.*

    e. Place doors approximately as shown, exact location not given.

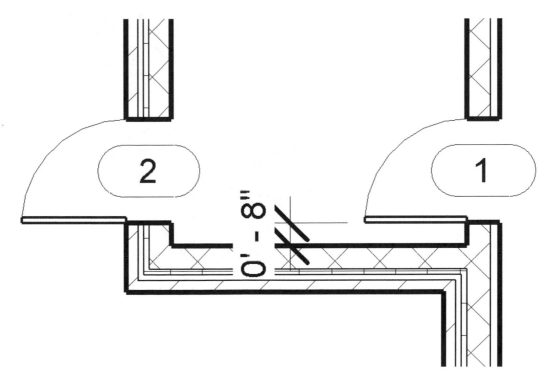

FIGURE 3-5.4 Doors in West stair shaft

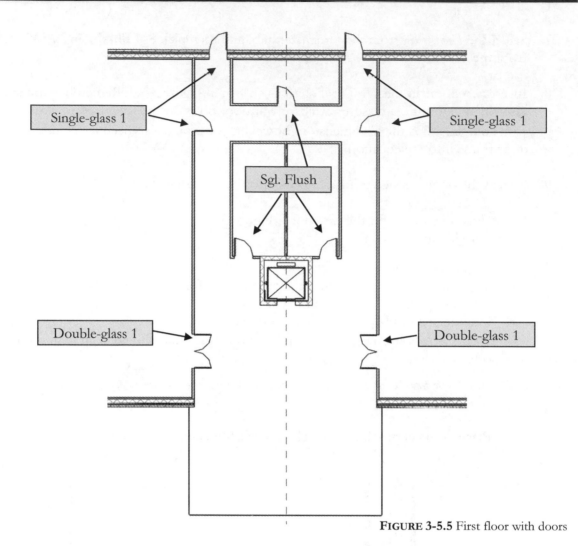

FIGURE 3-5.5 First floor with doors

FIGURE 3-5.6 Door type properties

Door Type Properties:

If you select *Edit Type* while the Door tool is active (selected), you can modify various properties related to the door.

You can easily add another standard door size to the *Family* as required. Click "Duplicate," and type a name (Figure 3-5.6).

Standard door sizes (and *Families*) can be added to your template file, so you don't have to load them for every new project.

Insert Windows

Adding windows to your project is very similar to adding doors. Like the doors, the template file you started from has one family preloaded into your project, the FIXED family. Looking at the *Type Selector* drop-down you will see the various sizes available for insertion. At this point you should also see the SIDELIGHT family that you loaded in the previous exercise. First, you will add a few interior borrowed lights using the sidelight family.

Interior Windows (Borrowed Lights)

11. With the *Window* tool selected, pick *Sidelights 1 :* **18″ x 84″** from the *Type Selector*.

12. On the West side of the atrium, insert the borrowed lights as shown in Figure 3-5.7; do not add the dimensions.

Make sure the borrowed light frames are flush with the atrium side of the wall. You can control that option by moving your cursor to the side of the wall you want the frame flush with before clicking to insert. After drawing the window, you can select the frame and use the flip icon (similar to doors and walls).

13. Repeat the previous steps to insert the borrowed lights on the East side of the atrium.

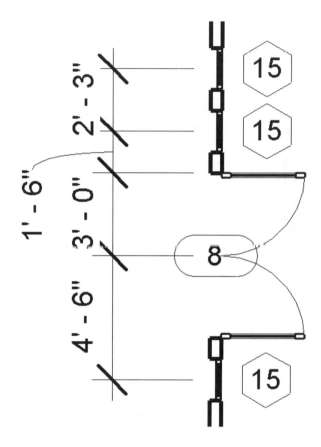

FIGURE 3-5.7 Sidelights added

Exterior Windows

14. Using the methods previously covered in this book, create a new window size in the *FIXED* family. Create *Fixed:* **32″ x 48″**. You are creating this new size to fit coursing in the plan view. The largest window (preloaded) that fits coursing in the plan view is 24″.

15. Adjust the sill height for your new window size to fit within coursing as well. Set the sill height for *Fixed:* **32″ x 48″** to be **3′-4″** (Figure 3-5.8).

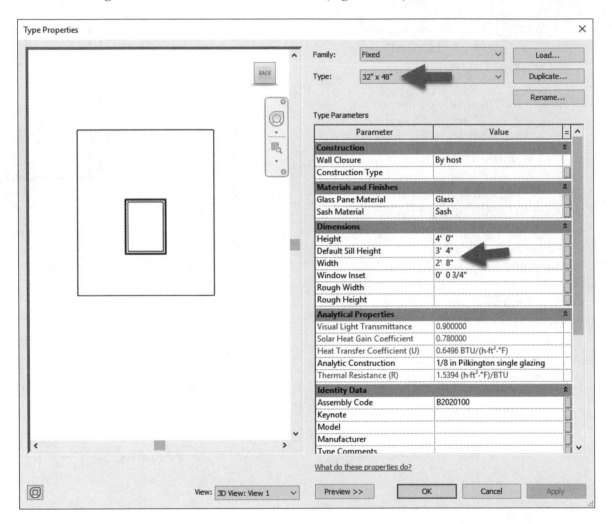

FIGURE 3-5.8 New window size with modified window sill height

16. Insert the window as shown in **Figure 3-5.9**. The window should be inserted with masonry coursing in mind.

> *NOTE: The dimensions displayed while inserting the window will not work as displayed for coursing because Revit is measuring from the center of the adjacent exterior wall. Thus, you will have to insert the window as close as possible and adjust its location, verifying with the* Measure *tool.*

Array Window

The *Array* tool allows you to quickly copy several objects that have the same distance between them. You will use *Array* to copy the windows:

17. Click the *Modify* tool and then select your window.

18. With the window selected, pick the **Array** tool from the *Modify* tab.

19. In the *Options Bar*, type **6** for the *Number* field.

20. Click the left mouse button at the midpoint of the window and move your mouse to the East until the dimension displayed is **8′-6″**.

21. You should now see the windows arrayed in the wall. 8′-6″ is not coursing, so select the **Activate Dimensions** button on the *Options Bar* and then enter **8′-8″** in the displayed dimension to adjust the window openings. This allows you to more accurately adjust the dimensions.

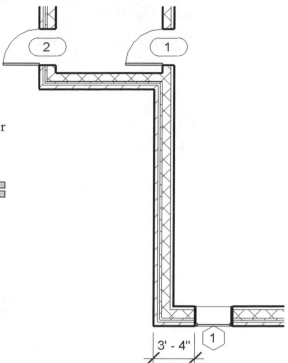

FIGURE 3-5.9 Exterior window

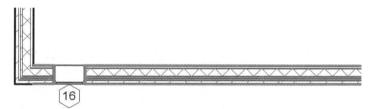

FIGURE 3-5.10 Window to be arrayed

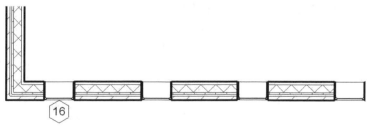

FIGURE 3-5.11 Window after array

22. Set the windows up on the three remaining walls of the first floor (Figure 3-5.12). Consider the following:

    a.   This would be a good use for the *Mirror* tool.

    b.   If you need to create a temporary wall for a mirror reflection axis, make sure the temporary wall is set to centerline.

    c.   You can use the *Reference Plane* to mirror the windows in the East West direction.

    d.   Use the *Measure* tool to verify accuracy.

    e.   Use the Ctrl key to select multiple windows.

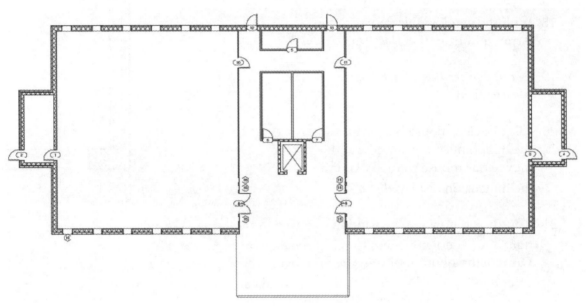

FIGURE 3-5.12 Completed window layout

Cleaning House:

As previously mentioned, you can view the various Families and types loaded into your project. The more Families and Types you have loaded the larger your project file is, whether or not you are using them in your project. Therefore, it is a good idea to get rid of any door, window, etc., that you know you will not need in the current project.

23. In the *Project Browser*, navigate to Families → Windows → Fixed. Right click on **36″ x 48″** and select **Delete**.

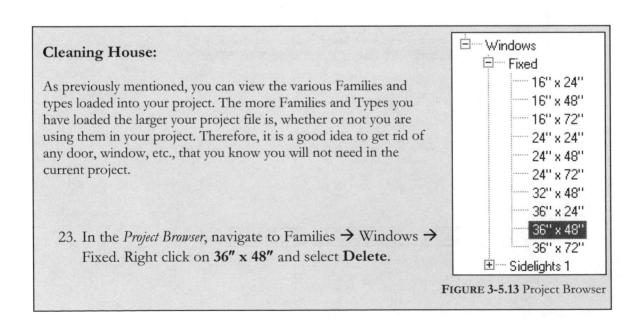

FIGURE 3-5.13 Project Browser

Self-Exam:

The following questions can be used as a way to check your knowledge of this lesson. The answers can be found at the bottom of this page.

1. The *Options Bar* allows you to set the height of a wall when first drawing it. (T/F)

2. It is not possible to draw a wall with the interior or exterior face of the core as the reference point. (T/F)

3. Elements cannot be moved accurately with the **Move** tool. (T/F)

4. The _____ tool, in the *Ribbon*, has to be selected in order to select an element in your project.

5. A wall has to be _____ to see its flip icons.

Review Questions:

The following questions may be assigned by your instructor as a way to assess your knowledge of this section. Your instructor has the answers to the review questions.

1. Revit comes with many predefined doors and windows. (T/F)

2. The length 3'-8" is a masonry dimension. (T/F)

3. You can delete unused families and types in the *Project Browser*. (T/F)

4. It is not possible to load families and types from the internet. (T/F)

5. It is not possible to select which side of the wall a window should be on while you are inserting it. (T/F)

6. What tool will break a wall into two smaller pieces? _____

7. The _____ tool allows you to match the surface of two adjacent walls.

8. Occasionally you use the Split Element command to turn one wall into two walls. (T/F)

9. You can use the _____ tool to copy an element multiple times in one step.

10. The template file has a few doors, windows and walls preloaded in it. (T/F)

Notes:

Lesson 4
FLOOR PLAN (Second and Third Floors):

In this lesson you will set up the upper two floors. This will mostly involve copying elements from the first floor with some modifications along the way. You will also adjust the floor-to-floor height and insert stairs into the stair shafts.

Exercise 4-1:
Copy Common Walls from First Floor

Setting Up the Second (and Third) Floor View

The first thing you need to do is make a few adjustments to the second (and third) floor settings. The default template you started your project from already has a second floor view setup in the project. The third floor has not been set up, so you will do that.

1. Open Exercise ex3-5.rvt and Save-As **ex4-1.rvt**.

2. In the *Project Browser*, double-click on the **Level 2** view under Floor Plans (Figure 4-1.1).

The current view is always bold in the *Project Browser*.

You should now see the second floor plan. Notice that the dark wall lines, shown in this view, exist at this level. The light gray lines are walls for the floor below (Figure 4-1.2).

You will turn off the view of the lower level and set the *Detail Level* to show more detail in the walls.

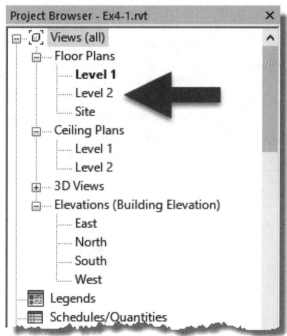

FIGURE 4-1.1 Project Browser; Level 2 view

3. Make sure nothing is selected and the *Properties Palette* is open (**type PP** to open it); when nothing is selected the *Properties Palette* shows the current view's properties.

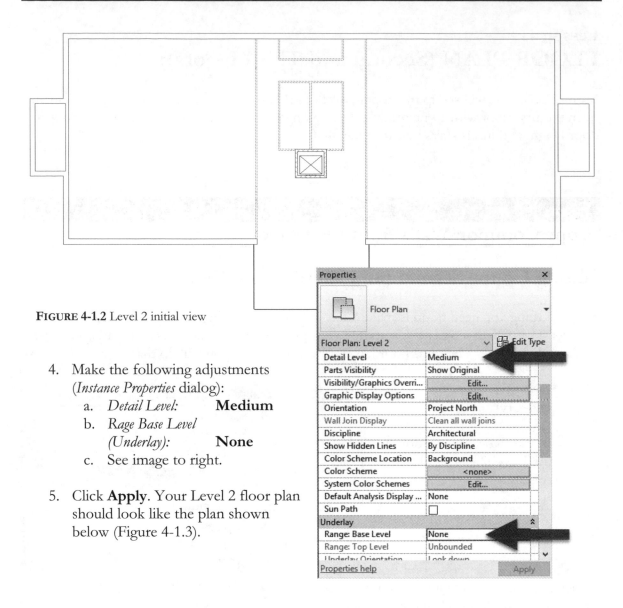

FIGURE 4-1.2 Level 2 initial view

4. Make the following adjustments (*Instance Properties* dialog):
   a. *Detail Level:* **Medium**
   b. *Rage Base Level (Underlay):* **None**
   c. See image to right.

5. Click **Apply**. Your Level 2 floor plan should look like the plan shown below (Figure 4-1.3).

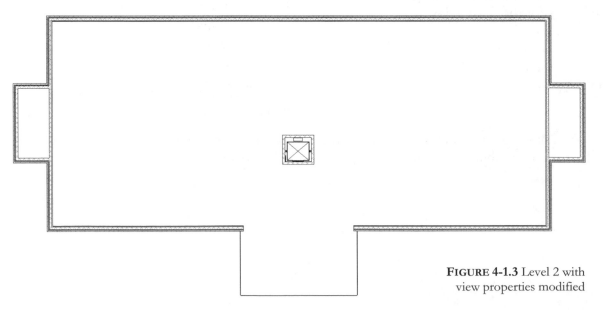

FIGURE 4-1.3 Level 2 with view properties modified

Because the walls and doors you will copy from the first floor are set up to extend to the floor above, you need to set up the third floor before you copy the walls from first to second (so the second floor walls have a floor to extend to).

Adding another floor is surprisingly simple. You switch to an elevation view and draw in a *Level* datum. By doing that Revit automatically sets up a Level 3 view in the *Project Browser*.

6. Double-click on one of the four elevation views listed under *Elevations* in the *Project Browser*. If you do not see your drawing in elevation, try another view and/or see the tip below.

7. With an elevation in the drawing window, select *Modify* and then select **Architecture → Datum → Level** on the *Ribbon*.

8. As you move your cursor near the Level 2 symbol you will see a dimension displayed, indicating the distance between Level 2 floor and Level 3 floor you are about to insert. For now, **set Level 3 to be 10′-0″ above Level 2** (Figure 4-1.4).

   a. Notice "Make Plan View" is checked on the Options Bar.

   b. Pick two points (left to right) to draw the *Level* datum.

   c. Make sure you see the "alignment" reference lines before picking the two points, so it aligns with the other levels

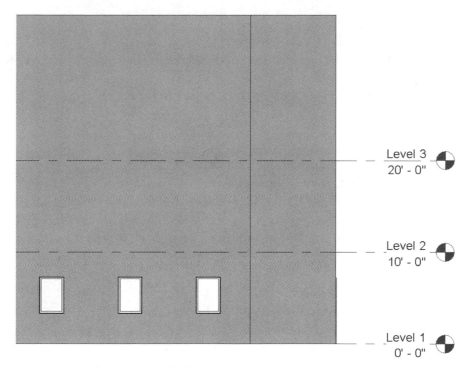

FIGURE 4-1.4 (Partial) South elevation

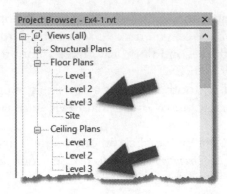

Notice that the Level 3 floor plan view was
automatically added to the *Project Browser*.
(See image to the right.)

TIP: *The default template has four Elevation tags shown in the plan view.
These tags represent what the four pre-setup views (under elevation) will see.
Therefore, you should start drawing your plan in the approximate center of the
four symbols. The symbols can be moved by dragging them with your mouse.
This is covered more thoroughly in Lesson 7.*

Next you will copy walls and doors from the first floor.

9. Switch to the **Level 1** view (see step 2).

10. Select all the interior walls (except the elevator shaft), doors and interior windows.

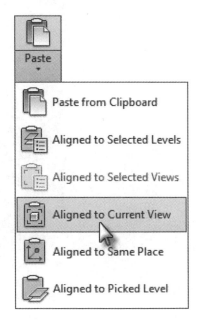

   TIP: *You will need to hold the Ctrl key to select multiple elements with
   multiple picks. You can drag a window(s) to select multiple objects at
   once.*

11. With the objects selected, pick **Modify | Walls** →
    Clipboard → **Copy to clipboard** from the *Contextual
    Tab* on the *Ribbon*, or **Ctrl + C** on the keyboard.

12. Click **Modify** on the *Ribbon* and then switch back to the
    Level 2 view.

13. Select **Modify** → **Clipboard** → **Paste** (down-arrow) →
    Aligned to Current View.

   FYI: *Paste aligned will make the new elements align with the original
   elements below but on the current level.*

Notice the walls, doors and interior windows are now copied to Level 2 (Figure 4-1.5). We still need to copy the exterior windows and the elevator door.

Also, notice that the new doors have different numbers (note: you will only have tags on level 2 if they were selected in the copy / paste) while the interior windows have the same number. Why is this? It relates to industry standards for architectural documentation. Each interior window that is the same size and configuration has the same type number throughout the project (this is a *Type Parameter*). Each door has a unique number because doors have so many variables such as locks, hinges, closer, panic bar, material, and fire rating (this is an *Instance Parameter*). To make doors easier to find, many architectural firms will make the door number the same as the room number the door opens into. You can change the door number by selecting the door tag and then clicking on the text. The door schedule will be updated automatically.

14. Using the same techniques described in the previous steps, copy the exterior windows and elevator door to Level 2.

   TIP: You will need to ungroup your windows (grouped with array) before copying them. Select one of the windows and pick the ungroup button on the Ribbon.

Why not draw these interior walls 36'-0" high like the exterior walls and elevator shaft?

Simulating real-world construction is ideal for several reasons. Mostly, you can be sure shafts align from floor to floor when the shaft is one continuous wall. Although the toilet and atrium interior walls align, they do not necessarily have to because they are separated by floor construction; this allows one floor to be modified later easily.

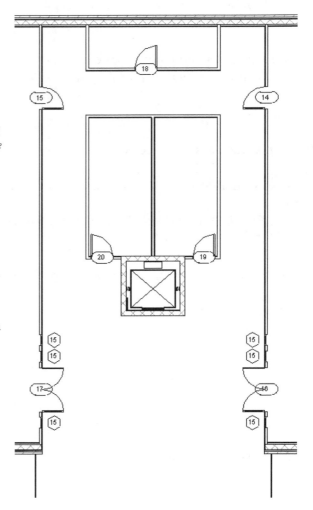

FIGURE 4-1.5 (Partial) Level 2 – walls added

Finally, you will copy several elements to Level 3. But first you need to change the height setting for the walls (on Level 2) before you paste them to Level 3 because there is nothing above (yet) to extend the walls to (e.g., roof or floor).

15. In the Level 2 view, select all the interior walls, doors and windows (except the elevator shaft).

You need to narrow your selection down to just the walls.

16. Select the **Filter** button on the *Ribbon*.

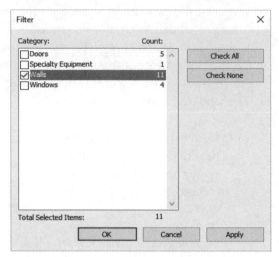

17. **Uncheck** all the items listed except *Walls* (Figure 4-1.6).

The list varies depending on what elements are in the selection set (Figure 4-1.6).

18. Click **OK**.

Now only the walls are selected.

FIGURE 4-1.6 Filter dialog

19. In the *Properties Palette*, change the *Top Constraint* to **Unconnected**, then **Apply**.

   FYI: If you pasted the walls without changing the Top Constraint *it would still be set to Level 3 but with a Top Offset of 10'-0".*

20. Select the elements again; now copy the selected elements to the *Clipboard*.

21. Switch to **Level 3** and make the *View Properties* changes listed in step 4 above (e.g., *Underlay* and *Detail Level*).

22. Paste (aligned to current view) the Level 2 elements to Level 3, including the exterior windows and elevator door.

23. Select all the Level 2 walls again and set the *Top Constraint* back to **Level 3**.

24. **Save your project.**

Exercise 4-2:
Additional Interior Walls

This short exercise will help reinforce the commands you have already learned. You will add walls and openings to your project.

Adding Walls

1. Add the interior walls and doors to **Level 1** as shown in Figure 4-2.1. Use the stud wall you used previously. Use the *Align* tool to align the walls, which are not dimensioned, with the adjacent walls previously drawn.

    FYI: Doors are not labeled to be single flush. Also, dimensions are to centerline of interior walls and to the finished face of the exterior walls.

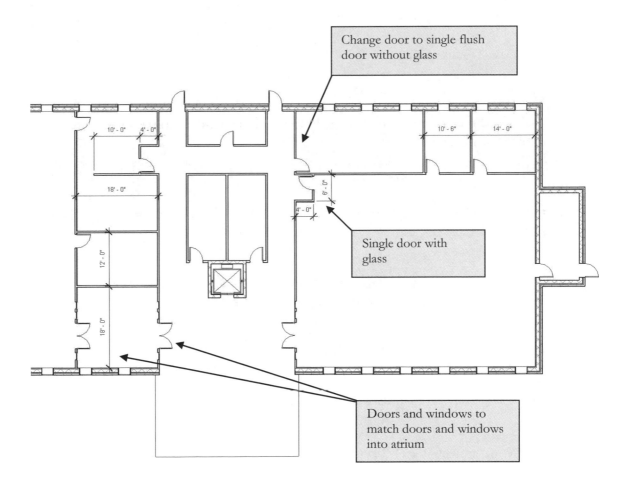

Change door to single flush door without glass

Single door with glass

Doors and windows to match doors and windows into atrium

FIGURE 4-2.1 Level 1 - Added walls

2. Similar to step 1, add the walls and doors shown in Figure 4-2.2 to **Level 3**.

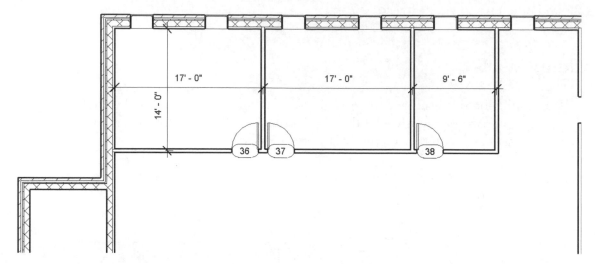

FIGURE 4-2.2 Level 3 - Added walls

3. Use the *Mirror* command to mirror the walls in Figure 4-2.2 to the other three corners of Level 3.

    TIP: Draw a "horizontal" reference plane.

4. Finally, modify the small office on the south, each side of building (Figure 4-2.3 and Figure 4-2.4).

    TIP: If you use the Trim tool (per the TIP in Figure 4-2.3), you will need to select "Delete Instance" to tell Revit to delete the door from the portion of wall that is being deleted.

    FYI: Your modifications to level 3 included adding a few executive offices to the top floor with the "good" views. You deleted the small office on the south side to make room for a reception desk at the main doors from the atrium. Ideally you would add windows to the interior walls of the executive offices to let borrowed light into the open office area. The center area will be open office area for executive assistants.

5. Save your project as **ex4-2.rvt**.

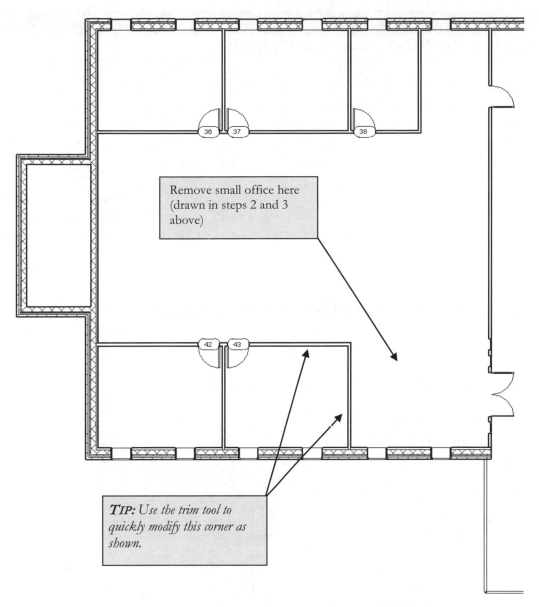

Remove small office here
(drawn in steps 2 and 3
above)

TIP: Use the trim tool to
quickly modify this corner as
shown.

FIGURE 4-2.3 Level 3 - Modify walls

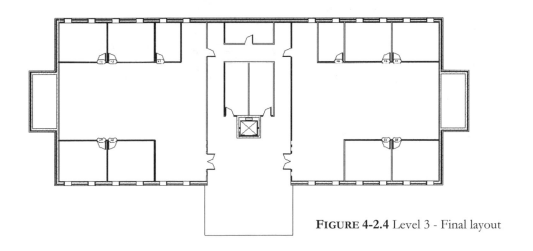

FIGURE 4-2.4 Level 3 - Final layout

Exercise 4-3:
Setting the Floor-to-Floor Height

You will modify the building's floor-to-floor height in this lesson. The reasons for doing this vary. Some examples might be to make the building shorter or taller to accommodate ductwork in the ceilings or the depth of the floor structure (the longer the span the deeper the structure). The default floor-to-floor height in the template file you started from is 10'-0", which is not typically feasible for commercial construction.

Don't forget to keep a backup of your files on a separate disk (i.e., Flash Drive, CD or DVD). Your project file should be about 3 MB when starting this exercise. Remember, your Revit project is one large file (not many small files). You do not want anything to happen to it!

Modify the Building's Floor-to-Floor Height

1. Open ex4-2.rvt, Save As **ex4-3.rvt**.

2. Open the **South** exterior elevation from the *Project Browser*.

Next you will change the floor-to–floor height to be 12'-0" for each level.

3. Select the Level 2 level datum, and then select the text displaying the elevation. You should now be able to type in a new number. Type **12** and then press **Enter** to see the changes. Notice the windows move because the sill height has not been changed (Figure 4-3.1).

FYI: The warning about overlapping walls will be resolved with the next step.

4. Change Level 3 to **24'-0"**.

5. **Save** your project.

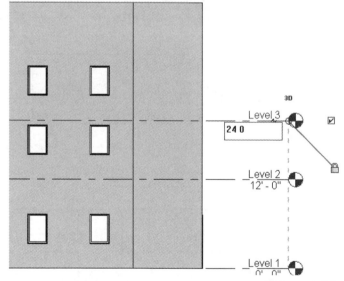

FIGURE 4-3.1
Exterior elevation: modifying Level 3 elevation

Self-Exam:

The following questions can be used as a way to check your knowledge of this lesson. The answers can be found at the bottom of this page.

1. The default settings for the floor plan view shows the walls for the floor below. (T/F)

2. It is not possible to add a new floor level while in an elevation view. (T/F)

3. You should start drawing your floor plan generally centered on the default elevation tags in a new project. (T/F)

4. You can use the *Align* tool to align one wall with another across a hallway from the other line. (T/F)

5. Changing the floor level height also changes a window's sill height. (T/F)

Review Questions:

The following questions may be assigned by your instructor as a way to assess your knowledge of this section. Your instructor has the answers to the review questions.

1. It is not possible to copy/paste objects from one floor to another and have them line-up (with the original objects). (T/F)

2. If a shaft wall is to be built from the lowest level to the roof, and not interrupted at each floor level, the wall should be drawn with that height (not separate walls on each floor level). (T/F)

3. Each Revit view is saved as a separate file on your hard drive. (T/F)

4. You select the part of the wall to be deleted when using the *Trim* tool. (T/F)

5. You can change the floor-to-floor height by changing the level datum (e.g., 24'-0" to 22'-0") in elevation. (T/F)

6. What parameter should be set to none, in the view's properties, if you do not want to see the walls from the floor below? _____

7. You use the _____ tool to create a new floor plan level when in an elevation view.

8. You can use the _____ tool to quickly select a certain type of element from a large group of selected elements.

9. If the Properties Palette is not open, type _____ to open it.

SELF-EXAM ANSWERS:
1 – T, 2 – F, 3 – T, 4 – T, 5 – F

Notes:

Lesson 5
Vertical Circulation:

In this chapter Revit's Stair, Railing and Ramp tools will be covered. This can be one of the most challenging features in Revit given the vast variety in the world when it comes to stairs and railings. These tools have some limitations and cannot accommodate all situations. In these cases, one might have to use In-Place families and model a highly customized stair and/or railing.

The first few "exercises" will provide a basic introduction to the stair and railing tools in Revit. There are no required steps for the law office project in these sections. However, it is highly recommended that this information be reviewed prior to completing the steps required later in this chapter.

Exercise 5-1:
Introduction to Stairs and Railings

Prior to getting into the details, this first section will provide a broad overview of Stairs and Railings in Revit.

This image is a photorealistic rendering of a stair and railing modeled in Revit. This example is from the author's *Interior Design Using Autodesk Revit* textbook. Notice the railing occurs on both sides of all three "runs" of stairs and along the [second] floor edge.

Basic Stair and Railing Terminology

First, let's define some terms related to stairs and railings in Revit. These terms generally correspond to real-world architecture/construction terminology. However, the main goal here is to define the terms in the context of Revit.

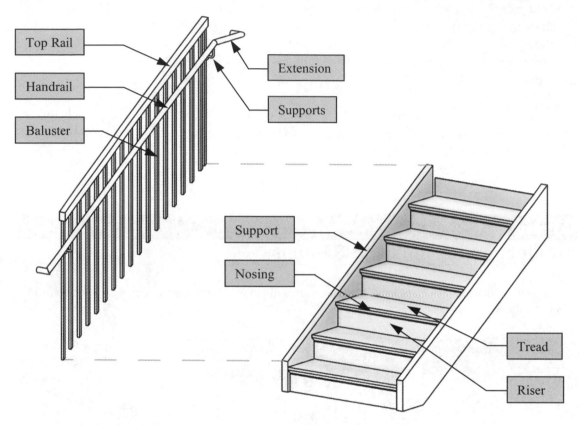

FIGURE 5-1.1a Basic Railing   **FIGURE 5-1.1b** Basic Stair

- Revit Stair Element
    - **Run**: a continuous section of stairs, consisting of Risers and Treads, between main floor levels and/or landings.
    - **Tread**: the flat horizontal part you step on.
    - **Riser**: the vertical portion which fills the gap between the treads. Not all stairs have risers.
    - **Nosing**: the outer edge, where the riser and tread come together.
    - **Landing**: intermediate horizontal surface (i.e. floor) between the main floor levels of a building. Building codes require a landing if a stair "run" rises more than a certain distance—which allows someone to have a safe place to rest.
    - **Supports** (aka Stringer): the structural elements supporting the treads and risers. For a commercial project this is typically a steel tube or C-channel. A stair typically has a stringer on each side of the stair. For wider stairs, one or more stringers may be required. They either span between floors (e.g. Level 1 up to Level 2) or may be anchored to an adjacent wall.

- o **Cut Marks**: in a 2D plan view, the stair is cut where it intersects the view's cut plane—an angled line is added to graphically cut off the stair. This allows space below the stair to be seen.
- o **Stair Path** (Annotation)
  - ▪ **Down/Up arrows**: in a 2D plan view, the arrow graphically indicates the direction the stairs ascend or descend. The arrow is typically used in conjunction with text (see next definition).
  - ▪ **Down/Up Text**: Text used in conjunction with arrows to make it clear which direction the stairs are going, which would be difficult to determine without looking at other views (e.g. sections) which a printed set of drawings may not contain (e.g. a small remodel project).
    - • **FYI**: The Arrow and Text are added automatically to a new stair but can be deleted. Thus the Stair Path tool exists, on the Annotation tab, if they need to be added back to the model.
- o **Tread Number** (Annotation): This tool, on the Annotation tab, places a sequence of numbers along a stair "run" indicating the total number of risers or treads.

- • Revit Railing Element
  - o **Balusters**: vertical elements extending from floor or stair up to "top rail." For commercial projects, building codes state that a 4" sphere cannot pass through a railing system. **FYI**: Balusters are not needed if glass panels are used.
  - o **Handrails**: continuous rounded element, attached to a wall or balusters with a "support," along a floor edge, a ramp or stair "run" which allows someone to place their hand on to prevent falling.
  - o **Guardrail**: A taller railing system, which may include a handrail, to prevent falling from a stair or floor edge. **FYI**: In the USA, when a guardrail is required it must be at least 42 inches tall.
  - o **Supports**: bracket used to attach handrail to wall or balusters.
  - o **Top Rails**: continuous rail at the top of the railing system—supported by balusters.

Plan View Representation of Stairs

In Revit plan views, Stairs and their hosted Railings are not shown as true 3D elements being cut by the view's cut plane. Rather, they are a hybrid 2.5D representation which follows traditional architectural graphic standards for floor plans. An example of this can be seen in Figure 5-1.2a. Notice all parts of the stair and railing extend to the Cut Mark. Contrast this with Figure 5-1.2b which is a true 3D view of the same stair; notice the railing is hardly visible and there is no cut mark. The cut plane for both examples is 3'-0" as seen in Figure 5-1.2c.

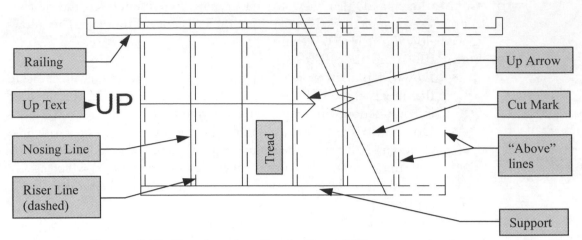

FIGURE 5-1.2a Plan view of a stair and railing – 2.5D representation

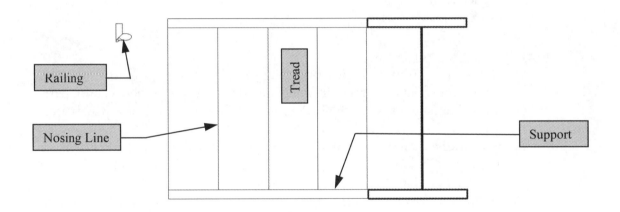

FIGURE 5-1.2b Plan view of a stair and railing – true 3D representation

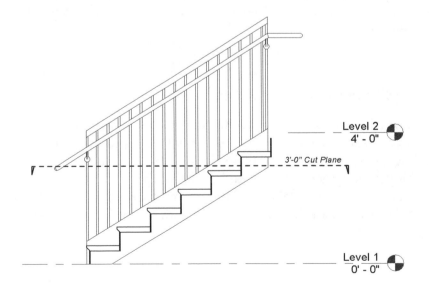

FIGURE 5-1.2c Stair in section with plan view's *cut plane* indicated

The Stair and Railing tools will be covered in more detail in the following "exercises" in this chapter. But, first, here is a high-level introduction to the tools available in Revit…

Stair Tools

Revit has a single stair tool: **Stair by Component**. Note that *Stair by Sketch* has been removed in Revit 2021.

Railing Tools

Revit has two main tools one can use to model railings:
- Sketch Path
- Place on Host

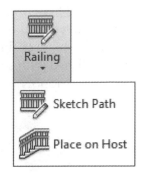

The **Sketch Path** tool allows a continuous section of railing to be added to the current level. The **Pick Host** command provides a quick way to add railings to a stair or railing, which is handy if they were deleted or not added during the stair/ramp creation.

In addition to these two Railing tools, railings can also be created while creating Stairs and Ramps. While in the Stair or Ramp command, selecting the **Railing** button on the Ribbon lets you specify which railing to use and how to position it (Figure 5-1.3). Revit will add a railing to both sides. Once created, one or both of the railings can be deleted. One tricky thing about the Pick Host tool is it only works if the stair or ramp does not have any railings hosted.

FIGURE 5-1.3
Specify how railing will be created

Basic Stair Types

Revit has three fundamentally different stair types which are needed to properly model the various types of stairs typically found in construction.

Stair Types:
- Assembled Stair
- Monolithic Stair (aka Cast-In-Place)
- Precast Stair

Assembled Stair

Revit's Assembled Stair represents the most common type of stair in construction: concrete pan and steel riser (Figure 5-1.4). A residential stair constructed of wood can also be accomplished with this option. With this type Revit provides separate settings for the treads, risers and supports.

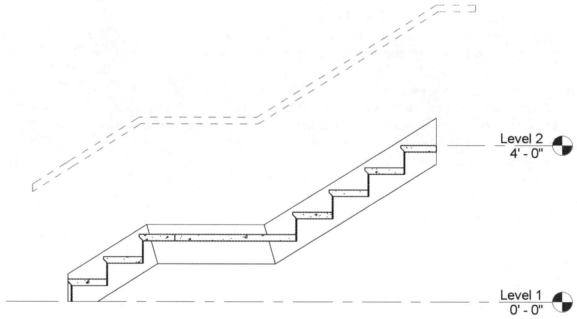

FIGURE 5-1.4 Assembled stair type example

Monolithic Stair (aka Cast-In-Place)

When a stair is constructed with cast-in-place concrete (CIP), the monolithic stair type should be used (Figure 5-1.5). This stair is often used outside the building (site design) or within the building whose primary structure is also CIP concrete. The properties for this type of stair vary a bit from the assembled stair type—for example, the minimum thickness and whether the bottom should slope or step can be specified. Structural designers can also add **Rebar** to this type of element.

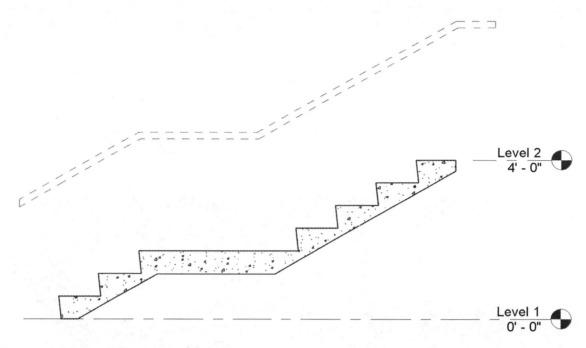

FIGURE 5-1.5 Monolithic stair type example

Precast Stair

When the various parts of a stair are manufactured off-site with concrete and then shipped to the job site, the Precast Stair type can be used to properly represent this special condition (Figure 5-1.6). Notice the special notches which are used to interconnect and support the individual parts. This stair type can also have Rebar added to it. **FYI:** This stair type can only be created using the *Stair by Component* command.

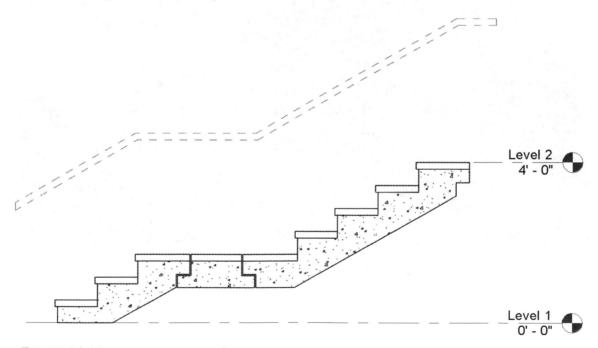

FIGURE 5-1.6 Precast stair type example

Because the parameters vary between stair types, it is often not a simple task to select a stair of one type (e.g. Assembled) and switch it to another (e.g. Monolithic) via the Type Selector. The stair needs to be modified or recreated from scratch to switch between types.

To see the difference between the **End with Riser** versus **End with Tread** options, compare Figure 5-1.2c, which is set to end with riser, with Figure 5-1.4 set to end with tread. In this example, the "end" is the top and each option is common in construction. The 'end with riser' option is used if the stair stops right at the floor edge. The 'end with tread' option is used when the stair ends a distance away from the main floor and something similar to a landing is added to fill in the gap.

This image is another view of the stair and railing design shown at the beginning of this section. The railings at the level 2 floor edge are separate elements from the railings hosted to the stair.

Graphic Controls

There are a number of ways to manage the visibility of Stairs and Railings in Revit. For the most part, these options apply to all stairs and railings regardless of which tool was used to create them.

Object Styles (project wide settings)

The highest level of graphic control in a Revit project is **Manage → Object Styles** (Figure 5-1.7). Here, one can specify the line thickness and color for the various parts of a Stair (same for Railings). For Stairs and Railings, these settings only apply if there are no view specific overrides or filters. *Tip:* Compare the terms listed to the previous images.

Visibility/Graphic Overrides (view specific settings)

The graphics of Stairs/Railings can also be controlled on a view by view basis via the **Visibility/Graphics Overrides** dialog (Figure 5-1.8); while in a view, type **VV**. For example, maybe a Demolition plan or Code plan should show less detail…this can be achieved just in those views.

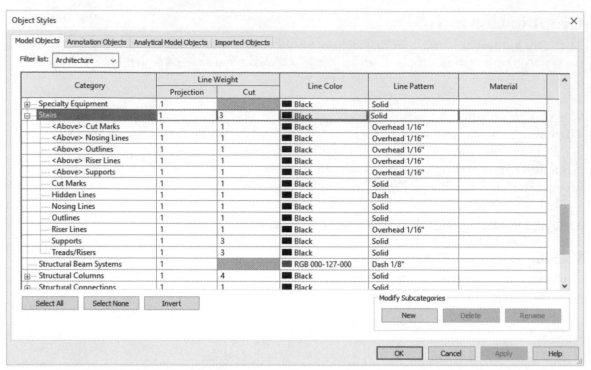

FIGURE 5-1.7 Object Styles dialog – <u>project-wide</u> graphics controls for stairs

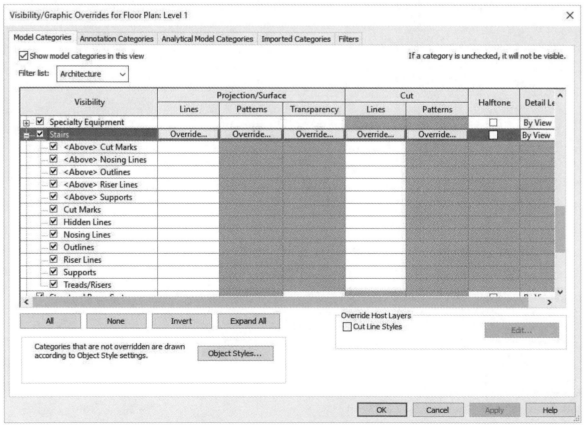

FIGURE 5-1.8 Visibility/Graphic Overrides dialog – <u>view specific</u> graphics controls for stairs

Calculate Tread/Riser Size

Although Revit automatically calculates the rise and tread dimensions for you, it is still a good idea to understand what is happening.

The **riser** is typically calculated to be as large as building codes will allow. Occasionally, a grand stair will have a smaller riser to create a more elegant stair.

Similarly, the **tread** is usually designed to be as small as allowable by building codes. This author worked on the design of a ski chalet where the treads were deeper than required by code for the comfort of those wearing ski boots in the building.

The largest riser and shortest tread create the steepest stair allowed. This takes up less floor space; see the next image. A stairway that is too steep is uncomfortable and unsafe.

Building codes vary by location; for this exercise, you will use 7" (max.) for the risers and 12" (min.) for the treads (11" tread plus a 1" nosing).

Codes usually require that each tread be the same size in a single run, likewise with risers.

To calculate the number and size of the risers:

> *Given:*
> Risers: **7"** max.
> Floor to floor height: **13'-4"**.

> *Calculate the number of risers:*
> 13'-4" divided by 7" (or 160" divided by 7") = 22.857

Seeing as each riser has to be the same size we will have to round off to a whole number. You cannot round down because that will make the riser larger than the allowed maximum (13'-4" / 22 = 7.3"). Therefore, you have to round up to 23. Thus: 13'-4" divided by 23 = 6.957.

So you need **23** risers that are **6 15/16"** each.

Multistory Stairs

If a stair extends multiple stories (i.e. levels) in a building, Revit has a feature which will support this in some cases (but not all). When a stair is selected, click the **Select Levels** command on the Ribbon while in an elevation view—from here you select levels to extend the stair to. This really only works with straight run or U-shaped stairs (aka switchback stair).

In the image below (Figure 5-1.9), there are triple-run (left) and double-run (right) stairs. As a side note, the triple-run stair might be used to minimize required floor space but requires enough vertical distance between floors as two of the runs overlap and can create headroom issues. When using the Multi-Story Stair feature in this case, the **double-run will work, but the triple-run will not**.

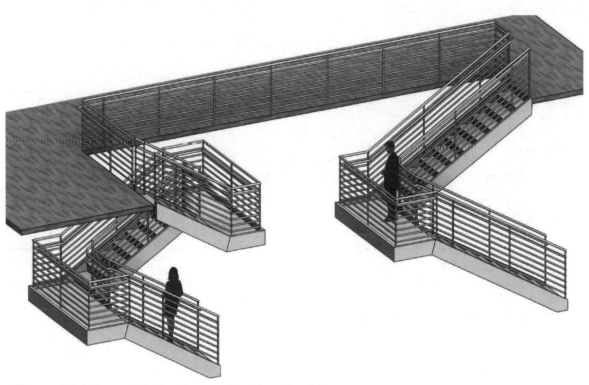

FIGURE 5-1.9 Two stair designs between level 1 and level 2

If the Multi-Story Stair feature is used on these two cases, we see the result in Figure 5-1.10. Because the stair is essentially arrayed vertically, and the triple-run starts and ends on the same side, we see a major issue. However, the double-run stair works correctly.

FIGURE 5-1.10 Stairs arrayed vertically using the Multistory Top Level parameter

Another problem with the new MS Story stair tool is it does not create half of the landings as shown in the image to the right (Figure 5-1.11). Often, in a stair shaft, the stair is all made of the same construction. The assumption with this tool, and the original stair tool, is that the primary floor construction, at each level, would extend into the stair shaft and define the landing at that location. While that may be done in some situations, it is not the norm.

Conclusion

This concludes the basic introduction to Stairs and Railings in Revit. The next sections will dive deeper into each of these tools.

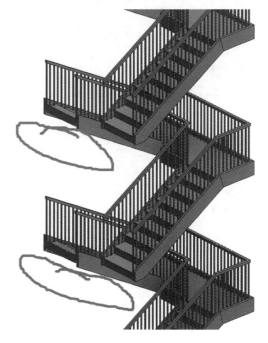

FIGURE 5-1.11 MS Stair missing landings

Exercise 5-2:
Stair by Component

Stair Types

When the *Stair by Component* command is selected, the **Type Selector** lists three options (Figure 5-2.1); this example is based on the Commercial-default template. At this point it is helpful to compare these three options with their corresponding entries in the **Project Browser** under Families - marked with #1 in Figure 5-2.2. The fourth item marked with #2, "stair," contains everything related to the older *Stair by Sketch* command. The remaining items are the "components" for this newer stair tool. These components are selected within the properties of the three main options highlighted here.

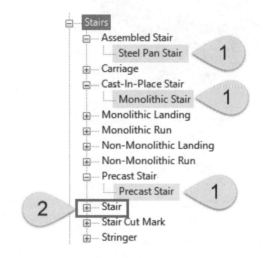

FIGURE 5-2.1 Type Selector; Stair by Component tool active

FIGURE 5-2.2 Family\Stairs in Project Browser

Stair Component Types

To see how a Stair by Component type, listed above, makes use of the **component types** take a look at the Type Properties for the "Steel Pan Stair" in Figure 5-2.3. The selected component types, which have been highlighted, can each be seen in the Project Browser as shown in Figure 5-2.4, also highlighted.

Unlike the Stair by Sketch, which contains all required properties in one Type Properties dialog (covered in the previous section), the Stair by Component has Type Properties for the entire stair (Steel Pan Stair in this example) and then Type Properties for the various components: Run Type, Landing Type, Supports and Cut Mark Type.

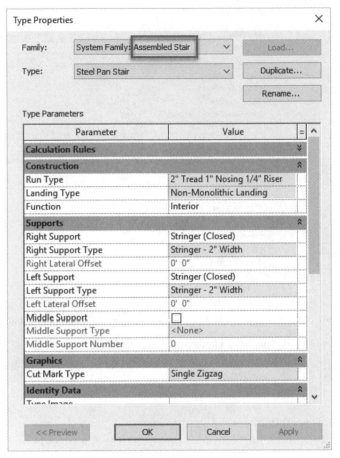

FIGURE 5-2.3 Steel Pan Stair type properties

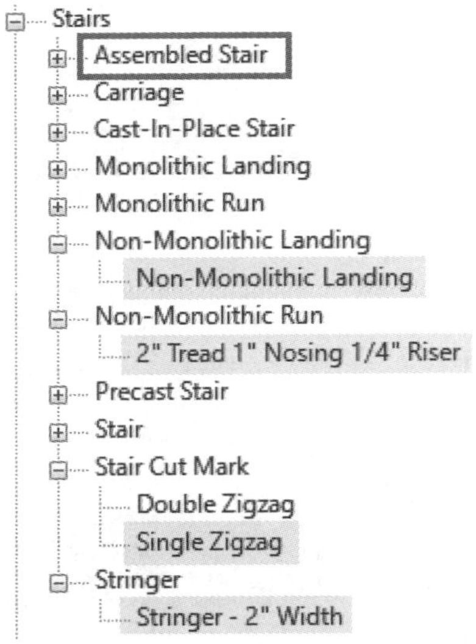

FIGURE 5-2.4 Component families used by Steel Pan Stair

> **Supports defined:** Both a **Carriage** and a **Stringer** are able to support the stair. In Revit, a carriage is below the riser and treads while a stringer runs alongside them. A *Stair by Component* stair can have three supports, Left, Right and Middle, as seen in Figure 5-2.4. All three supports can be carriage-style but only the Left/Right can be stringers. **FYI:** There is some legacy terminology in Revit that is a bit confusing; a stringer is called "Closed" and a carriage is called "Open."

Examples of parameters associated with component types can be seen in Figure 5-2.5; here we see the *Run Type* named **2" Tread 1" Nosing ¼" Riser** and the *Right Support Type* named **Stringer – 2" Width**.

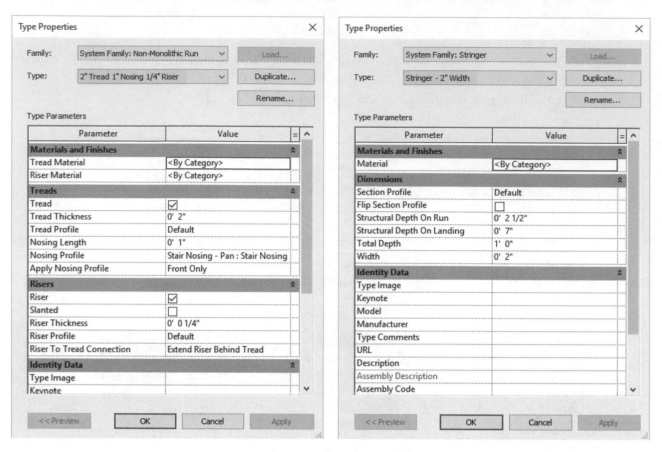

FIGURE 5-2.5 Type properties for two of the stair component types associated with 'Steel Pan Stair'

Notice in the image above that things like material and tread profile can be selected. This entire **Run Type** can be duplicated and modified when required—for example, a project has a simple utilitarian stair and a more formal open stair. Also notice that a profile can be selected for the supports: **Section Profile**. Thus, we can define a channel-shape which is fairly common in commercial construction. Remember, with the Stair by Sketch, the support can only be rectangular in shape.

All these settings will make more sense once you start using this tool. However, at this point you should have the basic understanding that there are properties for the entire stair system and then separate properties (i.e. types) for the individual components of the stair.

Creating Stairs using Stair by Component

While creating a stair with Stair by Component, the 3D "components" are created along the way. Using this tool, the individual sketch lines are not seen or editable (e.g., boundary and riser) like they are with the Stair by Sketch tool. This makes it easier to see what the stair will look like before clicking Finish Edit Mode. Also, because there are not sketch lines, it is possible to create a stair with multiple switch backs.

Here are the basic steps for the Stair tool:

- Start at the lowest level
- Select **Stair** tool
- Verify stair style in **Type Selector**
- Verify Base and Top Level settings in **Properties**
- Adjust settings on **Options Bar**
    - Location Line
    - Offset (horizontal)
    - Stair Width
    - Landing creation
    - See Figure 5-2.6 below
- **Pick points** to define one or more runs
- *Optional:* Add **Landing** to top or bottom if needed
    - select *Landing* on Ribbon while in Stair by Component tool
- *Optional:* Select and delete individual **Support** elements
    - For example, if there is a door at an intermediate landing
    - Use the *Support* option to replace if needed

The image below shows the **Options Bar** while the Stair by Component tool is active (Figure 5-2.6). Be sure to select the best option for Location Line. If in a stair shaft, one of the exterior support options is likely the more convenient option. Note that these settings can vary per run within the same stair instance.

Location Line: Run: Center   Offset: 0' 0"   Actual Run Width: 3' 0"   ☑ Automatic Landing

FIGURE 5-2.6 Options bar for Stair by Component tool

Picking the points on-screen is similar to the Stair by Sketch tool. Two clicks could create a continuous single-run from floor to floor, or, as seen in the image below, multiple clicks creates multiple runs. All along, Revit indicates how many risers remain before reaching the next level.

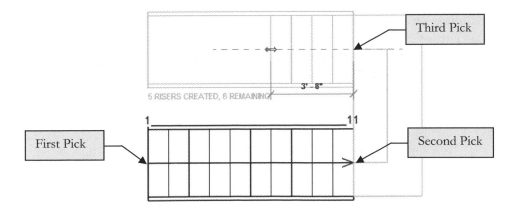

FIGURE 5-2.7 Creating stair with Stair by Component tool

Once the components are created, and while still in the Stair by Component tool, individual components can be selected.

Stair Landings

In the first image (Figure 5-2.8) the intermediate landing is selected. Notice there are six grips. There is one grip on each side and one grip at each stair run.

The next image (Figure 5-2.9) shows how the landing can be modified when adjusting the grips by clicking and dragging. In this example, the three circled grips were moved to change the size and shape of the intermediate landing.

Landings are only automatically added between runs when **Automatic Landing** is checked on the Options Bar. If additional landing construction is required at the top or bottom of the stair, select the **Landing** option (see Ribbon image below) and then sketch a closed loop which is touching the adjacent stair run.

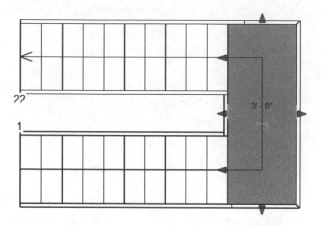

FIGURE 5-2.8 Stair by Component edit mode with landing selected

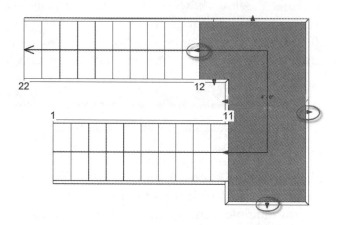

FIGURE 5-2.9 Landing adjusted using grips

The result is a landing with added supports on all three sides as seen in the upper part of Figure 5-2.10. Selecting the end support allows you to delete it. Notice how the side supports adjust from mitered to flush cut in the lower image once the support has been deleted.

When a landing is selected, its **Relative Height** and thickness are listed in the Properties Palette. Adjusting the elevation will automatically adjust the runs, plus the number entered will be changed to the closest riser position. However, the runs are not adjustable if any of them have been converted to a sketch (more on this in a moment).

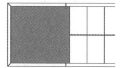

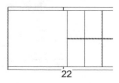

FIGURE 5-2.10 Added landing and support deleted

Non-rectilinear Landings:

The intermediate landings automatically created by Revit can be modified. While in Stair Edit mode, select the landing component as shown in the first image in Figure 5-2.11. Click the **Convert** command on the Ribbon.

At this point, a message appears indicating the conversion is irreversible. Click **Close** to complete the conversion. The landing portion of the stair is now a Sketch-based stair element.

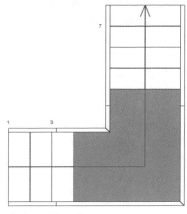

Original Landing

With the landing still selected, click the **Edit Sketch** command on the Ribbon. The lines which define the perimeter of the landing, in plan-view, are now editable. Use **Draw** and **Modify** tools, such as Line, Circle, Arc, Split, Trim, Offset, etc. to adjust the footprint of the landing. An example of a modified landing can be seen in the second image to the right.

When finished editing the landing sketch, click the **green checkmark** to exit sketch mode. If the sketch is not a clean outline with any overlapping or crossing lines Revit will create the landing. Unlike other sketches, like floors, there cannot be multiple closed loops (e.g. in a floor, a loop within a loop would create a hole in the floor). If a sketch error is shown, Revit will highlight the problem area. Simply click Continue and fix the problem.

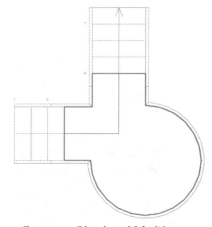

Convert to Sketch and Modify

Click the green checkmark one more time to finish the entire stair instance; Revit will recreate any hosted railings as well. This can be seen in the last image to the right.

If the landing needs to be modified again, select the stair and click Edit Stair and then select the landing component and click Edit Sketch.

Custom Landing Result

FIGURE 5-2.11 Custom Stair Landing

Landing Properties:

A landing can have its main properties defined by the Run, which is the default setting, or independently.

Selecting a landing and then looking at its Type Properties we see **Same as Run** is selected (Figure 5-2.12). This tells Revit to inherit several properties from the stair run that the landing is connected to within the *Stair by Component* instance.

TIP: A component within the stair can be selected without entering edit mode. Simply hover your cursor over the component and then tap the **Tab** key and click to select once the component highlights.

When **Same as Run** is unchecked, several properties appear as shown in Figure 5-2.13.

Landing Material:

In addition to controlling the various tread properties, notice that a **material** parameter is also now available when **Same as Run** is unchecked. If the landing requires a different floor finish, uncheck the **Same as Run** option and specify the material using the **Tread Material** parameter.

An example of the nosing and material matching the run can be seen in Figure 5-2.14.

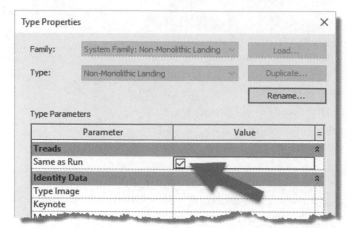

FIGURE 5-2.12 Landing type properties; 'Same as Run' checked

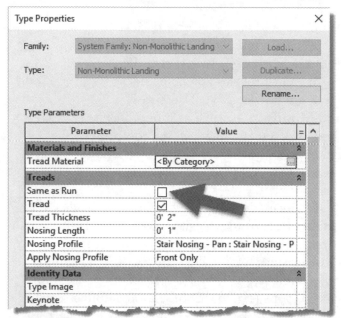

FIGURE 5-2.13 Landing type properties; 'Same as Run' unchecked

FIGURE 5-2.14 Landing nosing and material example

Stair Supports

When a support (aka Stringer) is selected, there are three ways in which the end can be cut: Horizontal, Perpendicular and Vertical (Figure 5-2.15). The default is Vertical. This setting is only adjustable at the ends of a top or bottom run. Notice in the image that the setting can vary for each side of the stair.

These options for supports are more limiting than those for the Stair by Sketch tool. Looking back at that section, we see it is possible to control whether the stringer continues to slope or is flat, and its vertical dimension. With the Stair by Component tool we can only extend the stringer if there is a landing to host it.

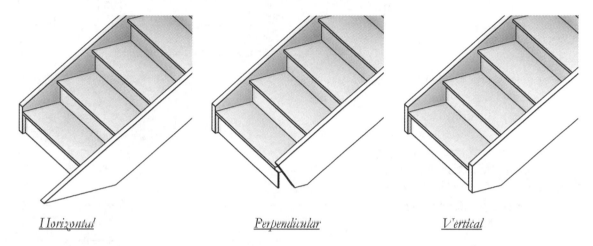

Horizontal   *Perpendicular*   *Vertical*

FIGURE 5-2.15 Three options for a support <u>end cut</u>: Horizontal, Perpendicular and Vertical, respectively

Revit generally does a good job at how supports transition. Sometimes there are issues as seen in the example below, on the left. This is not really Revit's fault. The issue is we did not give Revit enough room to make an appropriate transition. Once the landing has been adjusted, we see the buildable result on the right. As we will see later, this is true for railings as well.

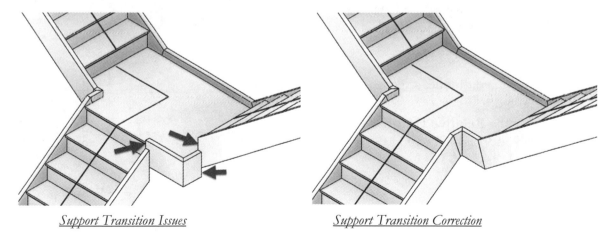

Support Transition Issues   *Support Transition Correction*

FIGURE 5-2.16 Dealing with supports which do not transition correctly

The example in the previous image (Figure 5-2.16) shows two runs coming off of the same landing, up to the next level. The steps to accomplish this will be covered in the next section on stair runs.

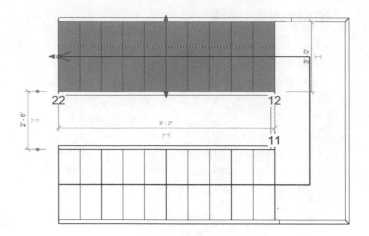

FIGURE 5-2.17 Stair run selected

Stair Runs

When an individual stair run is selected there are a few adjustment options per the grips as seen in Figure 5-2.17.

Dragging the side grips modifies the width of the stair – just for the selected run. Each run can be a different width.

Dragging the top end (or bottom end) grip will depress the number of steps in the selected run. The opposite run will have the same number of steps added (or removed); see Fig. 5-2.18.

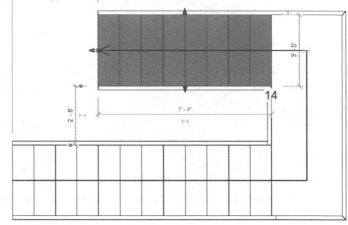

FIGURE 5-2.18 Stair run grip adjusted

A selected run can also be **Nudged** using the Arrow keys. When nudging the stair away from the landing, Revit will automatically extend the landing to meet up with the new stair position.

Additional runs can be modeled within the same stair instance as shown in the previous section. This may be for a grand stair or access to an adjacent roof level. Here is how this is done:

- While in edit mode, select the landing (Fig. 5-2.19)
- Copy the **Relative Height** to the clipboard; highlight and press **CTRL + C**
- Select the **Run** command on the *Ribbon* (Step #1, Fig. 5-2.20)

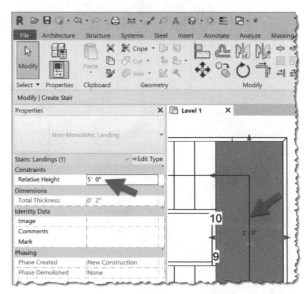

FIGURE 5-2.19 Height listed for selected landing

- Switch the *Properties Palette* drop-down to **Runs** (Step #2, Fig. 5-2.20)
- Paste the landing height into **Relative Base Height**; **CTRL + V** (Step #3, Fig. 5-2.20)
- Draw the stair run (Figure 5-2.21)

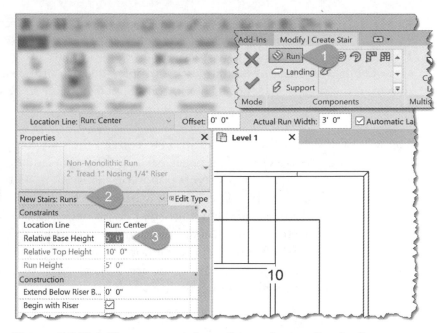

FIGURE 5-2.20 Adding an extra stair run from an intermediate landing

The result is a new run joined to the landing as seen in the two images that follow. Notice the run arrow is not able to connect multiple stair runs. Thus, make sure to model the desired primary path first. Also, if the secondary stair run needs to be repositioned, it will become un-joined with the landing. However, the landing can be adjusted and once it touches the run it will automatically join with it.

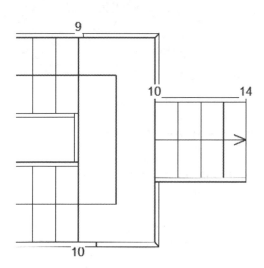

FIGURE 5-2.21 Extra run added to stair

FIGURE 5-2.22 3D view of modified stair

Stair Limitations

There are a few limitations with *Stair by Component* compared to *Stair by Sketch* when it comes to modeling a stair run or landing. For example, the edge of the stair cannot be angled as shown in the previous section. The bottom tread is not able to be extended out and rounded either.

The way to deal with these limitations is to convert the stair component to a sketch-based run or landing. As covered a few pages back for a custom landing, with a stair run selected, while in edit mode, click the **Convert** option on the Ribbon. As the following message indicates, this is irreversible. The main drawbacks are not being able to drag grips to resize the components, but runs and landings still join properly. Thus, the drawbacks are minimal when compared to the need to model the stair correctly. Once a component is converted, select it and click **Edit Sketch** on the Ribbon to modify a run or landing. At this point you are two "edit modes" deep—meaning the green checkmark, for finish, must be clicked twice to fully complete the stair modification.

Stair Annotation

There are a few annotation tools related to Stair by Component to be aware of.

The **Stair Path** tool, found on the Annotation tab, adds the UP or DN text and an arrow. Keep in mind that every tool on this tab is 2D and view specific. When a stair is initially created, the Stair Path annotation is automatically added to every plan-view. When subsequent plan views are created (or copied without detailing), stair arrow annotations are not automatically added to those.

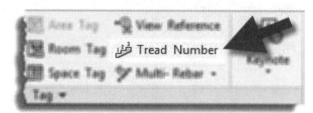

The Stair Path element can be selected and then repositioned or even deleted in each view. If the annotation is deleted, or a new plan-view is created, use the Stair Path tool to add it to a stair instance.

The visibility of the text and arrows can be controlled separately and per view. Figure 5-2.23 shows the **Visibility/Graphic Overrides** dialog with the Stair Paths category expanded on the Annotation Categories tab.

The text and arrow can be moved separately. The text, "UP" or "DN," can be moved in any direction. The arrow can only be moved perpendicular to the run of the stair.

The **Tread Number** tool, also found on the Annotation tab, enumerates each tread for a selected run. This tool only works with *Stair by Component* elements; it will not work with a stair created with *Stair by Sketch*. However, if a run is converted to a sketch within a *Stair by Component* stair this command will still work.

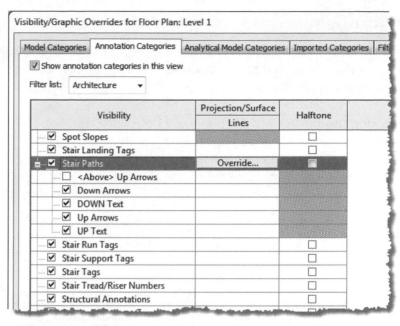

FIGURE 5-2.23 Stair path category

This tool is simple to use. To add tread numbers to a stair, simply start the command and then select a stair run within a plan or section view. Each separate run needs to be selected, even within one Stair by Component instance. The numbers only appear in the current view.

An example of the results can be seen in Figure 5-2.24 for both section and plan views.

Once placed, select it to access several formatting options via the Properties Palette. The Tread Number element is selected in Figure 5-2.24; notice the dashed line running through the numbers. When not selected, there is no visible line.

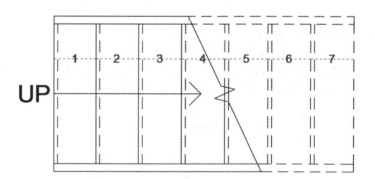

Plan View

Section View

FIGURE 5-2.24 Tread Number tool results

Warnings

If a stair has a problem, for example a stair instance does not reach a floor level, the **Show Related Warning** button will appear whenever that stair is selected.

Clicking this button will open a dialog describing the problem as shown in the image below (Figure 5-2.25). This can happen if a stair is created and there are "remaining" risers listed prior to clicking finish. Also, if a stair type is switched from non-monolithic to monolithic this can appear.

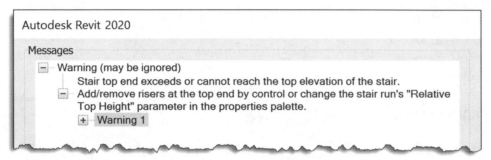

FIGURE 5-2.25 Warning message for selected stair

To see a full list of warnings for the entire model, select **Manage → Warnings** (Figure 5-2.26). This should be done occasionally to ensure there are no major issues with the model. Keeping this list free of major errors can improve the performance of the model as Revit is not constantly checking this situation.

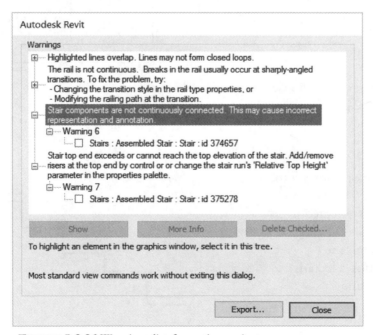

FIGURE 5-2.26 Warnings list for entire project

Exercise 5-3:
Introduction to Railings

This "exercise" will provide a basic introduction to the railing tool in Revit. There are no required steps for the law office project in this section. However, it is highly recommended that this information be reviewed prior to completing the steps required later in this chapter.

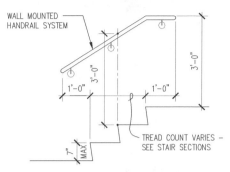

Railing Types

When the *Railing* command is selected, the railing Types in the current project are listed in the **Type Selector**. Compare this list to the same items listed in the **Project Browser** under Families\Railings\Railings. The items listed in the Project Browser can be Duplicated or Deleted.

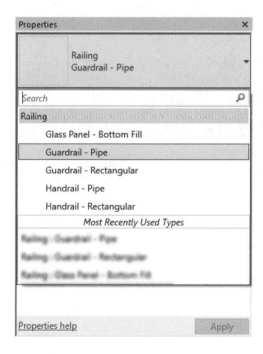

FIGURE 5-3.1 Type Selector; Railing tool active

FIGURE 5-3.2 Family\Railing in Project Browser

Railing Component Types

Similar to the way Stairs work, Railings have several high-level properties (i.e. Type Properties). These can be seen in Figure 5-3.3. Notice the option to select a Top Rail and Handrail type. Also, there is an option for two handrails on the same railing. This allows a Handrail to be placed on both sides of a railing.

The options available in the Top Rail and Handrail component drop-down lists is based on the Types created in the current project. These options can be seen in the Project Browser as shown in Figure 5-3.4. These Types can be Duplicated or Deleted via the right-click menu.

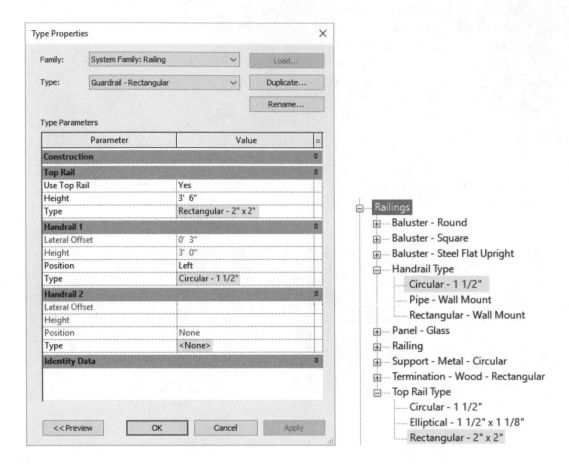

FIGURE 5-3.3 Guardrail – Rectangular railing properties

FIGURE 5-3.4 Component families used by railing

Conceptually, take note that a Top Rail and Handrail are more complex than just selecting a profile, as this will help you understand Revit's internal structure for these element types. The properties associated with the Top Rail and Handrail selected in the previous image can be seen in Figure 5-3.5.

In Revit, a Profile is a simple 2D closed outline—a series of connected lines and/or arcs. An example can be seen in Figure 5-3.7. Similar to other families, profiles can be parameterized. Thus, this simple circle is able to represent multiple diameters. The Project Browser shown in Figure 5-3.6 shows two sizes (i.e. types) for this profile. Just like doors, these types can be duplicated and new sizes created without the need to create a new profile family. However, creating a new profile is easy: **New → Family → Profile.rft** template.

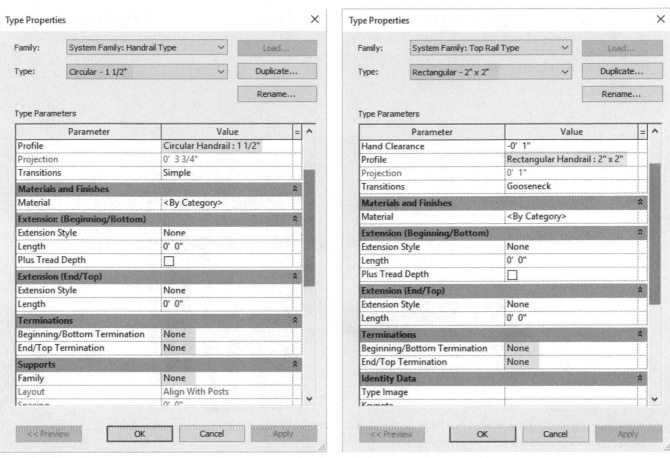

FIGURE 5-3.5 Type properties for two of the railing component types associated with "Guardrail – Rectangular"

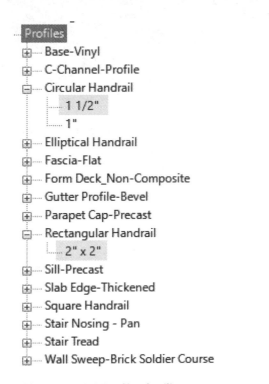

FIGURE 5-3.6 Profiles families used by handrail & top rail types

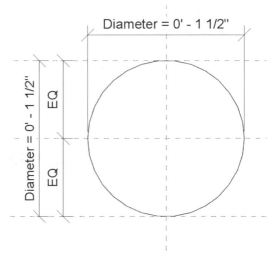

FIGURE 5-3.7 Profiles Circular handrail profile as seen in the family editor

In Figure 5-3.5, notice both Handrails and Top Rails have an option called **Terminations**. Handrails also have an option for **Supports**. In this example they are all set to "None." Figure 5-3.8 shows that a termination (aka escutcheon) occurs at the end of a Handrail (or Top Rail) and a Support is what holds the Handrail in place. Each of these items are separate families loaded into the current project.

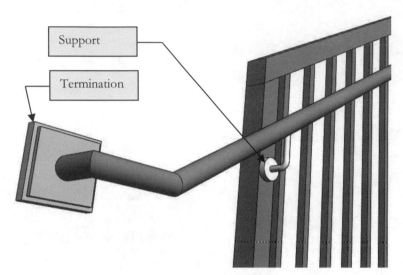

FIGURE 5-3.8 Termination and Supports example

Now that the various components and nested families in a railing have been covered, we need to back up to the beginning and look at how the rest of the railing is defined. This is the pattern, either horizontal, vertical or both, that fills that area from the stair/floor up to the Top Rail. You

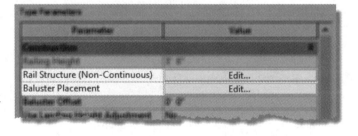

are about to see two of the most complex dialogs in Revit! We will compare the settings between the two guardrail types Pipe and Rectangular to help understand how the various options and settings work.

Rail Structure (Non-Continuous)

When a railing has horizontal/parallel rails, in addition to the Top Rail and Handrail, they are defined in the **Edit Rails (Non-Continuous)** dialog shown in Figure 5-3.10. To access this dialog, select a railing, click Edit Type, and then click Edit next to Rail Structure (Non-Continuous).

Looking at Figure 5-3.9 we notice the "Pipe" railing only has vertical elements. Thus, the Edit Rails dialog is empty. Contrast this with the "Rectangular" railing type which is mainly composed of horizontal/parallel rails.

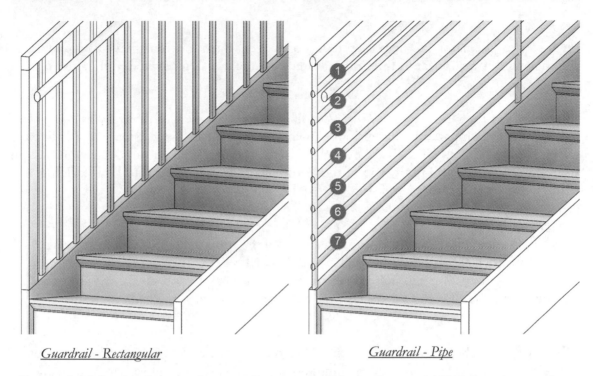

Guardrail - Rectangular   *Guardrail - Pipe*

FIGURE 5-3.9 Graphically comparing two railing types found in the Commercial-Default project template

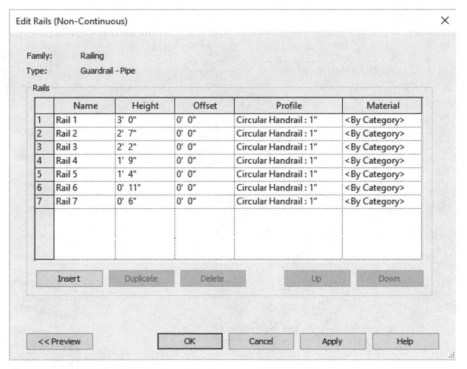

FIGURE 5-3.10 Defining horizontal elements within a railing

The numbers added to the "Rectangular" railing example, in Figure 5-3.9, correspond to the numbered rows in Figure 5-3.10. Notice each row has five settings as defined below.

Name

Each row can be named to help keep things organized. Each row must have a unique name. These named rows appear in the Edit Baluster Placement dialog covered in the next section.

Height

The Height of the rail is from the floor/landing or the front edge of the tread as shown in Figure 5-3.11. Compare the dimensions shown in this image to those listed in the Edit Rails dialog (Figure 5-3.10). Note that the height is to the top of the rail in each case. Dimensions from the tread edge and to the top of the Top Rail and Handrail relate to how building codes define stair, guardrail and handrail requirements.

Offset

This allows each rail to be offset horizontally from the main railing if needed. In previous versions of Revit this is how a handrail was defined. There are many creative ways this option can be used. Keep in mind there are two other "offset" options for a railing: the instance parameter *Tread/Stringer Offset* and the type parameter *Baluster Offset*.

Profile

This defines the shape continuously extruded along the handrail. Speaking of continuously, it is unclear why the dialog says "Non-Continuous"! Load and/or create additional Profiles in the current project for more options. Note that this could include something that looks like a stair stringer to maintain a consistent look when the railing at a stair transitions to a railing at an open floor edge (Figure 5-3.12).

Material

This defines the material applied to the shape created by the extruded profile.

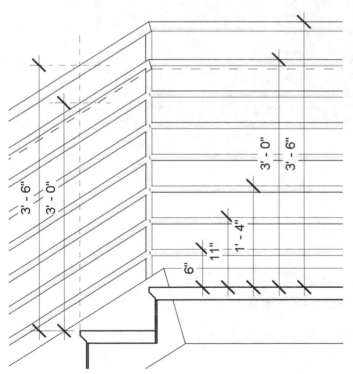

FIGURE 5-3.11 Railing rail dimensions

FIGURE 5-3.12
Railing with stringer

In the Edit Rails dialog, new rows can be inserted, and current rows can be duplicated or deleted. The order of the rows does not matter as the Height parameter dictates its position, but having them in the same relative order is helpful—thus the Up and Down buttons. Finally, expanding the preview option allows you to see the results without closing the dialog or clicking apply. However, clicking Apply will update the model without the need to close the dialog. This shows the changes in your model rather than a generic sample preview.

Baluster Placement

Where the *Rail Structure* dialog defines all the horizontal/parallel elements (except for top and handrail) the **Edit Baluster Placement** dialog defines all the vertical elements within a railing system. This dialog, as seen in Figure 5-3.13, has two main sections: **Main Pattern** and **Posts** as pointed out.

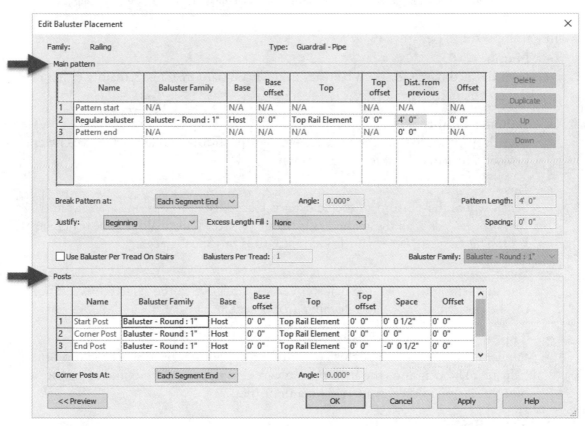

FIGURE 5-3.13 Edit Baluster Placement dialog; Guardrail – Pipe (4'-0" main pattern spacing)

Main Pattern
In the example above, for Guardrail – Pipe, there is not a "main pattern" per se. This is because the "main pattern" is defined by horizontal/parallel bars. Thus, in this example the Main Pattern is the "structural" supports for the railing system which are spaced 4'-0" (note the highlighted value in Figure 5-3.13). **FYI:** The 1" diameter pipe at 4'-0" on center may or may not be adequate to develop the code required lateral resistance at the top edge of the railing.

Comparing the two guardrail examples (Figure 5-3.9) we see the "Pipe" type has the "main pattern" spaced at **4'-0"** (Figure 5-3.13) whereas the "Rectangular" option spacing is only **4"** (Figure 5-3.14).

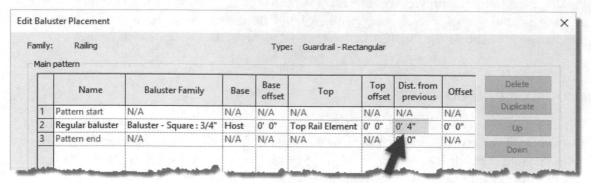

FIGURE 5-3.14 Edit Baluster Placement dialog; Guardrail – Rectangular (4" main pattern spacing)

Main Pattern settings per row:

- **Name**

  User defined name for each baluster in the main pattern. This is mainly to help keep things organized.

- **Baluster Family**

  This drop-down lists all the baluster families (post or panel) loaded in the current project. Note that *None* is also an option.

- **Base**

  Select where the bottom of the baluster is placed. Built-in options are *Host* and *Top Rail Element*. Also, all the named rails (from the Edit Rails dialog) appear here as well. **FYI:** The host can be a stair, floor, ramp, roof or toposurface.

- **Base Offset**

  Use this setting to move the baluster vertically relative to the selected *Base*. The value can be positive (for up) and negative (for down).

- **Top**

  Similar to the *Base* settings, the Top setting specifies the position of the top of the baluster. The height of the baluster can vary depending on the context, e.g. hosted on a stair (sloped) versus hosted on a floor (flat).

- **Top Offset**

  Use this setting to move the baluster vertically relative to the selected *Top*. The value can be positive (for up) and negative (for down).

- **Dist. From previous**

  Determines the spacing between the balusters.

- **Offset**

  Horizontally offset the baluster relative to the sketched path. The value can be positive and negative.

Both of these examples are simple and easy to understand. The example shown below depicts a more complex baluster pattern. Note that the Base and Top options selected relate to named rows in the Edit Rails dialog. This entire "main pattern" is repeated along the railing length. Note the **Justify** and **Excess Length Fill** settings as well.

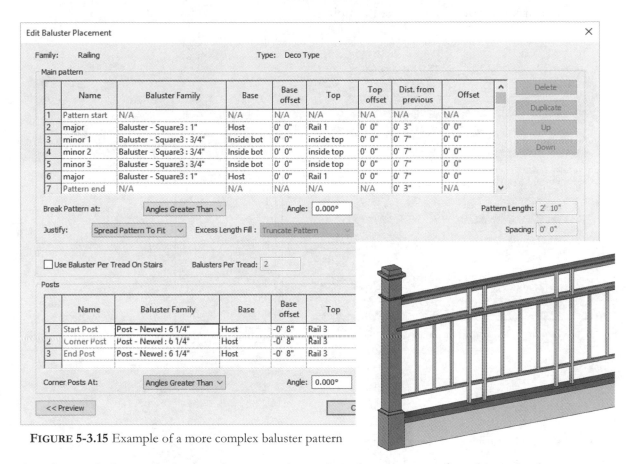

FIGURE 5-3.15 Example of a more complex baluster pattern

Another variation on **baluster placement** is to place them **per tread** as shown in the example below Figure 5-3.16. In this residential-type application there are two balusters per tread. Notice the more elaborate newel post family selected as a starting post.

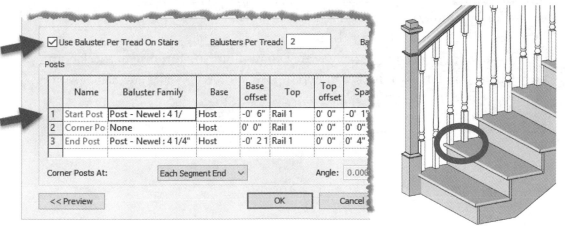

FIGURE 5-3.16 Baluster per tread

A **baluster family** is unique in that its top and bottom must be able to slope, at stairs, or be flat at a floor or landing. An example of the Baluster – Square family can be seen in the image below   shown in the family editor (Figure 5-3.17). Notice the **Top Cut Angle** and **Bottom Cut Angle** parameters. All of the required parameters are provided in the **Baluster – Post** family template file. If a new baluster family needs to be created, either copy (the RFA file) or start from the provided template.

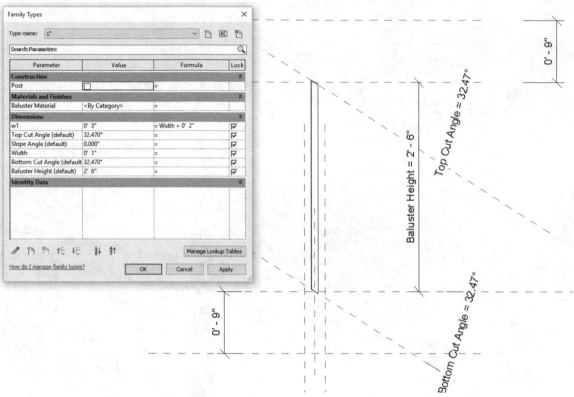

FIGURE 5-3.17 Baluster example in the family editor

For more information on Stairs and Railings in Revit:
- Visit the Autodesk Help website
- Visit www.revitforum.org, www.augi.com or https://forums.autodesk.com
- Check out **Tim Waldock's** in-depth posts on Revit's stairs and railings on his blog RevitCat: http://revitcat.blogspot.com/

Posts

The Posts section is more straightforward.

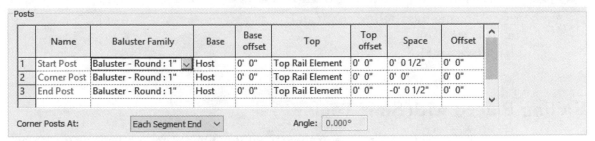

	Name	Baluster Family	Base	Base offset	Top	Top offset	Space	Offset	
1	Start Post	Baluster - Round : 1"	Host	0' 0"	Top Rail Element	0' 0"	0' 0 1/2"	0' 0"	
2	Corner Post	Baluster - Round : 1"	Host	0' 0"	Top Rail Element	0' 0"	0' 0"	0' 0"	
3	End Post	Baluster - Round : 1"	Host	0' 0"	Top Rail Element	0' 0"	-0' 0 1/2"	0' 0"	

Corner Posts At: Each Segment End   Angle: 0.000°

FIGURE 5-3.18 Posts section of Edit Baluster Placement dialog

Posts settings per row:

- **Name**
  There are only three built-in options named First, Corner and End.

- **Baluster Family**
  This drop-down lists all the baluster families (post or panel) loaded in the current project. Note that *None* and *Default* are also options.

- **Base**
  Select where the bottom of the post is placed. Built-in options are *Host* and *Top Rail Element*. Also, all the named rails (from the Edit Rails dialog) appear here as well.
  FYI: The host can be a stair, floor, ramp, roof or toposurface.

- **Base Offset**
  Use this setting to move the post vertically relative to the selected *Base*. The value can be positive (for up) and negative (for down).

- **Top**
  Similar to the *Base* settings, the Top setting specifies the position of the top of the post. The height of the post can vary depending on the context; e.g. hosted on a stair (sloped) versus hosted on a floor (flat).

- **Top Offset**
  Use this setting to move the post vertically relative to the selected *Top*. The value can be positive (for up) and negative (for down).

- **Space**
  Reposition the post along the length of the railing, if needed. One example might be to align the edge of the post with the vertical cut edge of the stringer below.

- **Offset**
  Horizontally offset the post relative to the sketched path. The value can be positive and negative.

Railing Tools

There are two main ways a railing can be added to a Revit project:

- In conjunction with the **Stair** or **Ramp** tools
- Separately, using the **Railing** commands
  a. Sketch Path
  b. Place on Host

Railing Placed with Stair Tool

When a new project is created using one of the default Revit templates, and a stair (or Ramp) is created, a railing is automatically placed on both sides. The image below (Figure 5-3.19) shows what the default results look like in a 3D view. These two railings are hosted by the stair and will update if the stair is modified. There are a number of things we can change to convey the design intent when needed.

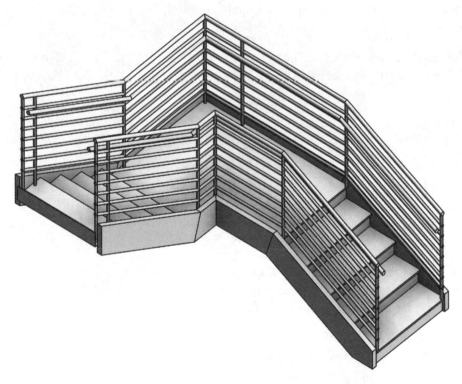

FIGURE 5-3.19 Default railing created with a new stair

If all railings have been deleted, use the **Railing → Place on Host** tool to quickly recreate the railings hosted by the stair. This tool only works if all railings have been deleted from the stair.

There are two settings related to railings we can adjust prior to creating the Stair (or Ramp). While creating the stair, click the **Railing** button on the Ribbon. This opens the Railing dialog shown in Figure 5-3.2. Here the railings type can be changed and its position relative to the stringer and treads. Note that both of these settings can be changed later.

Given the railing is hosted by the stair there are only a few things that can be done to modify the path.

The railing system can be continued at the top or bottom (or both) of the stair as shown in the image below (Figure 5-3.21).

FIGURE 5-3.20 Railing options within stair tool

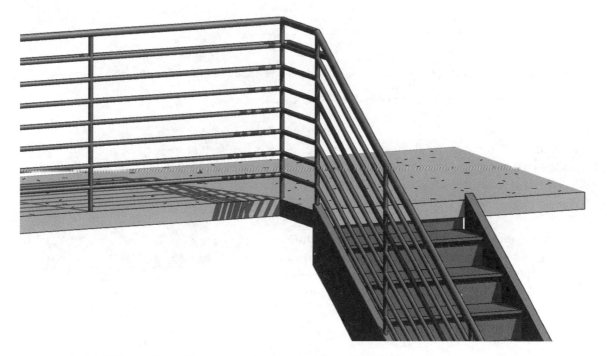

FIGURE 5-3.21 Extending railing at top of stair – along open floor edge

To do this (after the stair and railings are created):
- Start at the top or bottom level of the stair (or 3D view)
- Select the railing
- Click **Edit Path** on the *Ribbon*
- Select one of the **Draw** options (i.e. Line, Arc) from the *Ribbon*
- **Sketch** lines from the end of the stair hosted railing
- **Select** the newly sketched line (one at a time)
- Set the **Slope** option to **Flat** on the *Options Bar*
- See Figure 5-3.22

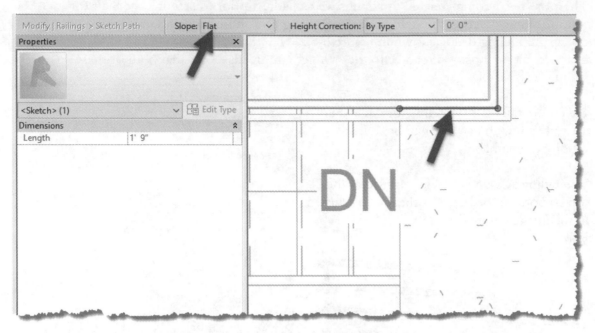

FIGURE 5-3.22 Extending railing at top of stair – along open floor edge

If the railing changes at the floor edge, for example the handrail stops or a stringer needs to be added to the railings as in Figure 5-3.12, then a separate railing element needs to be added—not part of the stair hosted railing. The trick in this case is adjusting the sketch lines so there is not odd overlap or gaps between the two railings.

When two sketch segments are selected, the Options Bar changes to show the **Rail Join** options. This can be changed if the transition between rails does not look as desired. This is similar to the **Edit Joins** tool on the Ribbon while in sketch mode.

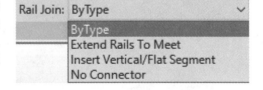

Railing → Sketch Path Tool

When a railing needs to be created apart from a stair or ramp, use the **Railing → Sketch Path** tool. When this command is selected, you simply sketch a path for the railing system to follow. If the railing has a Handrail, the side it is placed on is determined by which direction the sketch is created. While in sketch mode, click the **Preview** option to see the railing as it is to be sketched (similar to how Stair by Component works).

A sketched railing can be hosted to a floor—even a sloped floor. While in sketch mode, select the **Pick New Host** option on the Ribbon and then select the floor. The default host for a sketched railing is the level associated with the current plan-view.

The next two topics, **Handrails** and **Top Rails**, are relatively new features associated with a Railing type in Revit. Prior to Revit 2013, these elements were defined within the Edit Rails dialog. Now they are their own sub-element and can be selected separately from the main railing in the model (tapping the Tab key may be required). It is helpful to know this in case you are working in older project files—even if they have been upgraded. These old format railings do not allow the handrail or top rail to be selected.

Handrail

A Handrail, in Revit, is a component used within the Railing type. It cannot be used on its own. Keep in mind that any changes made to a Handrail type will change it wherever it is used. It is easy to edit a Handrail type while working in the context of a specific railing type, not realizing the same Handrail type is used in another railing type. In some cases, the Handrail type needs to be duplicated so changes do not affect other railing types.

Here are the main options related to a handrail type:

- Profile
- Material
- Extension (top and bottom)
- Terminations
- Supports (including spacing)

A few of these options have been adjusted to produce the results shown in the image below (Figure 5-3.23). The steps required to accomplish this will be covered next.

FIGURE 5-3.23 Handrail with several parameter adjustments

Railing Type Settings:

First, we notice the Railing type is where the Handrail is specified (Figure 5-3.24). In this case, two railing types are used, one for **Handrail 1** and one for **Handrail 2**. Creating multiple Handrail types is required if you want to have a different profile or position them at a different height as shown in this example.

For a center railing on a wide stair, Handrail 1 and 2 might be the same type but with different positions (one Left and one Right). This would place a handrail, at the same height, on both sides of the railing.

In this case, for two railing heights, the **Position** is the same.

This double-handrail example is not uncommon in elementary-grade schools. The normal higher rail is for adults and the lower for students.

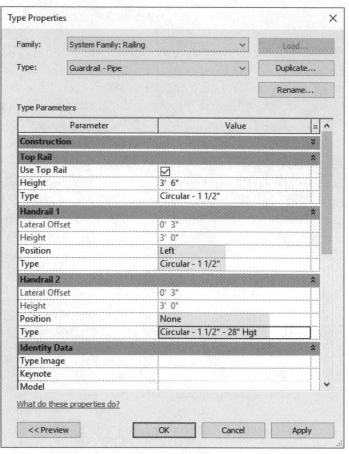

FIGURE 5-3.24 Handrail type and position

Creating a duplicate handrail is accomplished by **right-clicking** on a handrail type in the Project Browser and selecting **Duplicate** (Figure 5-3.25). Once a new type is created, the various properties covered on the next few pages can be adjusted as required.

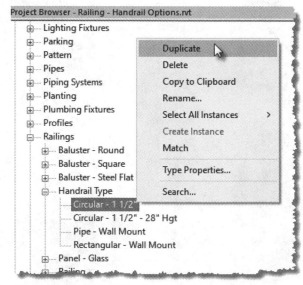

FIGURE 5-3.25 Duplicating a handrail type

Handrail 1 Properties:
The upper handrail, which is the default handrail as seen in Figure 5-3.9, has been modified per the following (see Figure 5-3.26).

- **Material:** Set to <u>Wood</u>

- **Extension Style:** Set to <u>Wall</u> (other options are None, Floor, Post). Default setting is None. This setting makes the handrail automatically return to the wall.

- **Length** (Bottom Extension): <u>1'-0"</u>. This will extend the length of the handrail from its default end, which aligns with the first riser. This is typically required to make the handrail comply with building codes.

- **Plus Tread Depth** (Bottom Extension): <u>Checked</u>. This makes the handrail continue along the slope the length of one tread. Checking this option makes the flat portion, controlled by the previous "Length" parameter, the correct height off the floor.

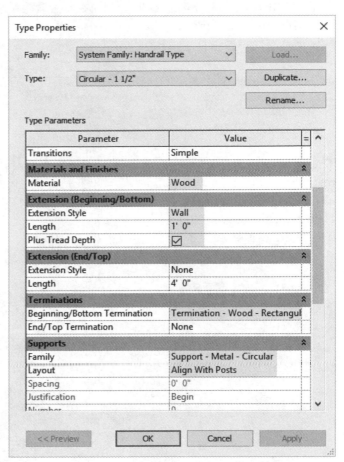

FIGURE 5-3.26 Upper handrail settings (i.e., Handrail 1)

- **Bottom Termination:** <u>Termination – Wood – Rectangular</u>. Places a family at the end of the handrail.

- **Family** (Supports): <u>Support – Metal – Circular</u>. Select the desired family for the handrail support.

- **Layout:** <u>Align with Posts</u>. Determines how the supports are placed/spaced. Other options are Fixed Distance, Fixed Number, Maximum Spacing and Minimum Spacing. Some of these options activate subsequent parameters such as Spacing and Justification.

The families listed for Terminations and Supports are created with a special subcategory designation Railing\Terminations and Railing\Supports.

Handrail 2 Properties:
The **lower handrail** is a duplicate of the previous handrail. Just the variations will be highlighted here (see Figure 5-3.27).

- **Height:** 2'-4"

- **Extension Style:** Set to <u>Floor</u>

The extension might more realistically just be set to wall like the upper handrail. However, this example helps to depict the options.

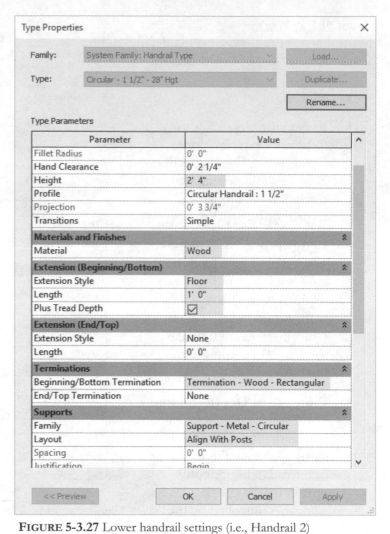

Type Properties   ✕

Family: System Family: Handrail Type   ⌄   Load...

Type: Circular - 1 1/2" - 28" Hgt   ⌄   Duplicate...

                                                          Rename...

Type Parameters

Parameter	Value
Fillet Radius	0' 0"
Hand Clearance	0' 2 1/4"
Height	2' 4"
Profile	Circular Handrail : 1 1/2"
Projection	0' 3 3/4"
Transitions	Simple
Materials and Finishes	≈
Material	Wood
Extension (Beginning/Bottom)	≈
Extension Style	Floor
Length	1' 0"
Plus Tread Depth	☑
Extension (End/Top)	≈
Extension Style	None
Length	0' 0"
Terminations	≈
Beginning/Bottom Termination	Termination - Wood - Rectangular
End/Top Termination	None
Supports	≈
Family	Support - Metal - Circular
Layout	Align With Posts
Spacing	0' 0"
Justification	Begin

        << Preview   OK   Cancel   Apply

FIGURE 5-3.27 Lower handrail settings (i.e., Handrail 2)

This stair shows a custom railing with both the Handrail and Top Rail employing a custom Extension Style.

Top Rail

The Top Rail is defined via a Railing's *Type Properties* as shown in Figure 5-3.28. Additional Top Rail types can be created by duplicating types listed in the *Project Browser*; **Families → Railings – Top Rail Type**. This is also where the height is defined.

The Top Rail has settings similar to the Handrail, but no supports, and can be adjusted separately from the Handrail. In the image below the Top Rail's **Extension Style** is set to **Post**. It can also be set to Wall, Floor or None just like the Handrail (Fig. 5-3.29).

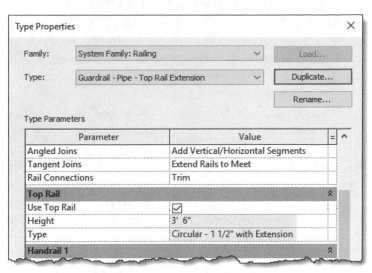

FIGURE 5-3.28 Top Rail defined in railing's type properties

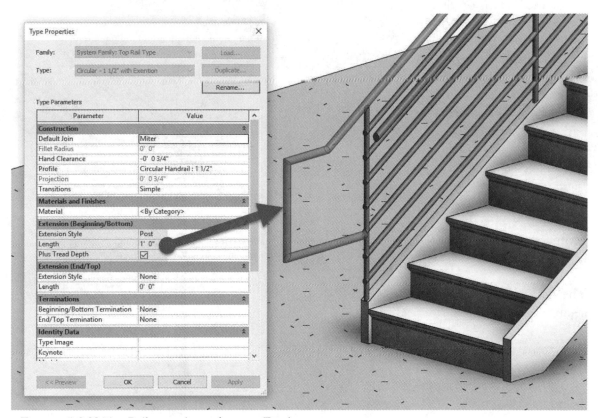

FIGURE 5-3.29 Top Rail extension style set to 'Post'

Remember that changes made to the Top Rail properties will affect all Railing types which use it. Notice the Top Rail type selected in Figure 5-3.28 is different than the Top Rail type specified in Figure 5-3.24 (also compare Top Rail pictures in the Handrail section).

Edit Rail (Top Rail or Handrail)

When a Handrail or Top Rail is selected, an option to edit the path of the railing appears on the Ribbon. This option only allows editing the path in a vertical plane aligned with the railing.

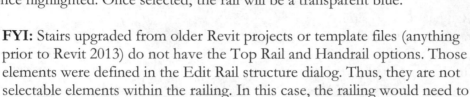

The Handrail or Top Rail can be selected by hovering the cursor over the Top Rail, being careful not to move the mouse, and then tapping the Tab key and clicking to select once highlighted. Once selected, the rail will be a transparent blue.

> **FYI:** Stairs upgraded from older Revit projects or template files (anything prior to Revit 2013) do not have the Top Rail and Handrail options. Those elements were defined in the Edit Rail structure dialog. Thus, they are not selectable elements within the railing. In this case, the railing would need to swap it out for a new supported type.

When the Extension Style options, Wall, Floor, Post and None, do not cover what you want, use the **Edit Rail** option. Clicking this from an elevation, section or 3D view allows a custom path to be defined.

The image to the right, Figure 5-3.30, shows the Top Rail path being modified. Notice the geometry appears as each line is sketched.

The custom result can be seen in Figure 5-3.31.

Unfortunately, at this time the railing cannot change planes/direction—to wrap around a column for example.

Selecting the Top Rail and clicking **Reset Rail** on the Ribbon will undo any changes.

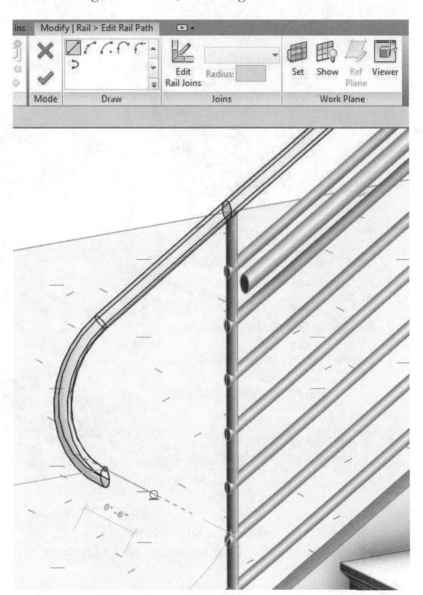

FIGURE 5-3.30 Editing the Top Rail path

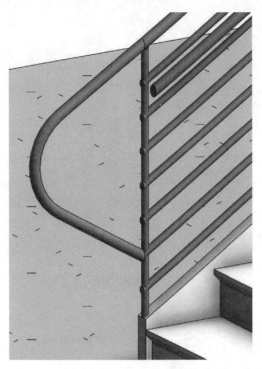

FIGURE 5-3.31 Custom Top Rail defined

FIGURE 5-3.32
Handrail only with supports at 6'-0" O.C. max.

Handrail Only

A railing or guardrail is not required when a stair, or ramp, is against a wall (Figure 5-3.32). In this case we just need a handrail without a railing. Unfortunately, this is not directly possible as there is no Handrail tool, just the Railing tool. Here is how this can be accomplished...

The first step is to **Duplicate** a railing—Guardrail – Rectangular is a good option as it requires fewer changes. In the new railing's Type Properties, set the **Top Rail** Use Top Rail to **unchecked** (Fig. 5-3.33). Next, click Edit for **Rail Structure** and Delete all rows. Finally, click Edit for **Baluster Placement** and set everything, Main pattern and posts, to **None** (Fig. 5-3.34).

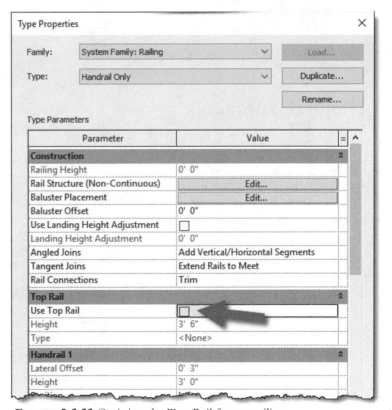

FIGURE 5-3.33 Omitting the Top Rail from a railing

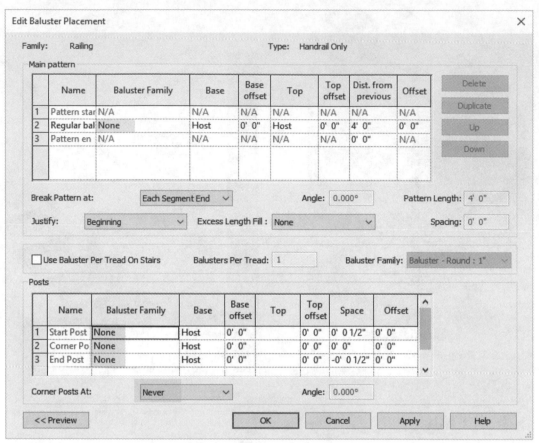

FIGURE 5-3.34 Omitting Balusters from a railing

Handrail Supports

Within the Handrail <u>Type</u> supports can be selected (Figure 5-3.35). The **Family** drop-down is based on special families currently loaded in the project file. The **Layout** option selected determines which of the remaining parameters are editable. In this example, Maximum Spacing causes the Spacing parameter to be active, but Justification and Number are inactive (i.e. greyed out).

The automatically placed supports do not always land in the ideal position. Fortunately, they can be repositioned and even deleted individually. To modify a support it

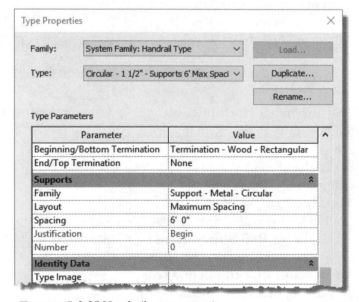

FIGURE 5-3.35 Handrail support settings

must be selected. To select a Support, hover the cursor over a support (don't move it) and then tap (not press) the Tab key until the Support highlights—and then click to select it. Select the pin icon to unpin that instance (Figure 5-3.36). It can now be moved or deleted as needed.

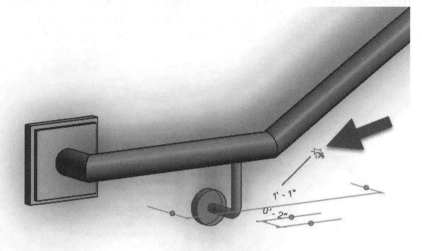

FIGURE 5-3.36 Support selected and un-pinned

Railing Hosting Opportunities

Revit has the ability to host railings to floors with modified points and, as shown below, the top of walls—even when they have been modified with Edit Profile. Railings can also be hosted by toposurfaces and roofs. Thus, a railing could be used to represent a fence on the site or lightning protection rods (using a highly customized railing type) on the roof.

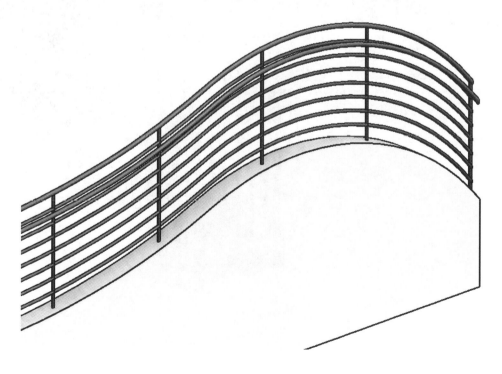

Sample Railing File

Make sure to examine the stair and railing sample file available with the provided Revit content; look in the **System Families** folder (click the *Imperial Library* shortcut on the left, in the Open dialog). While in this Revit project file, select a stair or railing and view its properties to see how it works. It is also possible to Copy/Paste selected examples into another open Revit project file.

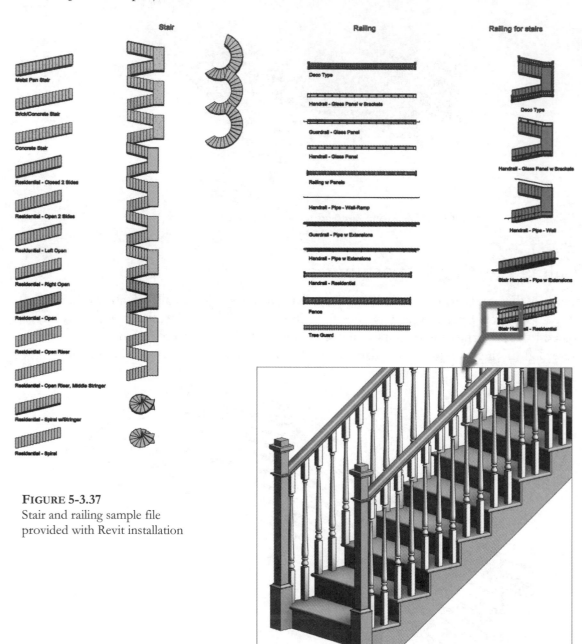

FIGURE 5-3.37
Stair and railing sample file
provided with Revit installation

Exercise 5-4:
Ramps and Sloped Floors

When two floors do not align vertically, or accessibility is required, a ramp must be provided. There are two main ways to model this in Revit. One is with the Ramp tool and the other is with the Floor tool. **For the most part the Ramp tool should never be used.** Rather, the Floor tool should be preferred as it generally works better as we shall see here.

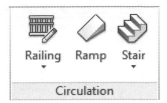

The Ramp Tool:

First, we will review how the ramp tool works and then talk about its limitations.

The ramp tool is closely related to the original *Stair by Sketch* tool in how it works. Starting at the lower level in a plan-view, when the Ramp tool is started, Sketch mode is entered and the **Run** option is selected on the Ribbon. First, verify the **Base Level** and **Top Level** properties are set correctly—this determines the required horizontal distance. Picking points on the screen will define the center of the ramp (Figure 5-4.1). Based on the **Ramp Max Slope** setting (via Edit Type) the required length, in plan, is listed. **Landings** are automatically added between gaps as shown in the two images below (Figure 5-4.1 and 5-4.2). The sketch lines can be modified if needed. For example, maybe one side of the ramp runs along an angled or curved wall.

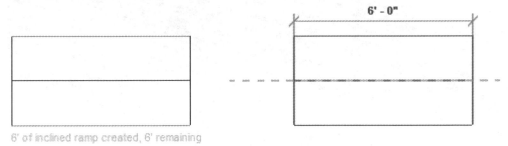

6' of inclined ramp created, 6' remaining

FIGURE 5-4.1 Ramp sketch mode; indicates length required based on vertical distance

12' of inclined ramp created, 0" remaining

FIGURE 5-4.2 Ramp sketch mode; sketch complete with landing

Once the sketch is complete and the Finish (green check mark) is selected, the ramp is created as seen in Figure 5-4.3. By default this will include a railing on each side of the ramp.

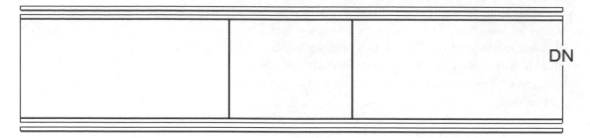

FIGURE 5-4.3 Ramp created including railings on each side

Looking at the ramp in 3D we notice the handrail is on the wrong side and the posts are half on, and half off, the ramp (Figure 5-4.4). Selecting the railings in plan (one at a time) and then clicking the **flip control-arrows** will move the handrail to the correct side. Also, with the railing selected, the **Tread/Stringer Offset** will reposition the railing.

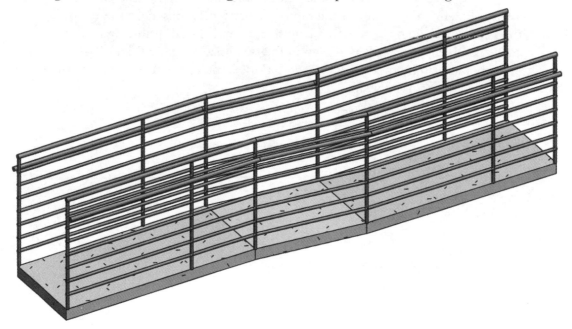

FIGURE 5-4.4 3D view of ramp

Most of the Instance (left) and Type (right) properties shown in Figure 5-4.5 are pretty straightforward. By default Revit set the **Base Level** to match the level the current plan-view is based on and the **Top Level** to the next Level above. This usually works as a Level should generally be created for any surface you can walk on in a building. In this example, the Top Level was changed to match the Base Level and the Top Offset was modified to define the total vertical distance.

The Type property **Shape** determines the shape (or profile) of the ramp in section as seen in Figure 5-4.6, set to **Thick**, and Figure 5-4.7 set to **Solid**. Also notice the Ramp Max Slope (1/x) setting. This defines the horizontal distance required for every 1" of vertical rise. This is typically set based on building code and accessibility code requirements.

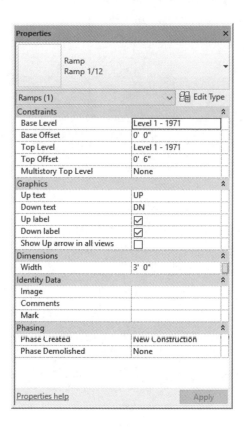 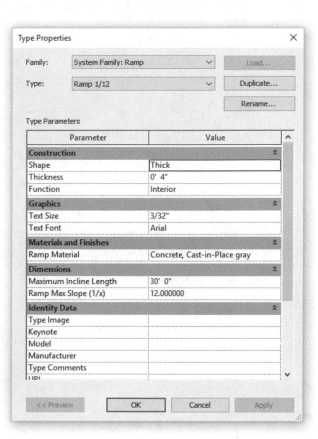

FIGURE 5-4.5 Ramp instance (left) and Type (right) properties

FIGURE 5-4.6 Ramp section with Shape properties set to Thick

FIGURE 5-4.7 Ramp section with Shape properties set to Solid

Now that the basics of the Ramp tool have been covered we will look at the limitations. The Ramp tool does not allow "Layers" of construction like the Floor tool does. Notice in Figure 5-4.5 that there is just one simple **Thick** property to define the thickness of the ramp. Thus, the ramp cannot look correct in both plan (Figure 5-4.8) and section (Figure 5-4.9), as there is only one material parameter for ramps. In previous versions of Revit, the **Spot Elevation** and **Spot Slope** annotation tools did not work on a Ramp element; they do now.

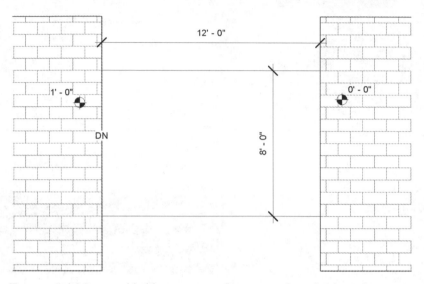

FIGURE 5-4.8 Ramp added between two floors – no floor finish or slope arrow

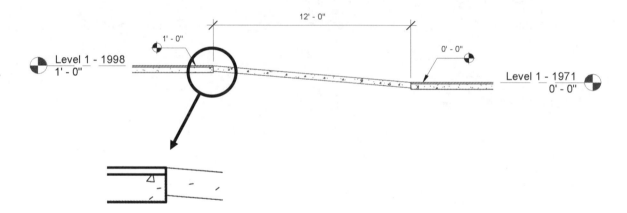

FIGURE 5-4.9 Ramp shown in section next to a concrete floor with a tile finish

The **Paint** tool can be used to apply a different material to the surface of the ramp, but the section and details still would not look correct. A Ramp cannot be joined to the adjacent floor construction as highlighted in the image above.

Sloped Floor:

An alternative to using the Ramp tool is to use the Floor tool and adjust it to slope.

First, we will look at the results of creating the same "ramp" as in the previous example. This Floor element can have multiple "Layers" of construction, including a separate surface finish. Additionally, we can apply a Spot Elevation and/or Spot Slope annotation as desired. The plan-view below has a Spot Slope element applied to indicate the slope and direction (Figure 5-4.10).

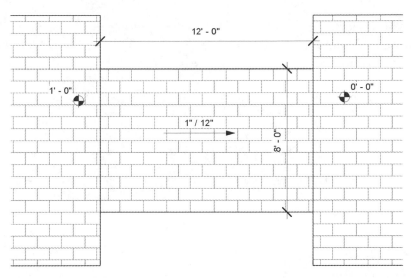

FIGURE 5-4.10 Sloped floor element added between two floors

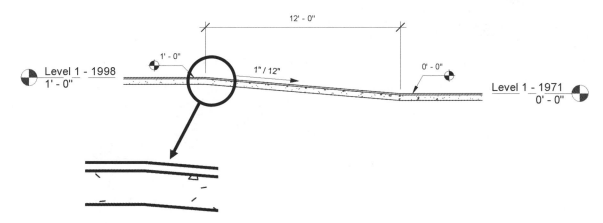

FIGURE 5-4.11 Sloped floor shown in section – joined to adjacent flat floor

As seen in the section above, the sloped floor joins nicely with the adjacent flat floor elements to create a continuous monolithic look that is common in this situation (Figure 5-4.11). Note that the Join command is used to achieve this look.

The only limitation to using a Floor rather than a Ramp element is a Railing cannot be added automatically. However, a Railing can be manually added and hosted to the floor.

Given the two examples, Ramp vs. Floor, it is generally preferred to use a Floor rather than a Ramp. Now we will look at how to create a sloped floor.

There are four ways to make a floor slope:
- Sketch line **defines slope**
- Sketch line **defines constant height** (preferred for existing conditions)
- **Slope Arrow** in sketch mode
- **Shape Editing** (generally avoid for simple ramps)

A sketch line, when selected, can be set to **Defines Slope** on the Options Bar (Figure 5-4.12). For a ramp, only one edge needs to be set to Defines Slope. Once Defines Slope is selected, the Slope parameter may be modified; e.g. 1"/12".

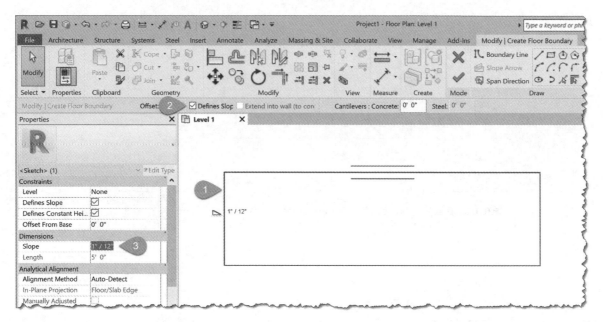

FIGURE 5-4.12 Using Defines Slope in floor sketch

When a sketch line is selected, while in sketch mode, there is a parameter called **Defines Constant Height** (Figure 5-4.13). Checking this box allows the **Offset From Base** to be modified. This value sets the specific height of that edit of the floor relative to the selected Level. This is helpful when documenting existing conditions and the slope is not known or not perfectly constructed per the original construction documents.

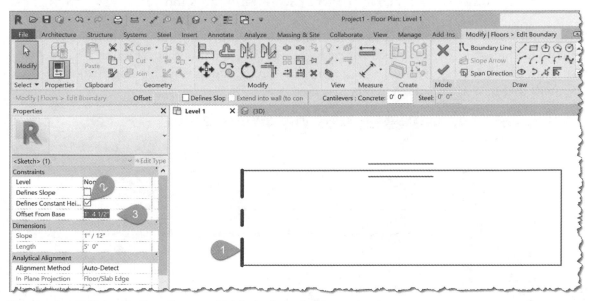

FIGURE 5-4.13 Using Defines Constant Height in floor sketch

The **Slope Arrow** option is easy to use while in sketch mode; simply select the Slope Arrow option on the Ribbon and point to point to define the direction the floor should slope (Figure 5-4.14). With the Slope Arrow selected, the Properties Palette lists the slope related settings. Changing **Specify** to "Slope" allows the Slope parameter to be set; e.g. 1"/12".

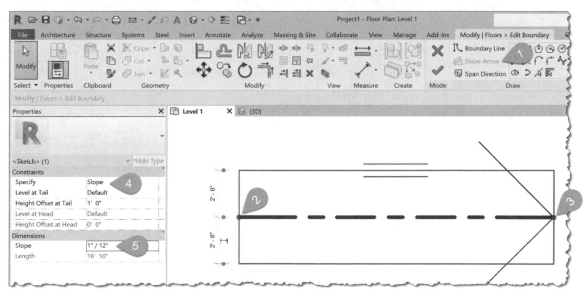

FIGURE 5-4.14 Using Slope Arrow in floor sketch

The **Shape Editing** feature is a little too complicated and should be avoided unless the surface warps or the bottom of the ramp needs to be flat. This feature is more for sloping a slab or roof around a floor/roof drain. To use this option, select the completed floor (not in sketch mode) and click the **Modify Sub Elements** tool on the Ribbon (Figure 5-4.15). Select one of the corner grips and then edit its offset value that appears next to the grip. Edit the grip for the other side of the ramp. Press Esc to finish the command.

If the bottom of the ramp needs to be flat, select the floor, go to Edit Type and then Edit Structure. Check the Varies option for each layer to have a flat bottom (Figure 5-4.15). The result can be seen in Figure 5-4.16. Note that the Varies setting will only affect floor instances that have their sub elements modified.

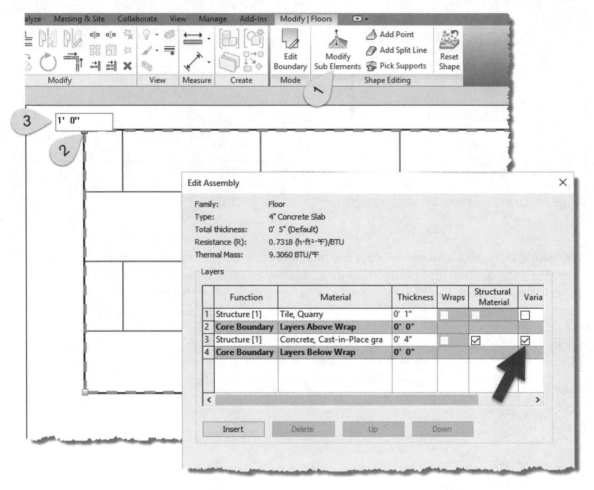

FIGURE 5-4.15 Modifying a floor's sub elements & floor structure dialog

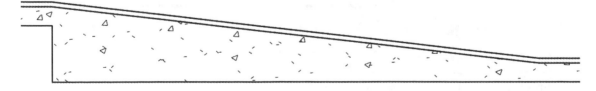

FIGURE 5-4.16 Floor with flat bottom

Exercise 5-5:
Elevators

Commercial buildings with more than one story often have an elevator for convenience and to comply with accessibility codes. This section will take a look at how they can be modeled within Revit.

Shaft Walls:

First, an elevator is contained within a vertical shaft defined by four walls as seen in the image to the right (Figure 5-5.1). These walls are modeled using **Revit's wall tool**. These walls should be modeled as they will be built. In this example the masonry walls are bearing walls (i.e. they are structural and will support the floors and roof) so a single wall extends from the footing to the roof. If the walls are built on each floor, then separate walls should be created at each level. Also, if different finishes are required at each floor, then separate walls are added adjacent to the shaft walls.

The dimensions of the shaft are defined by the elevator manufacturer. The width and depth are derived from the "car" size and required tracks/rails. The height can vary depending on the type of elevator—this often includes a pit (an area below the lowest stop) and an overrun (space above the highest stop).

Shaft Openings:

It is important that the hole in the floor align vertically. The best way to ensure this is to use the **Shaft Opening tool**. This tool is a vertical extrusion which cuts each floor it comes in contact with. This is preferred over editing the sketch of each floor to define the opening—which would potentially be wrong at one or more floors. Depending on building construction, this shaft is either on the inside or the outside of the shaft walls. If inside, the floors will extend into the shaft walls.

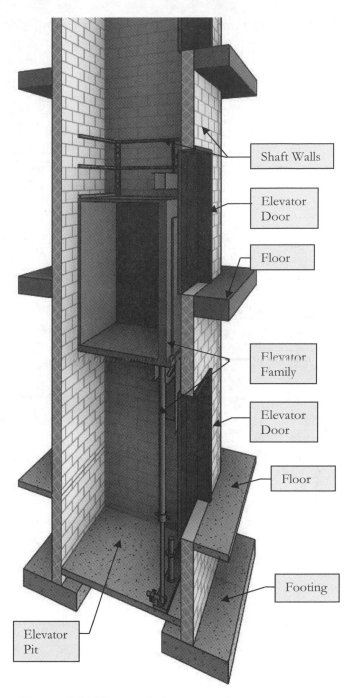

FIGURE 5-5.1 Elevator shaft cut away view

Shaft Walls

Elevator Door

Floor

Elevator Family

Elevator Door

Floor

Footing

Elevator Pit

Elevator Family:

The installed version of Revit does not come with any sample elevator content (except in the Metric Library folder).

The value in manufactured content is that it is the correct size and often has helpful features—for example, the family used on the previous page includes a box which defines the required shaft size, including pit depth and overrun. However, in the floor plan-views it does not follow the standard architectural graphics conventions for an elevator (whereas the generic one shown above does).

Another option is to create custom content. It is generic and parametric so it looks correct in plans and sections and is manually adjustable based on the elevator selected. There is a good example of a custom elevator family, including call buttons and indicator lights, in a thread at this location: http://www.revitforum.org/architecture-general-revit-questions/12415-elevators-what-do-you-do.html.

Yet another option is **DigiPara Elevatorarchitect**, which is a free Revit add-in. This tool includes several elevator manufacturers such as ThyssenKrupp, OTIS, KONE, Schindler and more. URL: http://www.digipara.com/.

Elevator Doors:

Elevators have several sets of doors. One set is associated with the elevator car while the rest are located on each floor the elevator stops at. The elevator cab door is represented in the elevator family. The doors at each floor are separate families individually placed at each floor.

Having separate elevator door families at each level will accommodate the various finishes at each level. **TIP:** when using multiple walls to represent the shaft wall and floor-specific finishes, use the join tool on those walls. This will cause the door family to cut through all walls.

Elevator Misc:

Elevators usually require an Elevator Equipment Room. This would be sized and positioned per the manufacturer's literature. If the elevator has a pit, it most likely requires a "pit ladder" for maintenance personnel to access the pit (this may have to be a custom family if one cannot be found).

Exercise 5-6:
Adding Utilitarian Stairs and Railings

In this exercise you will create a utilitarian stair in a stair shaft; this is a stair required for egress, with simple materials, and not intended to be primary circulation.

The Utilitarian Stair

The back stair is required for egress and is often referred to as a utilitarian stair due to the simple and durable finishes.

Utilitarian stairs are not readily visible and are primarily meant for emergency egress; their existence is primarily dictated by code. They certainly may be used as a means of vertical circulation within the building.

The finishes and light fixtures are simple, often painted CMU walls and surface mounted lights.

In stairs, building codes often require a gate at the ground level and at the upper level, if the stair continues to the roof. The reason for this is to help guide people in an emergency. Each floor in the utilitarian stair shaft looks the same, so it would be easy for a person, especially in a panicked state, to pass the ground level exit and continue down one or two more floors to the basement, extending their time in a dangerous situation. If the stair is used for normal, day-to-day, vertical circulation, some building officials will allow the gate to be placed on a magnetic hold-open which allows the gate to close in the event of a fire alarm system activation.

Utilitarian Stair Example
Durable, average cost stair with simple design

Most commercial codes also require that a sphere 4″ or larger not be able to pass through the guardrail. This has a large impact on design but creates a safe environment for the users; looking at the photo to the right, it is not difficult to see how older codes allowed for unsafe designs!

Next you will add stairs to your East and West stair shafts. Revit provides a powerful *Stair* tool that allows you to design stairs quickly with various constraints predefined (i.e., 7″ maximum riser).

Type Parameters

Before you draw the stair it will be helpful to review the options available in the stair family.

1. Open ex4-3.rvt and Save As **ex5-6.rvt**.

2. From the *Project Browser*, expand the Families → Stairs → Stair (i.e., click the plus sign next to these labels).

3. Right-click on the stair type **7″ max riser 11″ tread** and select the **Type Properties** option from the pop-up menu.

You should now see the options shown in Figure 5-6.1.

Take a couple minutes to see what options are available. You will quickly review a few below.

- <u>Tread</u>: Depth of tread in the plan view.

- <u>Nosing Length</u> (Depth): Treads are typically 12″ deep (usually code min.) and 1″ of that depth overlaps the next tread. This overlap is called the nosing.

- <u>Riser</u>: This provides Revit with the maximum dimension allowed (by code, or if you want it, less). The actual dimension will depend on the floor-to-floor height.

- <u>Stringer dimensions</u>: These dimensions usually vary per stair depending on the stair width, run and materials, to name a few. A structural engineer would provide this information after designing the stair.

- <u>Cost</u>: Estimating placeholder.

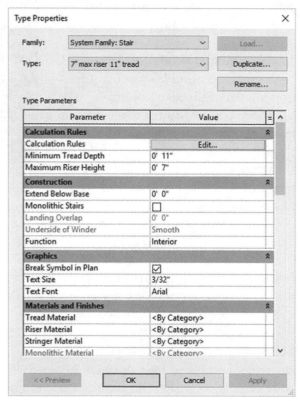

FIGURE 5-6.1 Stair type properties

Drawing the Stairs in Plan

You will be drawing a standard switch-back stair. At first, when using Revit to draw stairs, it may be helpful to figure out the number of risers and landings. That information will be helpful when drawing the stair. As you become more familiar with the *Stair* tool you will not need to do those calculations to draw a stair. Dividing the floor-to-floor height of 12'-0" by 7" we get 20.57. Obviously you cannot have a fraction of a riser so you need to round up to 21 (rounding down would make the riser higher than 7"). Therefore, 12'-0" divided by 21 equals 6.86". Thus you have 21 risers that are 6.86" high. Additionally, most codes would require a landing in a stair rising 12'-0".

4. Make sure you are in the **Level 1** floor plan view.

5. **Zoom in** to the West stair shaft.

6. Click on the **Architecture** tab on the *Ribbon*.

7. Select the **Stair** tool (circulation panel). Stairs

8. On the *Options Bar*, set the *Location Line* to **Exterior Support: Right**.

9. Also on the *Options Bar*, set the *Actual Run Width* to **3'-6"**.

Location Line: Exterior Support: Right ▼ Offset: 0' 0" Actual Run Width: 3' 6" ☑ Automatic Landing

10. Position the cursor approximately as shown in **Figure 5-6.3**; you are selecting the start point for the first step. Make sure you are snapping to the wall with *nearest*.

11. Pick the remaining points as shown in Figures 5-6.4, 5-6.5 and 5-6.6.

12. Hold the **Ctrl** key and then select the two runs of stairs (but not the landing or railings).

13. Click the **green check mark** to finish the stairs (Figure 5-6.2).

FIGURE 5-6.2 Stair Contextual Tab

14. Switch to **Level 2** and repeat the previous steps to add stairs from Level 2 up to Level 3.

Notice as you draw the stairs, Revit will display the number of risers drawn and the number of risers remaining to be drawn to reach the next level. If you click *Finish Stairs* before drawing all the required risers, Revit will display an error message. You can leave the problem to be resolved later.

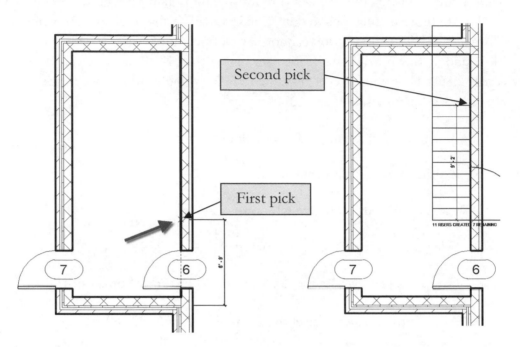

FIGURE 5-6.3 1st pick   **FIGURE 5-6.4** 2nd pick

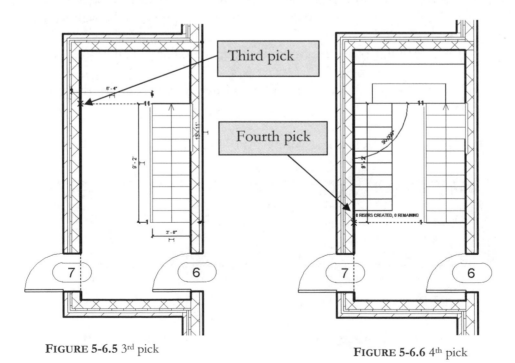

FIGURE 5-6.5 3rd pick   **FIGURE 5-6.6** 4th pick

FYI: *Revit has drawn the intermediate landings between levels. However, the landings at the main floor levels have not been created but could be using the* Landing *option within the* Stair *tool. Some projects extend the primary floor structure into the stair shaft to act as the landing for that level and also support the stair. In a later lesson you will draw a floor system that extends into the stair shaft.*

15. Repeat these steps for the East stair shaft.

16. **Add doors** to the <u>second</u> and <u>third floors</u> for both the East and West stair shafts (interior doors only!).

17. **Save** your project.

The final stairs still need some work, but with just a few steps you have a nice placeholder. Using the information presented earlier in this chapter, the stair can be further developed as the design progresses. Revit will add railings to both sides of the stair by default (this can be turned off on the *Ribbon* while in the Stair tool). The railings are separate elements which are hosted to the stair, much like a door is hosted to a wall. You can select a railing and delete it without deleting the stair. But if the stair gets deleted, the railings must go.

Exclusive Bonus Chapter:
Consider working through the **multistory stair** tutorial provided online with this book; see the inside front cover for instructions. The result of the exercise is the single Revit stair element shown to the right. If in a classroom setting, your instructor may assign this and possibly have you go back and apply this to this project.

Self-Exam:

The following questions can be used as a way to check your knowledge of this lesson. The answers can be found at the bottom of this page.

1. Stair landings are added automatically. (T/F)

2. A stair sample file can be downloaded from SDC Publications. (T/F)

3. Stairs and railings are shown as true 3D objects in plan views. (T/F)

4. A stair's railing cannot be deleted. (T/F)

5. Stair types are listed in the *Project Browser* under *Families*. (T/F)

Review Questions:

The following questions may be assigned by your instructor as a way to assess your knowledge of this section. Your instructor has the answers to the review questions.

1. Handrail supports can be moved or deleted. (T/F)

2. The riser is typically calculated to be as large as the building codes will allow. (T/F)

3. Stairs have multiple sub-categories which can be used to control stair visibility. (T/F)

4. To change a stair, select it and then click _____ on the *Ribbon*.

5. A railing extending past the end of a stair run must always slope. (T/F)

6. To extend the stair's stringer you draw _____ lines.

7. Posts, within a railing instance, can be unpinned and repositioned. (T/F)

8. After starting the *Stair* tool, the first thing you need to do is set the _____ level and the _____ level.

9. Railings can be modified to be shorter than the length of the stair. (T/F)

10. The riser height is based on the floor to floor height. (T/F)

Lesson 6
ROOF:

This lesson will look at some of the powerful options and tools for designing a roof for your building. You will also add skylights.

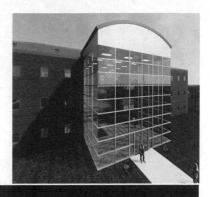

Exercise 6-1:
Hip Roof

The first step is to create a floor plan view at the roof level, at the top of your exterior masonry wall. This will create a working plane, called a level datum, for the *Roof* tool.

Add Level Datum

1. Open ex5-7.rvt and **Save As ex6-1.rvt**.

2. Open the **South** elevation view.

3. Select **Architecture → Datum → Level** tool from the *Ribbon*.

4. Draw a *Level* datum at the top of the exterior wall, at elevation 36'-0"; see the wall properties. Draw the datum "line" so both ends align with the other datum "lines" below it.

Next, you will rename the level datum. By default Revit will name the level based on the previous level created, plus 1. Thus, the new level should be named "Level 4." You will change this name so you (and others working on the same project) know it is the top of the masonry view and not another floor in the building.

5. Press **Esc** or select **Modify** from the *Ribbon*.

6. Now select the level datum you just drew (click on the line).

7. With the level datum selected, click on the [level name] text to rename the level label.

8. Change the label to **T.O. Masonry** (Figure 6-1.1).

   FYI: *T.O. means "Top Of."*

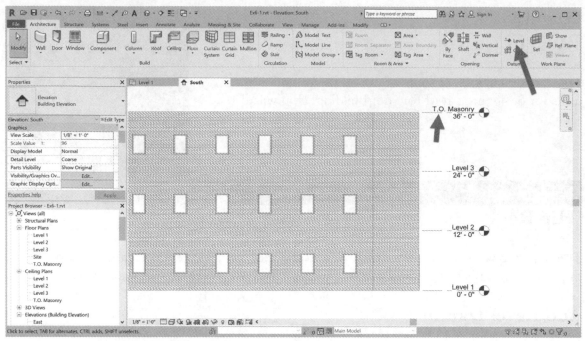

FIGURE 6-1.1 Renamed level datum

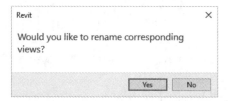

9. Click **Yes** when prompted to rename corresponding views (Figure 6-1.2).

FIGURE 6-1.2 Rename prompt

These steps are the same you used to add the third floor. Notice the "T.O. Masonry" label is now listed in the *Floor Plans* section of the *Project Browser*. You can delete the T.O. Masonry ceiling plan if you'd like as it will not be needed; simply right-click and select delete from the pop-up menu.

Add a Hip Roof

10. Open the newly created **T. O. Masonry** Floor Plan view.

11. In Properties, while nothing is selected, set the (Underlay) **Range: Base Level** to **Level 3**.

12. Select **Architecture → Build → Roof** (down-arrow).

13. Select **Roof by Footprint** from the drop-down list.

Before you start the roof, you will change the slope (pitch) of the roof.

14. Click the **Properties Filter** drop down and select New <Sketch> as pointed out in Figure 6-1.3.

15. Make sure **Defines Slope** is checked and then change the **Slope** to **6″/12″**. Click **Apply**.

This will make the roof pitch 6/12, which means for every 12″ horizontally the roof will *rise* 6″ vertically.

16. You are now prompted to select exterior walls to define the footprint. Select ONLY the wall segments that define the 120′-0″ x 60′-0″ portion of the building (Figure 6-1.4). Pick the exterior side of the walls.

You will notice in Figure 6-1.4 that there are three sections along the perimeter of the rectangle that are open because no wall is available to pick. You will need to draw three lines to close the "footprint."

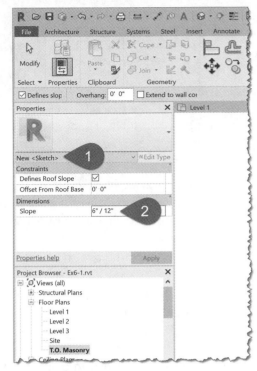

FIGURE 6-1.3 Properties Palette

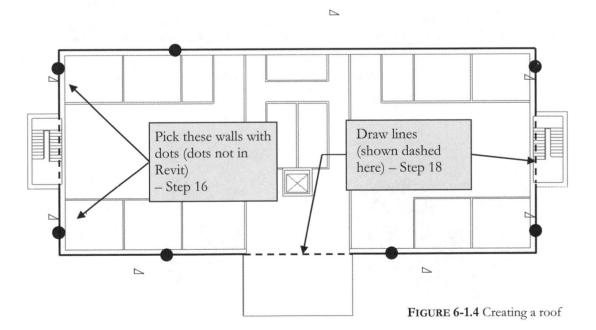

FIGURE 6-1.4 Creating a roof

17. Select the **Line** tool from the *Ribbon* (Figure 6-1.5).

FIGURE 6-1.5 Ribbon: Roof tool active

18. Draw three lines to create a complete rectangle, making sure you use the snaps to accurately snap to the endpoint of the lines already present: one line across the atrium and the other two at the stair shafts (Figure 6-1.4).

19. Now click the **green check mark** from the *Ribbon*.

20. Click **NO** when prompted to attach the highlighted walls to the roof (Figure 6-1.6).

You will now see a portion of the roof in your plan view. The cutting plane is 4′-0″ above the "floor" level, so you are seeing the roof thickness in section at 4′-0″ above the T.O. Masonry level.

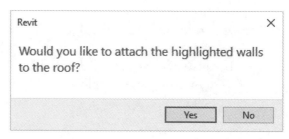

FIGURE 6-1.6 Prompt

- Switch to an elevation view to see the roof, South elevation, shown in Figure 6-1.7.

- You can also switch to the default 3D view to see the roof in isometric view via the *Quick Access Toolbar*.

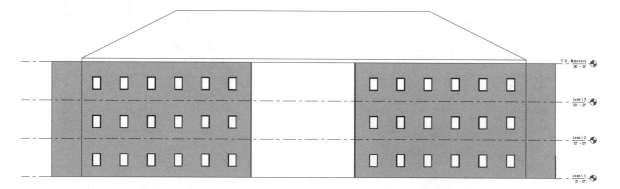

FIGURE 6-1.7 South elevation

After looking at the roof you have created, switch back to the plan view: **T.O. Masonry**. You will now add a roof over the East stair shaft.

21. **Zoom in** on the East stair shaft.

22. Select the **Roof** tool and click *"Roof by Footprint."*

23. With **Defines slope** checked in the *Options Bar* (Figure 6-1.5), pick the three exterior walls at the stair shaft.

24. Uncheck **Defines Slope**, and then select the **Line** tool and draw a line as shown in Figure 6-1.8 to close the footprint. Be sure to use snaps to accurately draw the enclosed area.

25. Pick the **Modify** button and then select the line you just drew.

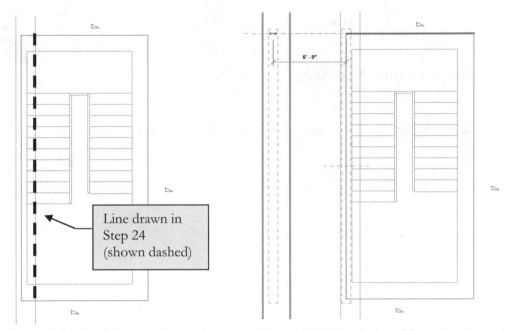

Line drawn in Step 24 (shown dashed)

6' - 0"

FIGURE 6-1.8 Roof footprint; East stair   **FIGURE 6-1.9** Modified roof footprint; East stair

26. Use the **Move** command to move the line **6'-0"** to the West (Figure 6-1.9).

27. Select the **green check mark** from the *Ribbon* to finish the roof.

28. Click **NO** when prompted to attach the highlighted walls to the roof (Figure 6-1.6).

29. Switch to the **South** elevation view (Figure 6-1.10).

30. Switch back to the **T.O. Masonry** view.

31. Select the roof element over the East stair shaft.

32. Switch to the *Default* **3D View** and adjust your view to look similar to Figure 6-1.11.

TIP: Drag on the ViewCube.

33. Select **Modify → Geometry → Join/Unjoin Roof**.

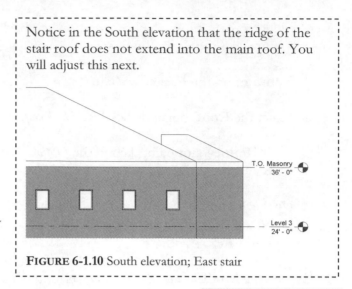

Notice in the South elevation that the ridge of the stair roof does not extend into the main roof. You will adjust this next.

FIGURE 6-1.10 South elevation; East stair

You will now select the two edges of the roofs that you want to come together.

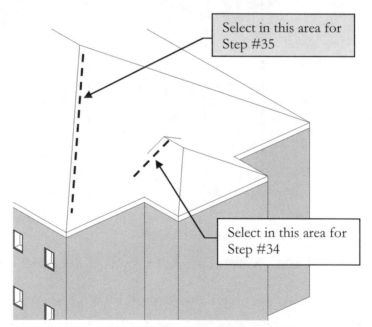

Select in this area for Step #35

Select in this area for Step #34

FIGURE 6-1.11 Default 3D View

34. Select the edge of the smaller roof; see Figure 6-1.11.

35. Select the edge of the larger roof; see Figure 6-1.11.

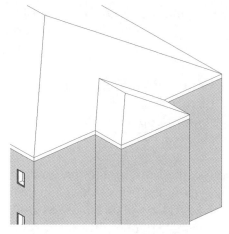

FIGURE 6-1.12 Joined roof

Your roof should now look similar to Figure 6-1.12. Take another look at the South elevation to see the revision.

36. Repeat the previous steps to create a roof over the West stair shaft.

TIP: You can also try mirroring about your Reference Plane.

Atrium Roof:

Next you will create a roof over the atrium area. We want a 4'-0" high aluminum panel above the curtain wall, thus pushing the atrium roof up higher. You will need to create a new wall type for the aluminum panels.

37. Switch to the **T.O. Masonry** view listed under *Floor Plans.*

38. Select the **Wall** tool and then select the *Basic Wall:* **Generic – 5"** type.

39. Click *Type Edit* on the *Properties Palette*, and then **Duplicate**.

40. Enter **exterior wall – aluminum** for the name.

41. Click **Edit** next to the *Structure* parameter. Add a new *Layer* using the *Insert* button. Set the exterior finish with the material set to **Aluminum** and then edit the thickness to be **5/8"** (Figure 6-1.13).

42. Draw three walls, so their exterior faces align with the exterior face of the curtain wall below. Be sure to use *Snaps* and set the wall height to **4'-0"** (Figure 6-1.14).

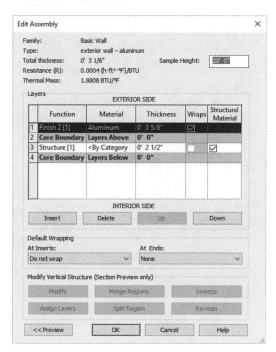

FIGURE 6-1.13 New wall structure

The walls running north-south need to extend far enough back into the main roof to avoid any holes.

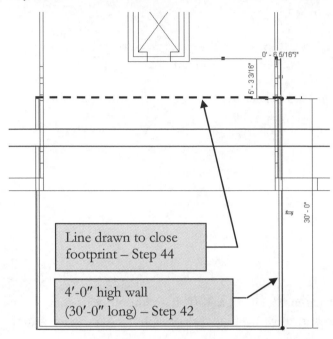

Line drawn to close footprint – Step 44

4'-0" high wall (30'-0" long) – Step 42

FIGURE 6-1.14 4'-0" high wall above curtain wall in atrium

43. Use the **Roof** tool and select the three walls just drawn using the footprint option; "Defines slope" checked.

44. Use the **Line** tool to draw a line to close the open side (Figure 6-1.14). This will create a closed rectangle to complete the roof; "Defines slope" unchecked.

45. Before finishing the roof, set the **Base offset from level** to **4'-0"** in the *Properties Palette*. This will place the roof on top of the 4'-0" high wall you just drew.

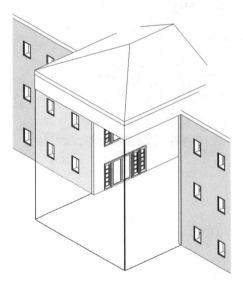

FIGURE 6-1.15 3D view

46. Select the **green check mark**.

Your 3D view should look similar to **Figure 6-1.15**. Like the stair roof, the atrium roof needs to be joined to the main roof.

47. Click **NO** when prompted to attach the highlighted walls to the roof (Figure 6-1.6).

48. Use the **Join/Unjoin Roof** tool to join the atrium roof to the main roof (similar to steps 32-34 in this section).

Look at the side elevations. If the roof does not extend all the way to the main roof, select the roof and pick *Edit Footprint* (on the *Ribbon*) to move the line further into the building. When finished it should look like Figure 6-1.16.

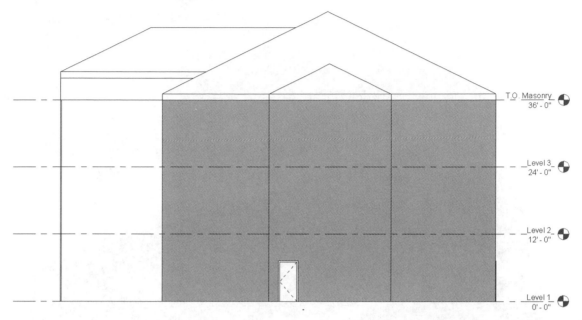

FIGURE 6-1.16 Atrium roof

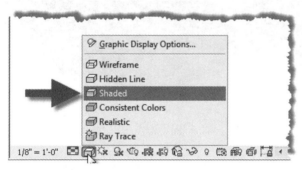

FIGURE 6-1.17 Setting visual style

Next you will take a quick look at the project, thus far, in an isometric view.

49. Click the *3D View* icon on the *Quick Access Toolbar*.

The 3D view can be improved by shading the surfaces.

50. Set the *Visual Style* to **Shaded** via the *View Control Bar* (Figure 6-1.17).

TIP: You can also turn on shadows from the View Control Bar (see image to right). This can make viewing and printing on larger, complex models much slower so use it sparingly.

The 3D Model should now be shaded (Figure 6-1.18).

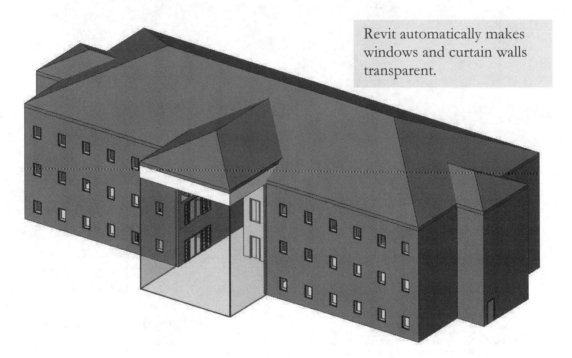

Revit automatically makes windows and curtain walls transparent.

FIGURE 6-1.18 Shaded model

Before moving on you will adjust the top constraint of the Level 3 interior walls to be tied to the new level.

51. Switch to the Level 3 Floor Plan view and select all the interior walls. Adjust their *Top Constraint* to the new level: *T.O. Masonry.*

52. **Save** your project.

Try adjusting the view; click and drag on the *ViewCube.*

Exercise 6-2:
Skylights

This short exercise covers inserting skylights in your roof. The process is much like inserting windows. In fact, Revit lists the skylight types with the window types, so you use the *Window* tool to insert skylights into your project. Technically, a skylight is a roof hosted window.

Inserting Skylights

You will place the skylights in an elevation view.

1. Load project file **ex6-1.rvt**.

2. Switch to the **South** elevation view.

3. Select the **Window** tool and load the *skylight* family (skylight.rfa) into the project (via the *Load family* button on the *Ribbon*).

Load Family

4. Select *Skylight:* **24″ x 27″** from the *Type Selector*.

You are now ready to place skylights on the roof. Revit will only look for roof elements when placing skylights, so you don't have to worry about a skylight ending up in a wall.

5. Roughly place four skylights as shown in Figure 6-2.1 on the roof.

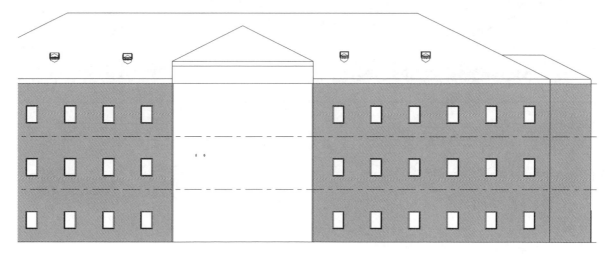

FIGURE 6-2.1 South elevation, skylights added

6. Press **Esc** or click the **Modify** tool to cancel the *Window* tool.

Next, you will want to align the skylights with each other.

7. Switch to the **West** elevation view.

8. Select one of the visible skylights.

You should now have the skylight selected and see the temporary dimensions that allow you to adjust the exact location of the element. Occasionally, the dimension does not go to the point on the model that you are interested in referencing from. Revit allows you to adjust where those temporary dimensions point to.

9. Click and drag the grip shown in Figure 6-2.2 (wait until it snaps) to the ridge of the main roof (Figure 6-2.3).

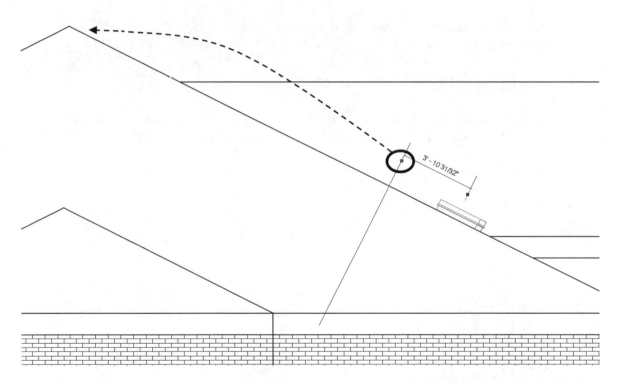

FIGURE 6-2.2 West elevation: default dimension shown when selecting skylight.

Click on the dimension text and change the text to **22'-8"**.

NOTE: *This will adjust the position of the skylight relative to the roof.*

10. Select the other skylight on the west elevation and adjust it to match the one you just revised.

11. Switch to the East elevation and repeat the above steps to adjust.

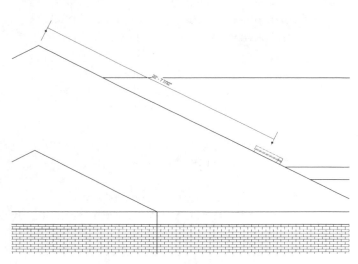

FIGURE 6-2.3 West elevation: default dimension shown when selecting skylight.

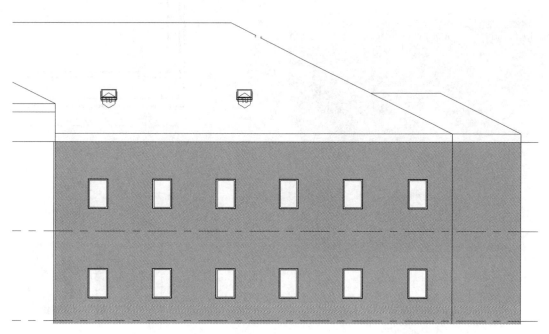

FIGURE 6-2.4 South elevation: with skylights added

Your drawing should look similar to the one above. Notice that the skylight tag is right on top of the skylight. You will adjust that next.

12. **Zoom in** on one of the skylights and **click** on the skylight tag (South view).

13. You should see a symbol appear near the bottom of the tag; drag on this symbol to move the tag down (Figure 6-2.5).

14. Position the skylight tag so the tag does not overlap the skylight (Figure 6-2.6).

15. Adjust the other skylight tags; as you reposition these tags you may see a reference line appear indicating the symbol will automatically align with an adjacent symbol.

Take a minute to look at your shaded 3D view and try changing the view so you can see through the skylight glass into the spaces below (Figure 6-2.7).

16. **Save** as **ex6-2.rvt**.

FIGURE 6-2.5 Enlarged skylight detail

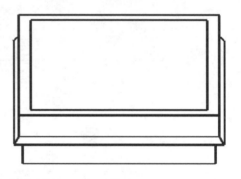

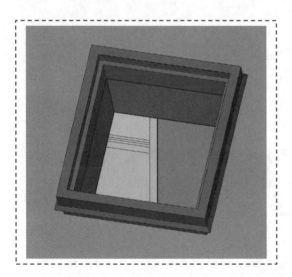

FIGURE 6-2.7 Shaded skylight view

FIGURE 6-2.6 Enlarged skylight detail – revised

Exercise 6-3:
Roof Design Options (Style, Pitch and Overhang)

In this lesson you will look at the various ways to use the *Roof* tool to draw the more common roof forms used in architecture today.

Start a New Revit Project

You will start a new project for this lesson so you can quickly compare the results of using the *Roof* tool.

1. Start a new project using the **default.rte** template (aka Imperial-Architecture).

2. Switch to the **North** elevation view and rename the level named *Level 2* to **T.O. Masonry**. This will be the reference point for your roof. Click **Yes** to rename corresponding views automatically.

    TIP: Just select the Level *datum and click on the level datum's text to rename.*

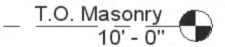

3. Switch to the *Level 1 Floor Plan* view.

Drawing the Buildings

4. Set the Level 1 *"Detail Level"* to **medium**, so the material hatching is visible within the walls.

    TIP: Use the View Control Bar at the bottom.

5. Use the *Wall* tool with the wall *Type* set to **"Exterior - Brick on Mtl. Stud,"** and draw a **40'-0" x 20'-0"** building (Figure 6-3.1).

    FYI: The default Wall height is OK; it should be 20'-0".

Be sure to draw the building within the elevation tags.

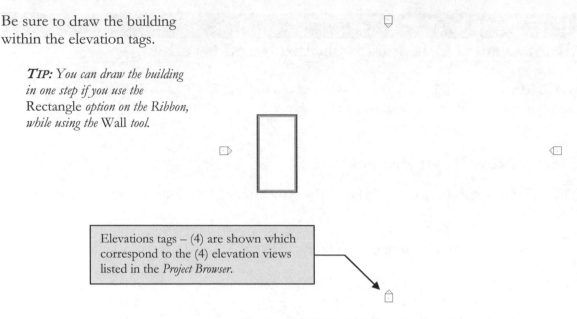

TIP: You can draw the building in one step if you use the Rectangle *option on the Ribbon, while using the* Wall *tool.*

Elevations tags – (4) are shown which correspond to the (4) elevation views listed in the *Project Browser.*

FIGURE 6-3.1 Bldg. and Elev. tags

You will copy the building so that you have a total of four buildings. You will draw a different type of roof on each one.

6. Drag a window around the walls to select them. Then use the **Array** command to set up four buildings **35'-0" O.C.** (Figure 6-3.2). See the *Array Tip* below.

TIP: Zoom in and make sure the brick is on the exterior side of the wall. If not, you can select each wall and click its flip *icon.*

ARRAY TIP: *Select the first building, select Array, and then, just like the Copy command, define a copy 35' to the right, then enter the number of copies.*

FIGURE 6-3.2 Four buildings

7. Select all of the buildings and click **Ungroup** from the *Ribbon.*

Ungroup

Hip Roof

The various roof forms are largely defined by the *"Defines slope"* setting. This is displayed in the *Options Bar* while the *Roof* tool is active. When a wall is selected and the *"Defines slope"* option is selected, the roof above that portion of wall slopes. You will see this more clearly in the examples below.

8. Switch to the **T.O. Masonry** *Floor Plan* view.

9. Select the **Architecture → Build → Roof** *(down-arrow)* → **Roof by Footprint** tool.

10. Set the overhang to **2'-0"** and make sure **Defines slope** is selected (checked) on the *Options Bar*.

☑ Defines slope | Overhang: 2' 0" | ☐ Extend to wall core

11. Select the four walls of the West building, clicking each wall one at a time.

    TIP: Make sure you select towards the exterior side of the wall, notice the review line before clicking.

12. Click **Finish Edit Mode** (i.e., the green check mark) on the *Ribbon* to finish the *Roof* tool.

13. Click **Yes** to attach the roof to the walls.

14. Switch to the **South** elevation (Figure 6-3.3).

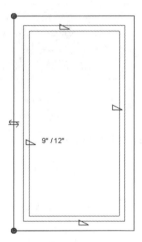

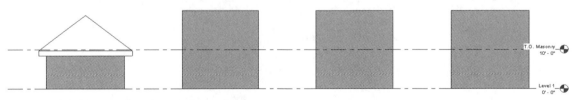

FIGURE 6-3.3 South elevation – Hip roof

You will notice that the default wall height is much higher than what we ultimately want. However, when the roof is drawn at the correct elevation and you attach the walls to the roof, the walls automatically adjust to stop under the roof object. Additionally, if the roof is raised or lowered later, the walls will follow; you can try this in the South elevation view by simply using the *Move* tool. **REMEMBER:** *You can make revisions in any view.*

15. Switch to the **3D** view using the icon on the *QAT* (Figure 6-3.4).

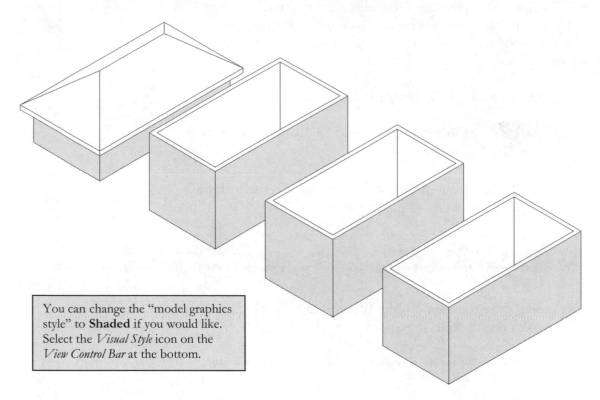

You can change the "model graphics style" to **Shaded** if you would like. Select the *Visual Style* icon on the *View Control Bar* at the bottom.

FIGURE 6-3.4 3D view – hip roof

Gable Roof

16. Switch back to the **T.O. Masonry** view (not the ceiling plan for this level).

17. Select the **Roof** tool and then **Roof by Footprint**.

18. Set the overhang to **2′-0″** and make sure **Defines slope** is selected (checked) on the *Options Bar*.

19. Only select the two long (40′-0″) walls.

20. **Uncheck** the **Defines slope** option.

21. Select the remaining two walls (Figure 6-3.5).

22. Pick the **green check mark** on the *Ribbon* to finish the roof.

23. Select **Yes** to attach the walls to the roof.

24. Switch to the **South** elevation view (Figure 6-3.6).

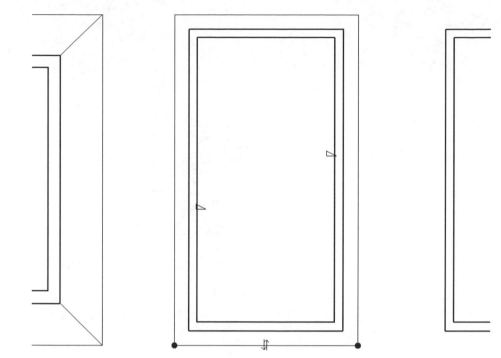

FIGURE 6-3.5 Gable – plan view

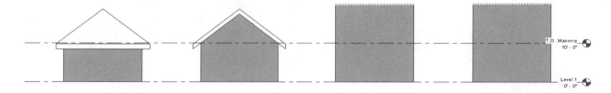

FIGURE 6-3.6 South elevation – gable roof

25. Switch to the **3D** view (Figure 6-3.7).

Notice the wall extends up to conform to the underside of the roof on the gable ends.

FYI: You may be wondering why the roofs look odd in the floor plan view. If you remember, each view has its own cut plane. The cut plane happens to be lower than the highest part of the roof – thus, the roof is shown cut at the cut plane. If you go to Properties Palette → View Range (while nothing is selected) and then adjust the cut plane to be higher than the highest point of the roof, then you will see the ridge line.

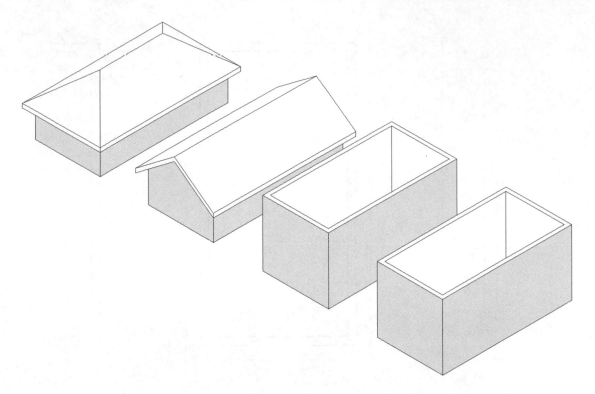

FIGURE 6-3.7 3D view – gable roof

Shed Roof

26. Switch back to the **T.O. Masonry** view.

27. Select the **Roof** tool, and then **Roof by Footprint**.

28. Check **Defines slope** on the *Options Bar*.

29. Set the overhang to **2′-0″** on the *Options Bar*.

30. Select the East wall (40′-0″ wall, right-hand side).

31. Uncheck **Defines slope** in the *Options Bar*.

32. Select the remaining three walls (Figure 6-3.8).

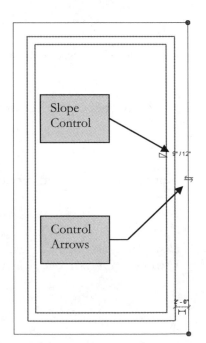

FIGURE 6-3.8 Selected walls

33. Set the **Slope**, or roof pitch, to 3/12 (Figure 6-3.9) on the *Properties Palette*.

34. Click **Apply** on the *Properties Palette*.

35. Pick the **green check mark** on the *Ribbon* to finish the roof.

36. Select **Yes** to attach the walls to the roof.

Revit ✕

Would you like to attach the highlighted walls to the roof?

[Yes] [No]

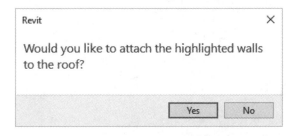

FIGURE 6-3.9
Properties for Roof tool

FYI: You can also change the slope of the roof by changing the Slope Control text (see Figure 6-3.8); just select the text and type a new number.

TIP: You can use the Control Arrows, while the roof line is still selected, to flip the orientation of the roof overhang if you accidentally selected the wrong side of the wall and the overhang is on the inside of the building.

37. Switch to the **South** elevation view (Figure 6-3.10).

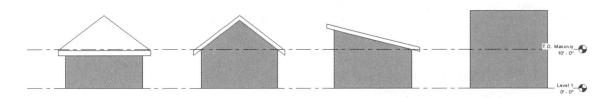

FIGURE 6-3.10 South elevation – shed roof

38. Switch to the *Default 3D* view (Figure 6-3.11).

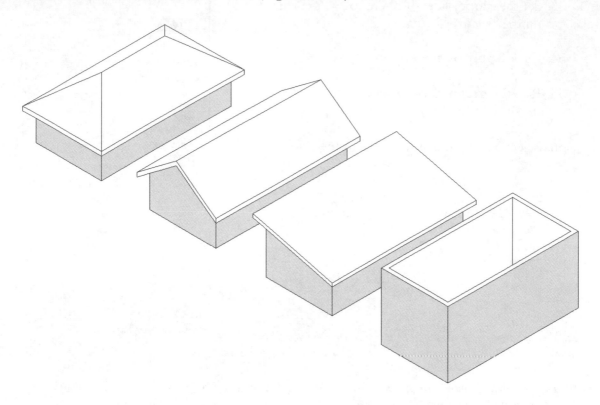

FIGURE 6-3.11 Default 3D view – shed roof

Once the roof is drawn, you can easily change the roof's overhang. You will try this on the shed roof. You will also make the roof slope in the opposite direction.

39. In **T.O. Masonry** view, select **Modify** from the *Ribbon*, and then select the shed roof.

40. Click **Edit Footprint** from the *Ribbon*.

Edit Footprint

41. Click on the East roof sketch-line to select it.

42. Uncheck **Defines slope** from the *Options Bar*.

43. Now select the West roofline and check **Defines slope**.

If you were to select the green check mark now, the shed roof would be sloping in the opposite direction. But, before you do that, you will adjust the roof overhang at the high side.

44. Click on the East roofline again to select it.

45. Change the overhang to **6'-0"** in the *Options Bar*.

Changing the overhang only affects the selected roofline.

46. Select the **green check mark**.

47. Switch to the South view to see the change (Figure 6-3.12).

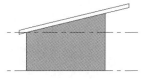

Thus you can see it is easier to edit an object than to delete it and start over. Just remember you have to be in sketch mode (i.e., *Edit Sketch*) to make changes to the roof. Also, when a sketch line is selected, its properties are displayed in the *Properties Palette*. That concludes the shed roof example.

FIGURE 6-3.12 South elevation – shed roof (revised)

Flat Roof

48. Switch back to the **T.O. Masonry** *Floor Plan* view.

49. Select **Architecture → Roof → Roof by Footprint**.

50. Set the overhang to **2'-0"** and make sure **Defines slope** is not selected (i.e., unchecked) in the *Options Bar*.

51. Select all four walls.

52. Pick the **green check mark**.

53. Select **Yes** to attach the walls to the roof.

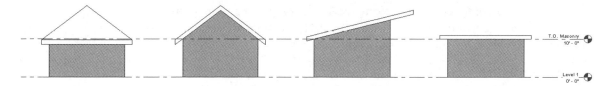

FIGURE 6-3.13 South elevation – flat roof

54. Switch to the South elevation view (Figure 6-3.13).

55. Also, take a look at the **Default 3D view** (Figure 6-3.14).

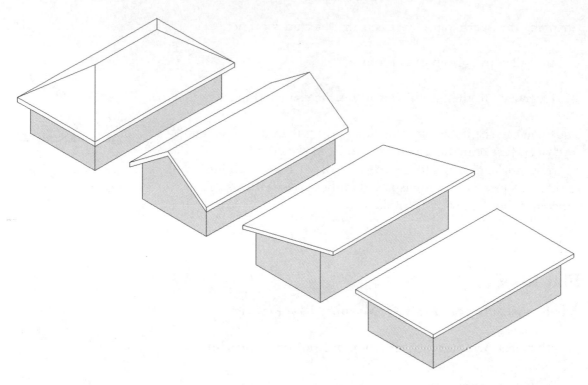

FIGURE 6-3.14 Default 3D view – flat roof

56. Save your project as **ex6-3.rvt**.

Want More?

Revit has additional tools and techniques available for creating more complex roof forms. However, that is beyond the scope of this book. If you want to learn more about roofs, or anything else, take a look at one of the following resources:

- Revit **Web Site:** www.autodesk.com
- Revit **Newsgroup** (potential answers to specific questions)
  www.augi.com; www.revitcity.com; www.autodesk.com; www.revitforum.org
- Revit **Blogs** information from individuals (some work for Autodesk and some don't)
  www.BIMchapters.blogspot.com, www.revitoped.com, revitclinic.typepad.com

Reference material: Roof position relative to wall

The remaining pages in this chapter are for reference only and do not need to be done to your model. You are encouraged to study this information so you become more familiar with how the *Roof* tool works.

The following examples use a brick and concrete wall example. The image below shows the *Structure* properties for said wall type. Notice the only item within the *Core Boundary* section is the *Masonry – Concrete Block* (i.e., CMU) which is 7⅝″ thick (nominally 8″). Keep this in mind as you read through the remaining material.

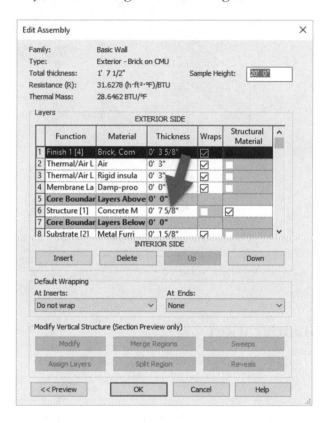

The following examples will show you how to control the position of the roof system relative to the wall below, both vertically and horizontally. The roof properties that dictate its position basically involve relationships between three things: the **Level Datum**, the exterior **Wall System** and the bottom edge of the **Roof System**. There are several other properties (e.g., pitch, construction, fascia, etc.) related to the roof that will not be mentioned at the moment so the reader may focus on a few basic principles.

The examples on this page show a sloped roof sketched with *Extend into wall (to core)* enabled and the *Overhang* set to 2'-0". Because *Extend into wall (to core)* was selected, the bottom edge of the roof is positioned relative to the *Core Boundary* of the exterior wall rather than the finished face of the wall. See the discussion about the wall's *Core Boundary* on the previous page.

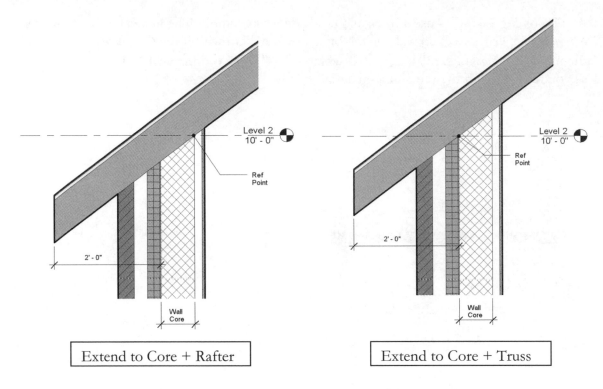

| Extend to Core + Rafter | Extend to Core + Truss |

Revit Roof Properties under Consideration

Extend Into Wall:
(To Core)
This option was *checked* on the *Options Bar* while sketching the roof.

Rafter Or Truss:
This option is an *Instance Parameter* of the roof object; the example on the above left is set to *Rafter* and the other is set to *Truss*.

NOTE: The Extend into Wall (to core) *option affects the relative relationship between the wall and the roof, as you will see by comparing this example with the one on the next page.*

Base Level:
Set to *Level 2*: By associating various objects to a level, it is possible to adjust the floor elevation (i.e., *Level Datum*) and have doors, windows, floors, furniture, roofs, etc., all move vertically with that level.

Base Offset:
From Level
Set to *0'-0"*: This can be a positive or negative number which will be maintained even if the level moves.

The examples on this page show a sloped roof sketched with *Extend into wall (to core)* NOT enabled and the *Overhang* set to 2'-0". Notice that the roof overhang is derived from the exterior face of the wall (compared to the *Core Boundary* face on the previous example when *Extend into wall* was enabled).

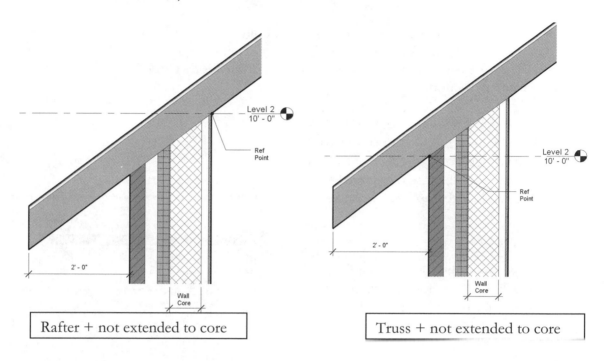

| Rafter + not extended to core | Truss + not extended to core |

Revit Roof Properties under Consideration

Extend Into Wall:
(To Core)
This option was *NOT checked* on the *Options Bar* while sketching the roof.

Rafter Or Truss:
This option is an *Instance Parameter* of the roof object; the example on the above left is set to *Rafter* and the other is set to *Truss*.

NOTE: The Extend into wall (to core) *option affects the relative relationship between the wall and the roof, as you will see by comparing this example with the one on the previous page.*

Base Level:
Set to *Level 2*. By associating various objects to a level, it is possible to adjust the floor elevation (i.e., *Level Datum*) and have doors, windows, floors, furniture, roofs, etc., all move vertically with that level.

Base Offset:
From Level
Set to *0'-0"*. This can be a positive or negative number which will be maintained even if the level moves.

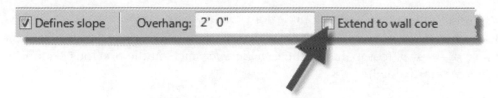

As you can see from the previous examples, you would most often want to have *Extend to wall (to core)* selected while sketching a roof because it would not typically make sense to position the roof based on the outside face of brick or the inside face of gypsum board for commercial construction.

Even though you may prefer to have *Extend to wall (to core)* selected, you might like to have a 2'-0" overhang relative to the face of the brick rather than the exterior face of concrete block. This can be accomplished in one of two ways:

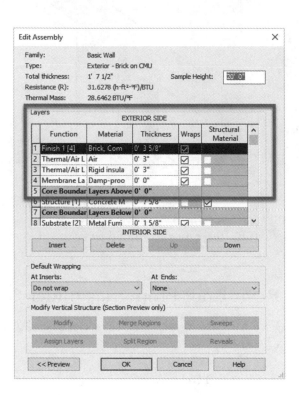

(A) You can modify the overhang, while sketching the roof, to include the wall thickness that occurs between the face of wall and face of core: $2'\text{-}0'' + 9\tfrac{5}{8}'' = 2'\text{-}9\tfrac{5}{8}''$. See the image to the right.

(B) The second option is to manually edit the sketch lines. You can add dimensions while in *Sketch* mode, select the sketch line to move, and then edit the dimension. The dimension can also be *Locked* to maintain the roof edge position relative to the wall. When you finish the sketch the dimensions are hidden.

Energy Truss

In addition to controlling the roof overhang, you might also want to control the roof properties to accommodate an energy truss with what is called an *energy heel*, which allows for more insulation to occur directly above the exterior wall.

To do this you would use the *Extend into wall (to core)* + *Truss* option described above and then set the *Base Offset from Level* to 1'-0" (for a 1'-0" energy heel). See the image and the *Properties Palette* shown on the next page.

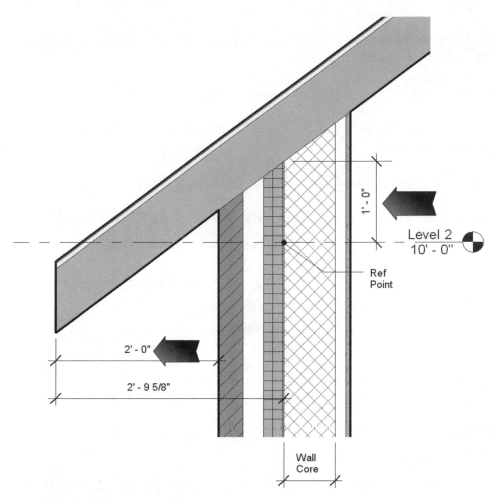

1' - 0"

Level 2
10' - 0"

Ref
Point

2' - 0"

2' - 9 5/8"

Wall
Core

Many other properties and techniques exist which one can use to develop the roof for a project, things like the *Rafter Cut* and *Fascia Depth* which control the fascia design. Also, you can apply a sweep to a roof edge to add a 1x fascia board around the building. These are intermediate to advanced concepts and will not be covered here.

This concludes the study of the *Roof* tool!

Self-Exam:

The following questions can be used as a way to check your knowledge of this lesson. The answers can be found at the bottom of this page.

1. You don't have to click *Finish Roof* when you are done defining a roof. (T/F)

2. The wall below the roof automatically conforms to the underside of the roof when you join the walls to the roof. (T/F)

3. The roof overhang setting is available from the *Options Bar*. (T/F)

4. To create a gable roof on a building with 4 walls, two of the walls should not have the _____ option checked.

5. Is it possible to change the reference point for a temporary dimension that is displayed while an object is selected? (Y/N)

Review Questions:

The following questions may be assigned by your instructor as a way to assess your knowledge of this section. Your instructor has the answers to the review questions.

1. When creating a roof using the "*roof by footprint*" option, you need to create a closed perimeter. (T/F)

2. Can the "*Defines Slope*" setting be changed after the roof is "finished"? (Y/N)

3. Skylights need to be rotated to align with the plane (pitch) of the roof. (T/F)

4. Skylights automatically make the glass transparent in shaded views. (T/F)

5. While using the **Roof** tool, you can use the _____ tool from the *Ribbon* to fill in the missing segments to close the perimeter.

6. You use the _____ parameter to adjust the vertical position of the roof relative to the current working plane (view).

7. A skylight cannot be repositioned in an elevation view. (T/F)

8. You need to use the _____ to flip the roofline when you pick the wrong side of the wall and the overhang is shown on the inside.

Lesson 7
FLOOR SYSTEMS and REFLECTED CEILING PLANS:

In this lesson you will learn to create floor structures and reflected ceiling plans.

Even though you currently have floor levels defined, you do not have an object that represents the mass of the floor systems. You will add floor systems with holes for stairs, elevators, and the atrium.

Ceiling systems allow you to specify the ceiling material by room and the height above the floor. Once the ceiling has been added it will show up in section views (sections are created later in this book).

Exercise 7-1:
Floor Systems

Similar to other Revit elements, you can select from a few pre-defined system families. You can also create new types.

Level 1, Slab on Grade

Sketching floors is a lot like sketching roofs (Lesson 5); you can select walls to define the perimeter and draw lines to fill in the blanks and add holes (cut-outs) in the floor element.

1. Open ex6-2.rvt and **Save As ex7-1.rvt**.

2. Switch to the **Level 1** floor plan view.

3. Select **Architecture → Build → Floor** *(down-arrow)* **→ Floor: Architectural**.

4. Click **Edit Type** on the *Properties Palette*.

5. Select type **Generic – 12″** from the *Type* drop-down list at the top.

6. Click **Duplicate** to start a new floor type.

7. Type **6″ Slab on Grade**, then **OK**.

8. Click the **Edit** button next to the *Structure* Parameter.

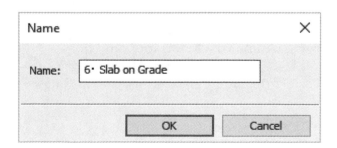

9. Change the material for the structure layer shown to **Concrete: cast-in-place concrete grey**, and change the thickness to **6″**.

10. Next you will add carpet on top of the slab. Click **Insert** and use the **Up/Down** buttons to position the *Layer* correctly (Figure 7-1.1).

11. Add another layer:

    a. *Function:* **Finish 1 [4]**
    b. *Material:* **Carpet (1)**
    c. *Thickness:* **1/4″**
       (Figure 7-1.1)

12. Click **OK** to close the open dialog boxes.

13. Select all the exterior walls on **Level 1**; this should include the curtain wall at the atrium and the stair shafts (Figure 7-1.3).

    TIP: Select the interior side of the wall; you can use the control arrows if needed.

14. Click the **green check mark**.

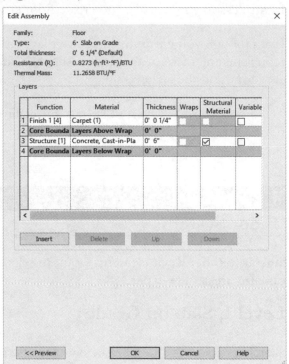

FIGURE 7-1.1 New Floor System

You will most likely get an error message. This is because the main exterior walls extend into the atrium (past the curtain wall). Because this is not a perfect corner (and it does not need to be), you can trim the "edge of slab" lines while in sketch mode to create a true corner, i.e., a closed line for the floor (Figure 7-1.2).

15. *(If you did not get an error, skip ahead to Step 18.)* Click **Continue**.

16. Use the **Trim** tool; select the two lines leading to the corner that needs to be trimmed. Do this for both sides of the atrium (Figure 7-1.3).

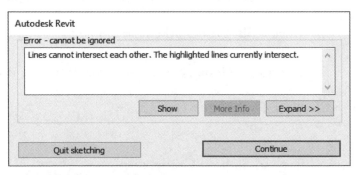

FIGURE 7-1.2 Floor error message

17. Click the **green check mark**.

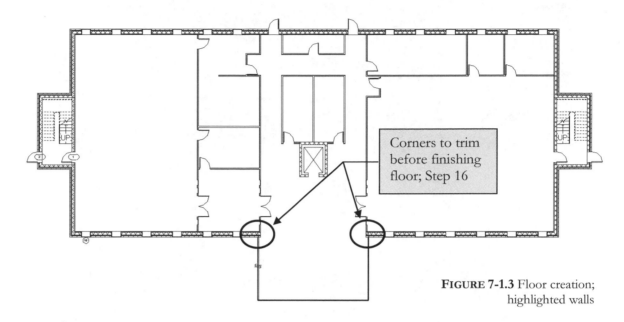

Corners to trim before finishing floor; Step 16

FIGURE 7-1.3 Floor creation; highlighted walls

You now have a floor at the first level. You should see a stipple pattern representing the floor area. You would most likely want to turn that pattern off for a floor plan. You will do that next.

18. Click **Modify** on the *Ribbon* to unselect the new floor.

19. Select **Edit**, next to the *Visibility/Graphic Overrides* parameter in the *Properties Palette* (Figure 7-1.4).

   TIP: Type VV to skip Steps 18 and 19!

20. In the *Visibility/Graphic Overrides* dialog, click the "cell" at the intersection of *Floors* (row) and *Surface / Patterns* (column); select **Override** (Figure 7-1.5).

   FYI: The word "Override" does not appear until you click within the "cell."

21. Uncheck **Visible**.

22. Click **OK** to close the dialogs.

The stipple pattern is no longer visible.

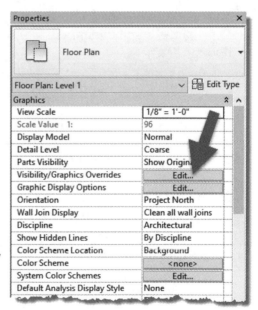

FIGURE 7-1.4 Level 1 view properties

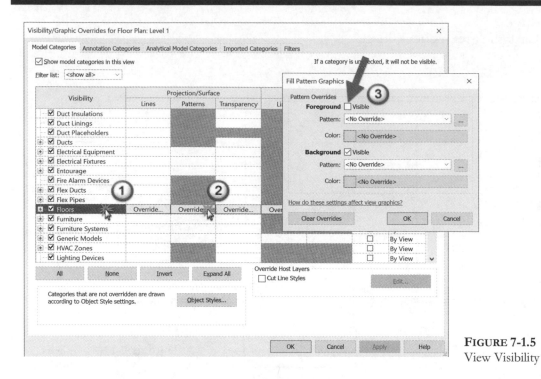

FIGURE 7-1.5
View Visibility

Levels 2 and 3, Conc. + Metal Deck + Bar Joists

23. Switch to **Level 2** view.

Next you need to load a profile to properly define the upper floors.

24. Use the **Load Family** tool to load the following: Profiles\Metal Deck\ **Form Deck_Non-Composite**. If prompted, select overwrite.

25. Activate the **Floor** tool and create a new floor type named **Steel Bar Joist 14″ – Carpet on Concrete**.

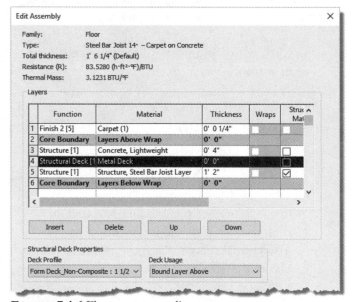

FIGURE 7-1.6 Floor system – edit structure

TIP: Use a similar floor type as a starting point (duplicate) when creating new floor types.

26. Adjust/add the layers shown in **Figure 7-1.6**.

   a. Be sure to set the *Structural Deck Properties* correctly near the bottom of the *Edit Assembly* dialog. Select the 1 1/2″ deck; this will appear in sections.

   NOTE: In an actual project you might not want to include the finishes, such as VCT and Carpet. These can be added separately to better control their locations.

Creating the second and third floors will be a little more involved than was the first floor. This is because the upper floors require several openings. For example, you need to define the openings for the elevator, the stair shafts and the atrium space. Revit makes the process very simple, however.

You should still be in the *Floor* tool.

27. On the *Options Bar*, check "**Extend into wall (to core).**"

FYI: *The "Extend into wall (to core)" option will extend the slab to your CMU (CMU is the core in our example) and go under the furring. Depending on the design, the floor may extend to the exterior face of the CMU, allowing the CMU to bear on the floor slab at each level. In this exercise you will select the interior side.*

28. Select the exterior walls indicated in Figure 7-1.7.

REMEMBER: To select the interior side of the wall, use the control arrows if needed.

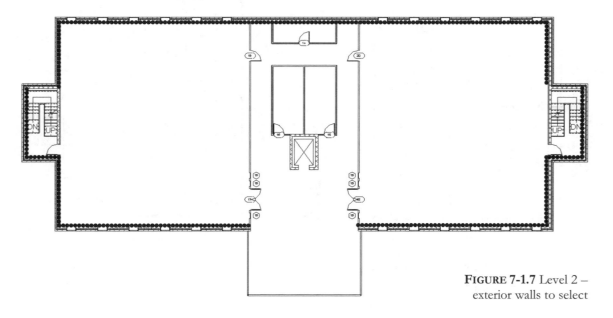

FIGURE 7-1.7 Level 2 – exterior walls to select

Next you will define the portion of floor that extends into the stair shaft to be the landing at this level. You will need to use the *Line* tool and the *Trim* tools to define this area.

29. Click on the **Line** tool from the *Draw* panel on the *Ribbon*.

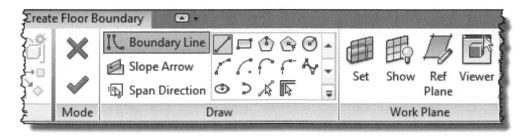

30. **Zoom In** to the West stair shaft.

31. Draw a <u>horizontal</u> line defining the edge of the landing; use Revit's snaps to accurately pick the top riser as shown in **Figure 7-1.8**.

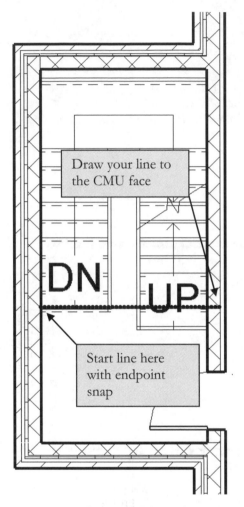

FIGURE 7-1.8 Level 2 – West stair

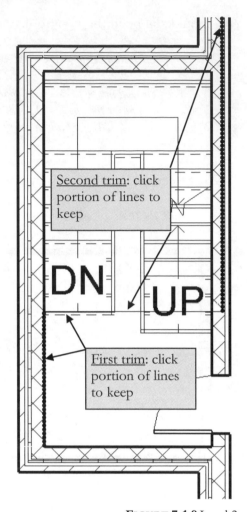

FIGURE 7-1.9 Level 2 – West stair trim lines

32. Select the **Trim** tool and trim the three lines referenced in **Figure 7-1.9**.

33. Repeat these steps for the East stair shaft.

34. Next you will pick the four walls at the elevator shaft, selecting the shaft side of the wall; be sure to use the **Pick Walls** feature from the *Draw* panel on the *Ribbon*.

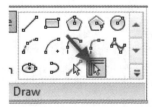

You are now ready to define the edge of the slab at the atrium.

35. Use the **Line** tool (from the *Draw* panel) to draw the edge of slab in the atrium (5 lines) as shown in **Figure 7-1.10**.

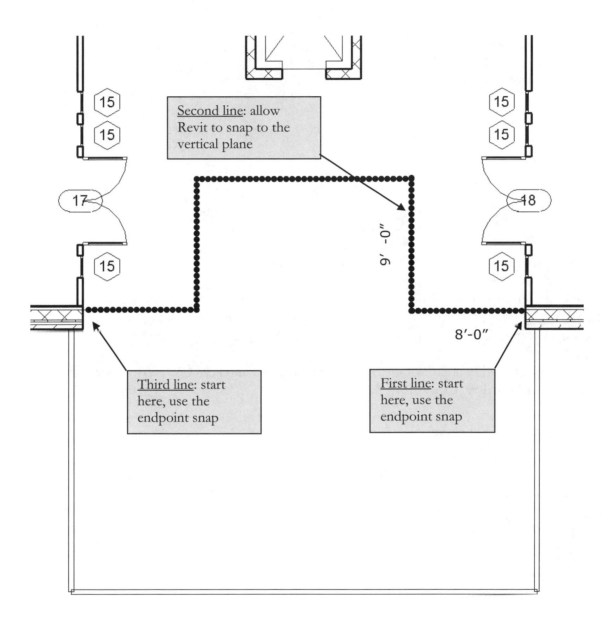

FIGURE 7-1.10 Atrium slab definition

36. Click the **green check mark** to finish the floor.

37. Click **Yes** to the prompt "*Would you like the walls that go up to this floor's level to attach to its bottom?*"

38. Click **Yes** for the prompt to join the walls that overlap the floor system (Figure 7-1.11).

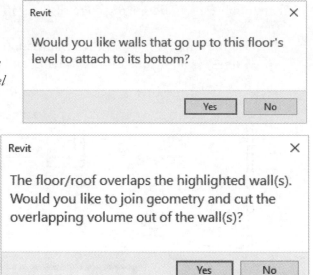

Revit	✕
Would you like walls that go up to this floor's level to attach to its bottom?	
	Yes No

Revit	✕
The floor/roof overlaps the highlighted wall(s). Would you like to join geometry and cut the overlapping volume out of the wall(s)?	
	Yes No

FIGURE 7-1.11 Finish floor prompts

39. Change Level 2's *Visibility* to turn off the floor pattern.

That completes the Level 2 floor system. Next you will copy the floor you just created to Level 3. You could switch to Level 3 and manually draw a floor following the same steps as for Level 2; however, because the two floors are identical, it would be faster to copy it.

40. Select the Level 2 floor element you just created and select **Modify Floors →
Clipboard → Copy** from the *Ribbon*.

TIP: *Selecting elements that overlap, like the exterior walls and the edge of slab (floor system), may require the use of the TAB key. The only way to select a floor element is by picking its edge. Revit temporarily highlights elements when you move your cursor over them. But, because the floor edge may not have an "exposed" edge to select (e.g., like we have in the atrium area), you will have to toggle through your selection options for your current cursor location. With the cursor positioned over the edge of the floor (probably with an exterior wall highlighted), press the TAB key to toggle through the available options. A tooltip will display the elements; when you see floor:floor-name, click the mouse to select it.*

41. Switch to **Level 3**.

42. Select *Modify → Clipboard → Paste (down arrow) →* **Aligned to Current View**.

43. Change Level 3's *Visibility* to turn off the floor pattern.

Explore your work by looking at the model in 3D (Figure 7-1.12).

Notice again how Revit automatically applies colors and patterns to surfaces to help you (and your client) better visualize your design with minimal effort. These colors and materials relate to the materials applied to each element. You will learn more about this in the rendering chapter.

44. Save your project as **ex7-1.rvt**.

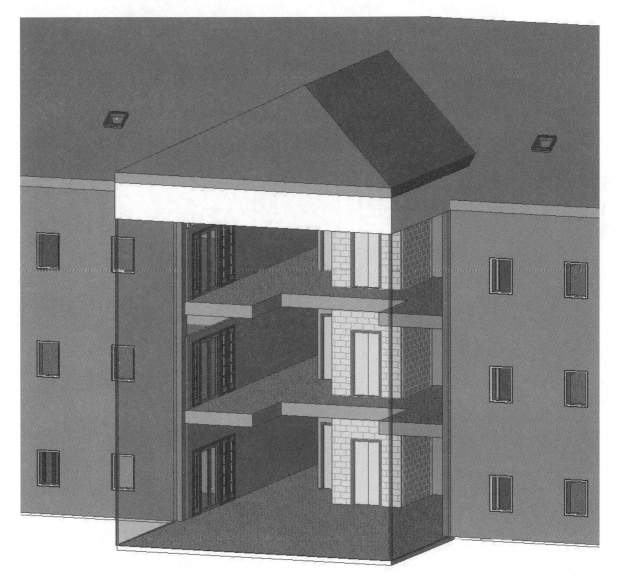

FIGURE 7-1.12 3D view with floors

Exercise 7-2:
Ceiling Systems (Susp. ACT and Gypsum Board)

This lesson will explore Revit's tools for drawing reflected ceiling plans. This will include drawing different types of ceiling systems.

Suspended Acoustical Ceiling Tile System

1. Open ex7-1.rvt and Save As **ex7-2.rvt**.

2. Switch to the **Level 1** <u>ceiling plan</u> view from the *Project Browser*.

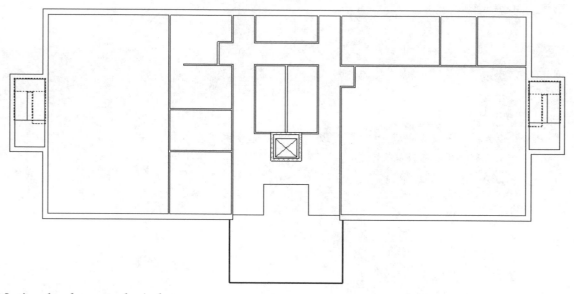

Notice the doors and windows are not visible in the ceiling plan views. The ceiling plan views have a cutting plane similar to floor plans. You can see this setting by selecting *Edit*, next to **View Range** in the *Properties Palette*.

The default value is 7'-6". You might increase this if, for example, you had 10'-0" ceilings and 8'-0" high doors. Otherwise, the doors would show because the 7'-6" cutting plane is below the door height (Figure 7-2.1).

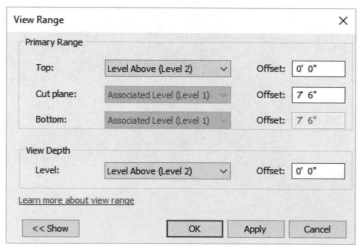

FIGURE 7-2.1 Properties: View Range settings

Click **<<Show** for more information…

3. Select **Architecture** → **Build** → **Ceiling**.

Ceiling

You have 4 ceiling types (by default) to select from (Figure 7-2.2).

4. Select *Compound Ceiling:* **2'x4' ACT System** from the *Type Selector*.

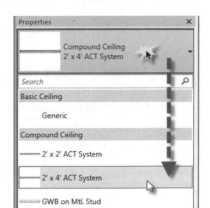

FIGURE 7-2.2 Type Selector

Next you will change the ceiling height. The default setting is 8'-0" above the current level. You will change the ceiling height to 9'-0" to make the large open office areas feel more spacious. This setting can be changed on a room by room basis.

5. Set the *Height Offset From Level* setting to **9'-0"** in the *Properties Palette* (Figure 7-2.3).

You are now ready to place ceiling grids. This process cannot get much easier, especially compared to traditional CAD programs.

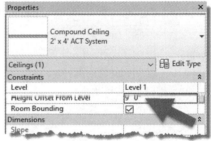

FIGURE 7-2.3 Properties Palette

6. Move your cursor anywhere within the large open office area in the West side of the building. You should see the perimeter of the room highlighted.

7. Pick within the large room; Revit places a grid in the room (Figure 7-2.4).

You now have a 2x4 ceiling grid at 9'-0" above the floor (Level 1 in this case).

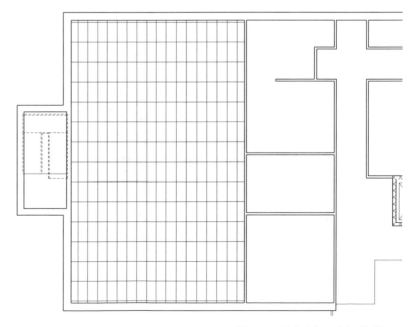

FIGURE 7-2.4 Level 1: Ceiling

When you place a ceiling grid, Revit centers the grid in the room. The general rule-of-thumb is you should try to avoid reducing the tile size by more than half at its perimeter. You can see in Figure 7-2.4 that the East and West sides look okay. However, the North and South sides are small slivers. You will adjust this next.

8. Select **Modify** from the *Ribbon*.

9. **Select** the ceiling grid (only one line in the ceiling grid will be highlighted).

10. Use the **Move** tool to move the grid 24″ to the North (Figure 7-2.5).

11. Place ceiling grids as shown in Figure 7-2.5.
    a. *Be sure to adjust the ceiling heights shown.*
    b. *Adjust the grids to avoid small tiles at the perimeter.*

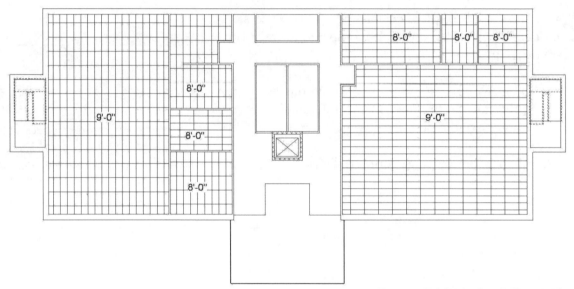

FIGURE 7-2.5 Level 1: Ceiling Grids

Modifying the Suspended Acoustical Ceiling Tile System

Making modifications to the grid is relatively easy. Next, you will adjust the ceiling height and rotate the grid.

12. Zoom in to the room in the upper right corner on *Level 1*.

13. Change the height to **8′-6″** in the *Properties Palette*, and then click **Apply**.

14. With the grid still selected, pick *Compound Ceiling:* **2′x2′ ACT System** from the *Type Selector* on the *Properties Palette*.

15. Again, with the grid still selected, use the *Rotate* tool to rotate the grid 45 degrees.

TIP: *When using the Rotate tool* 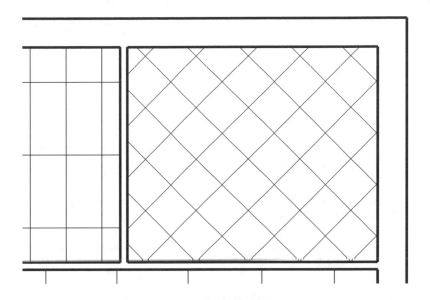 *you need to pick two points. The first point is your reference line. The second point is the number of degrees off that reference line. In this example, try picking your first point to the right as a horizontal line. Then move the cursor counterclockwise until 45 degrees is displayed. You can also type the angle instead of picking a second point.*

16. Your drawing should look similar to **Figure 7-2.6**.

FIGURE 7-2.6 Level 1: Modified Ceiling

The image below shows a camera view, looking South, of the large open office area, showing the ceiling. You will learn how to create camera views later in the text.

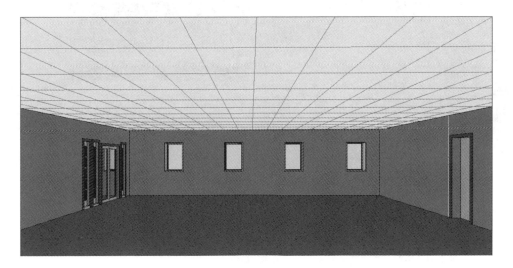

Next, you will look at drawing gypsum board (or drywall) ceiling systems. The process is identical to placing the grid system. Additionally, you will create a new ceiling type.

Gypsum Board Ceiling System

You will create a new ceiling type for a gypsum board ceiling. To better identify the areas that have a gypsum board ceiling, you will set the ceiling type to have a stipple pattern. This will provide a nice graphical representation for the gypsum board ceiling areas. First we will create a new material to define the stipple pattern which can be applied to our gypsum board ceiling type.

17. From the *Manage* tab, select **Materials**.
    This is the list of materials you select from when assigning a material to each layer in a wall system, etc.

18. Select *Gypsum Wall Board*, **right-click** on it and select **Duplicate** and then enter the name **Gypsum Ceiling Board**.

19. In the *Surface Pattern* area, pick the down-arrow and select **Gypsum-Plaster** from the list, and then click **OK** (Figure 7-2.7).

    FYI: *The appearance asset should also be duplicated but will not be done now; see chapter 11.*

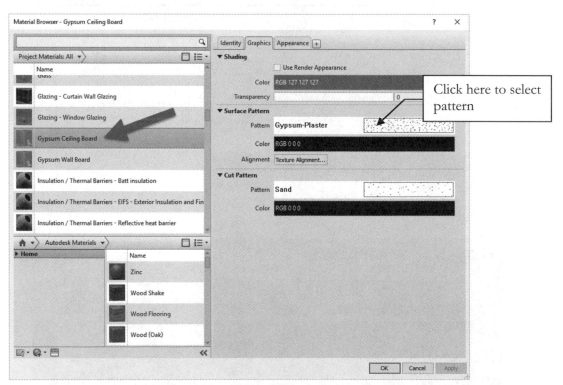

FIGURE 7-2.7 Materials dialog

The *Surface Pattern* setting is what will add the stipple pattern to the gypsum board ceiling areas. With this set to *none*, the ceiling has no pattern (like the basic ceiling type).

Thus, if you wanted Carpet 1 finish to never have the stipple hatch pattern, you could change the surface pattern to none via the *Materials* dialog and not have to change each view's visibility override.

20. Select **Architecture → Build → Ceiling**.

21. Click **Edit Type** on the *Properties Palette*.

22. Set the *Type* drop-down to **GWB on Mtl. Stud**.

   FYI: You are selecting this because it is similar to the ceiling you will be creating.

23. Click *Duplicate* and type **Gypsum Ceiling Board** for the name.

24. Select **Edit** next to the *Structure* parameter.

25. Set the values as follows (Figure 7-2.8):
   a. **1½″ Mtl. Stud**
   b. **¾″ Mtl. Stud**
   c. **Gypsum Ceiling Board**
      (This is the material you created in Step 18.)

26. Click **OK** two times.

FYI: The ceiling assembly you just created represents a typical suspended gypsum board ceiling system. The Metal Studs are perpendicular to each other and suspended by wires, similar to an ACT (acoustical ceiling tile) system. You are now ready to draw a gypsum board ceiling.

27. Make sure **Gypsum Ceiling Board** is selected in the *Type Selector* on the *Properties Palette*.

28. Set the ceiling height to **8′-0″**.

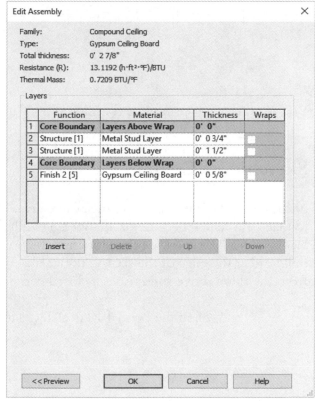

FIGURE 7-2.8 New ceiling – Edit assembly

29. Pick the two bathrooms on Level 1, which are the two rooms North of the elevator (Figure 7-2.9).

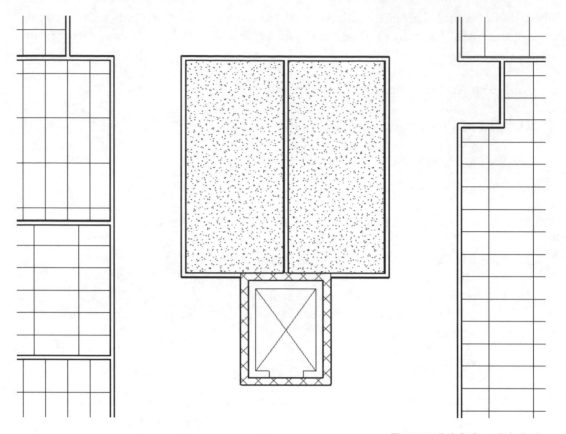

FIGURE 7-2.9 Gyp. Bd. Ceiling

You now have a gypsum board ceiling at 8'-0" above the finished floor in the toilet rooms.

Sketching a Ceiling

Next, you will draw a ceiling in the atrium area. However, you cannot simply pick the room to place the ceiling because of the opening in the floor. You will need to sketch the ceiling just like you sketched the floor system in the previous exercise. First, you will need to draw a bulkhead at the edge of the second floor slab. A bulkhead is a portion of wall that hangs from the floor above and creates a closed perimeter for a ceiling system to tie into.

30. While still in the Level 1 Reflected Ceiling Plan view, select the **Wall** tool.

31. Set the wall properties to **Interior - 4 7/8″ Partition (1-hr)** and the *Base Offset* to **9′-6″**. *(This will put the bottom of the wall to 9′-6″ above the current floor level, Level 1 in this case.)* (Figure 7-2.10)

32. Set the *Top Constraint* to **Up to level: Level 2** (Figure 7-2.10).

TIP: The next time you draw a wall you will have to change the Base Offset back to 0′-0″ or your wall will be 9′-6″ off the floor.

33. **Draw the bulkhead**; make sure you snap to the edge of the slab. Also, make sure the wall is under the floor system, not out in the opening. Do this by drawing the wall either from right to left or left to right depending on how you have the *Location Line* set (Figure 7-2.11).

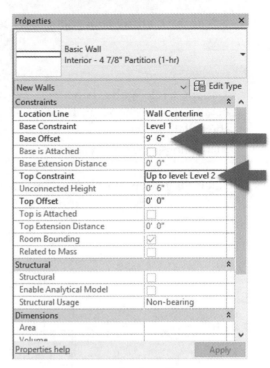

FIGURE 7-2.10 Bulkhead (wall) properties

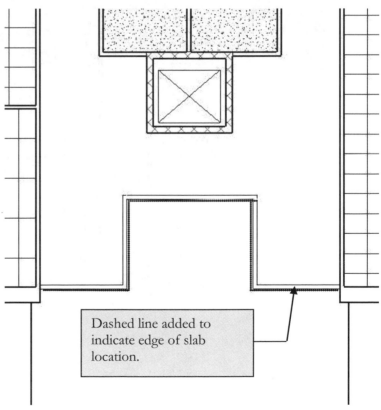

Dashed line added to indicate edge of slab location.

FIGURE 7-2.11 Bulkhead drawn

34. Select the **Ceiling** tool and then click **Sketch Ceiling** from the *Ribbon*.

Sketch
Ceiling

35. Use the **Pick Walls**, **Line**, and **Trim** tools to sketch a line at the perimeter of the ceiling area as shown in **Figure 7-2.12**. You will also need to sketch a line around the toilet/elevator area to define the area within the larger area that will not receive the ceiling pattern: **2′x2′ ACT ceiling, with the ceiling height set to 9′-6″**.

36. Click the **green check mark** and save project as **ex7-2.rvt**.

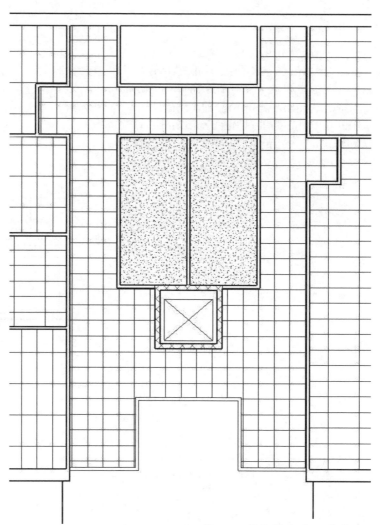

FIGURE 7-2.12 Atrium Ceiling

Exercise 7-3:
Placing Fixtures (Lights and Diffusers)

In this exercise, you will learn to load and place light and mechanical fixtures in your reflected ceiling plans.

Loading Families

Before placing fixtures, you need to load them into your project.

1. Select **Architecture → Build → Component** from the *Ribbon*.

2. Select **Load Family** on the *Ribbon* (Figure 7-3.1).

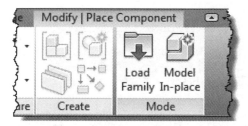

FIGURE 7-3.1 Component tool active; Ribbon

3. Double-click the *Lighting\Architectural\Internal* folder, and then double-click **Troffer Light- 2x4 Parabolic.rfa** (Figure 7-3.2).

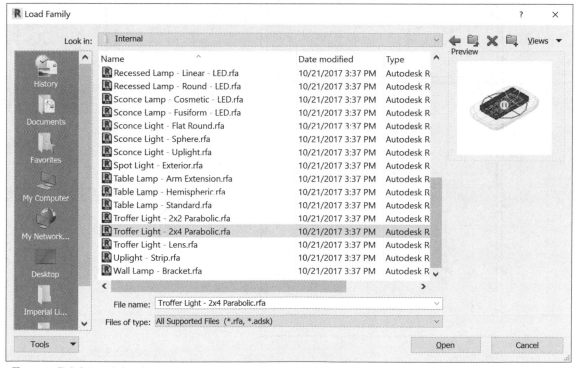

FIGURE 7-3.2 Load family

Before placing the light fixture you will load the other families first.

4. Select **Load Family** again, and browse to the *Mechanical\Architectural\Air-Side Components\Air Terminals* folder.

5. Select **Square Supply Diffuser** and **Square Return Register** (you can select both while holding the Ctrl key) and then **Open**.

    TIP: You can hold the Ctrl key to select and load multiple components from the same folder at once.

Placing Families

You are now ready to place the elements in your ceiling plans.

6. With **Component** still selected from the *Ribbon*, pick *Troffer Light – 2′x 4′ Parabolic:* **2′x4′ (2 Lamp) – 277V** from the *Type Selector*.

7. On **Level 1 RCP**, place fixtures as shown in **Figure 7-3.3**.

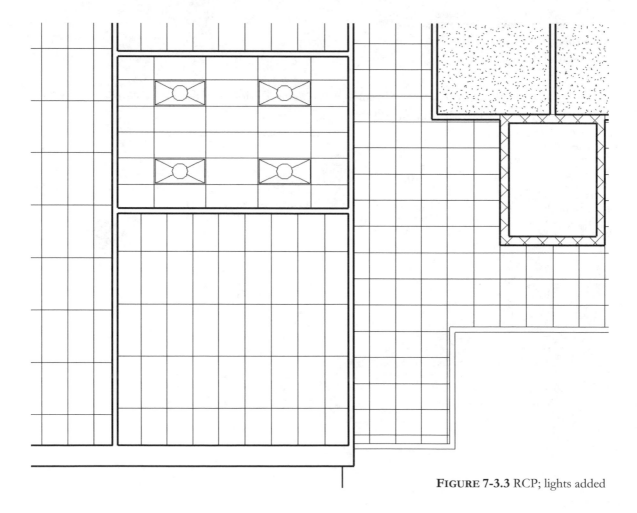

FIGURE 7-3.3 RCP; lights added

TIP: You can use array to quickly place several lights.

You may have to use the *Move* command (or better, the *Align* tool) to move the fixture so it fits perfectly in the ACT grid.

8. Now place another **2x4 light fixture** as shown in **Figure 7-3.4**.

Notice the fixture does not automatically orientate itself with the ceiling grid. There may be an occasion when you want this.

Also, notice the light fixture hides a portion of the ceiling grid. This is nice because the grid does not extend through a light fixture.

TIP: You may press the Space bar while placing the light fixture to rotate it.

9. Use **Rotate** and **Move** to rotate the fixture to align with the grid (Figure 7-3.5).

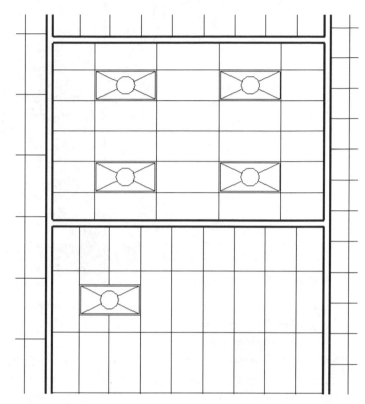

FIGURE 7-3.4 RCP; lights added

10. Once you have one fixture rotated, it is easier to use the *Copy* tool and the snaps to add rotated light fixtures. **Copy** the light fixture to match the layout in **Figure 7-3.5**.

TIP: You can check "multiple" on the Options Bar to quickly copy several lights at once.

11. Select *Square Supply Diffuser:* **24"x24"** from the *Type Selector*.

12. Place the diffusers as shown in Figure 7-3.6.

13. Select *Square Return Register:* **24"x24"** from the *Type Selector*.

14. Place the registers as shown in Figure 7-3.6.

15. Save your project as **ex7-3.rvt**.

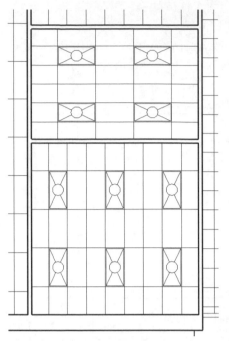

FIGURE 7-3.5 RCP; rotated lights

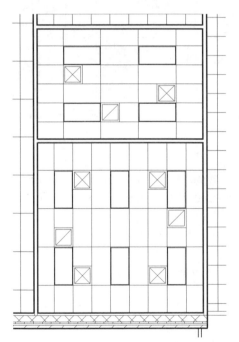

FIGURE 7-3.6 RCP; mechanical

Reflected Ceiling Plan Symbols:

Revit provides many of the industry standard symbols necessary in drawing reflected ceiling plans (RCP). As shown in Figure 7-3.6, supply air is represented with an X and return air has a diagonal line. It is typical to have an RCP symbol legend showing each symbol and material pattern and list what each one represents.

Component Properties

If you want to adjust the properties of a component, such as a light fixture, you can browse to it in the *Project Browser* and right click on it *(notice the right click menu also has the option to select all instances of the item in the drawing)* and select *Properties*. You will see the dialog below for the 2x4 (2 Lamps).

You can also click duplicate and add more sizes (e.g., 4′x4′ light fixture).

You can also select an inserted component and review the *Properties Palette* for additional properties for that particular instance.

It should be pointed out that, in professional practice, the mechanical and electrical design is often created in a separate model and then linked into the architectural model.

Exercise 7-4:
Annotations

This short section will look at adding notes to your RCP. A later chapter will cover annotations in much more detail.

Adding Annotations

1. Select **Annotate** → **Text** → **Text** from the *Ribbon*.

2. Pick **3/32" Arial** from the *Type Selector*.

3. Select the **Leader** option pointed out in **Figure 7-4.1**.

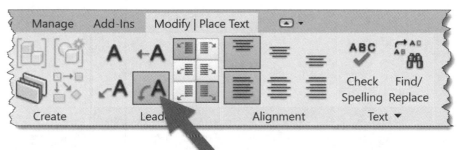

FIGURE 7-4.1 Text; Ribbon

Next, you will add a note indicating that the atrium area is open to the floor above (i.e., no floor or ceiling here). First you will draw a leader, and then Revit will allow you to type the text.

4. Add one of the leaders (i.e., arrows) shown in Figure 7-4.2.

5. Add the note "**OPEN TO ABOVE**" shown in **Figure 7-4.2**.

    a. Do the following to add the right-hand arrow:
        i. Click **Modify**;
        ii. Click the text to select it;
        iii. Click the right-hand arrow icon on the *Ribbon*.

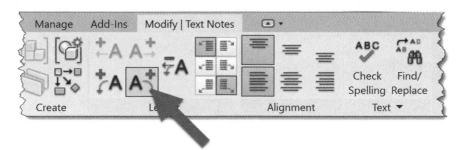

Notice in the image of the *Ribbon* above, with the text selected, the ability to remove leaders is available (i.e., *Remove Last*). The arrows are removed in the order they were added.

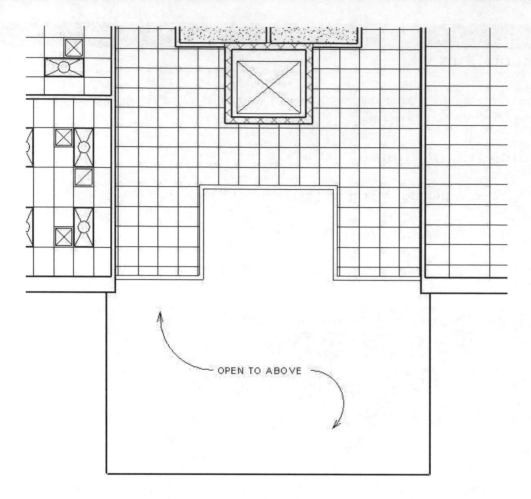

FIGURE 7-4.2 Text with leaders

Adding Text Styles to Your Project

You can add additional text styles to your project. Some firms prefer a font that has a hand lettering look and others prefer something like the Arial font. These preferences can be saved in the firm's template file so they are consistent and always available. You will add a new text style next.

6. Click on the **Text** tool.

7. Select **Edit Type** on the *Properties Palette*.

8. Select **Duplicate** and enter **1/4″ Outline Text** (Figure 7-4.3).

9. Next, make the following adjustments to the *Type Properties* (Figure 7-4.4):

    a. *Text Font:* **Swis721 BdOul BT**
    b. *Text Size:* **1/4″**

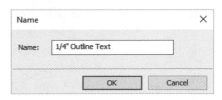

NOTE: You can use any Windows True-Type font. If you do not have this font, select another that best matches (see Figure 7-4.5 below).

FIGURE 7-4.3 New text name

The text size you entered in step 9 is the size of the text when printed. If you change the scale of the drawing, the text size will automatically change, so the text is always the correct size when printing. It is best to set the drawing to the correct scale first, as changing the drawing scale can create a lot of work: repositioning resized text that may be overlapping something or too big for a room.

10. Select **OK** to close the open dialog boxes.

You should now have the new text style available in the *Type Selector* on the *Properties Palette*.

11. Use the new text style to create the text shown below (Figure 7-4.5).

12. Erase the sample text (unless your instructor tells you otherwise).

13. Save as **ex7-4.rvt**.

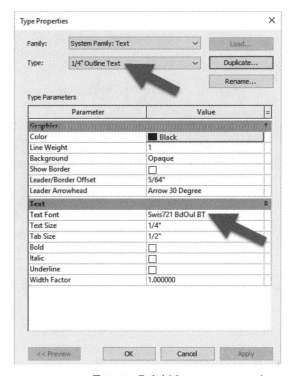

FIGURE 7-4.4 New text properties

TEST TEXT IN REVIT

FIGURE 7-4.5 New text style sample

Self-Exam:

The following questions can be used as a way to check your knowledge of this lesson. The answers can be found at the bottom of this page.

1. You must pick Walls to define floor areas. (T/F)

2. Use the Ctrl key to cycle through the selection options. (T/F)

3. When you add a floor object in the plan view, the floor does not show up right away in the other views, i.e., 3D, Sections, etc. (T/F)

4. You use the _____ tool if you need to add a new product, like exterior plaster, so you can add it to wall types and other systems.

5. You have _____ different types of leader options with the *Text* tool.

Review Questions:

The following questions may be assigned by your instructor as a way to assess your knowledge of this section. Your instructor has the answers to the review questions.

1. It is not possible to create new text styles. (T/F)

2. You can add additional diffuser sizes to the Family as required. (T/F)

3. The light fixtures automatically turn to align with the ceiling grid. (T/F)

4. You can adjust the ceiling height room by room. (T/F)

5. Use the _____ button to add additional elements, for insertion, into the current project (e.g., ceiling: linear box family).

6. Leaders can only be removed in the order in which they were originally drawn. (T/F)

7. Use the _____ tool if the ceiling grid needs to be at an angle.

8. Use the _____ tool to adjust the ceiling grid location if a ceiling tile is less than half its normal size.

9. The Visibility/Graphics Override dialog can control the visibility of an object's surface pattern, i.e., the stipple for the gypsum board ceiling. (T/F)

10. What is the current size of your project (after completing Exercise 7-4)?

    _____ MB.

Lesson 8
INTERIOR and EXTERIOR ELEVATIONS:

This lesson will cover interior and exterior elevations. The default template you started with already has the four main exterior elevations set up for you. You will investigate how Revit generates elevations and the role the elevation tag plays in that process. You will also be introduced to a feature called *Design Options*.

Exercise 8-1:
Creating and Viewing Exterior Elevations

Here you will look at setting up an exterior elevation and how to control some of the various options.

Setting Up an Exterior Elevation

Even though you already have the main exterior elevations set, you will go through the steps necessary to set one up. Many projects have more than four exterior elevations, so all exterior surfaces are elevated.

1. Open your project, ex7-4.rvt, and **Save As ex8-1.rvt**.

2. Switch to your **Level 1** *Floor Plan* view.

3. Select **View → Create → Elevation**.

   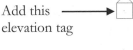

   TIP: Select the main part of the split-button, not the down arrow.

4. Place the temporary elevation tag in the plan view as shown in Figure 8-1.1.

   NOTE: As you move the cursor around the screen, the elevation tag automatically turns to point at the building.

You now have an elevation added to the *Project Browser* in the *Elevations* grouping.

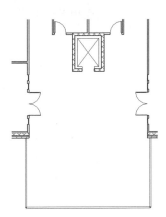

Add this ⟶ elevation tag

FIGURE 8-1.1 Added elevation tag

After placing an elevation tag, you should rename the elevation label in the *Project Browser*.

5. In the *Project Browser*, under *Elevations*, select the elevation label that was just added; it should be "Elevation 1 – a."

6. Right-click on the view label and select **Rename**.

7. Type: **South Temp**

The name should be fairly descriptive, so you can tell where the elevation is just by the label. This will be essential on a large project that has several exterior elevations and even more interior elevations.

8. Double-click on **South Temp** in the *Project Browser*.

The elevation may not look correct right away. You will adjust this in the next step. Notice, though, that an elevation was created simply by placing an elevation tag in the plan view.

9. Switch back to your **Level 1** floor plan view.

Next you will study the options associated with the elevation tag. This, in part, controls what is seen in the elevation.

10. The elevation tag has two parts: the pointing triangle (pointer) and the square center (body). Each part will highlight as you move the cursor over it. **Select the square center part**.

You should now see the symbol shown on the right (Figure 8-1.2).

View direction boxes:
The checked box indicates which way the elevation tag is looking. You can check (or uncheck) the other boxes. Each checked box relates to an elevation view in the *Project Browser*.

Rotation control:
Allows you to look perpendicular to an angled wall in plan, for example.

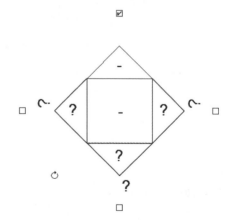

FIGURE 8-1.2 Selected elevation tag

Adjusting elevation tags:

While selected, you simply drag the tag to move it.

11. Press the **Esc** key to unselect the elevation tag.

12. Select the "pointing" portion of the elevation tag.

13. In the *Properties Palette*, set *Far Clipping* to "Clip without line" and then click **OK**.

Your elevation tag should look similar to Figure 8-1.3.

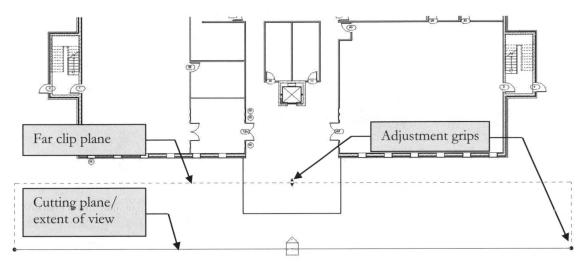

FIGURE 8-1.3 Selected elevation tag

The elevation tag, as selected in Figure 8-1.3, has several features for controlling how the elevation looks. Here is a quick explanation:

- **Cutting plane/extent of view line**: This controls how much of the 3D model is elevated from left to right (i.e., the width of the elevation).

- **Far clip plane**: This controls how far into the 3D model the elevation can see.

- **Adjustment grips**: You can drag this with the mouse to control the features mentioned above.

14. Select the view label **South Temp** in the *Project Browser*; now look at the *Properties Palette* settings for the view (see notes on the next page).

You have several options in the *Properties Palette* (Figure 8-1.4). Notice the options under the *Extents* heading; below is an examination of three of these settings:

- **Crop View**: This crops the width and height of the view in elevation. *Adjusting the width of the cropping window in elevation also adjusts the "extent of view" control in the plan view.*

- **Crop Region Visible**: This displays a rectangle in the elevation view indicating the extent of the cropping window (described above).
  When selected in elevation view, the rectangle can be adjusted with the adjustment grips.

- **Far Clipping**: If this is turned "off," Revit will draw everything visible in the 3D model *(within the "extent of view").*

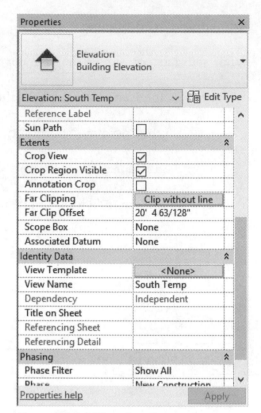

FIGURE 8-1.4 Elevation view:
South Temp - Properties

You will manipulate some of these controls next.

15. With the elevation tag still selected (as in Figure 8-1.3), drag the "cutting plane/extent of view" line up into the atrium as shown in Figure 8-1.5.

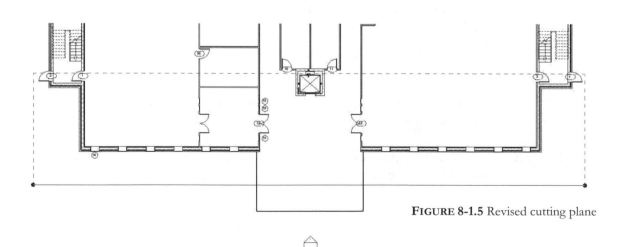

FIGURE 8-1.5 Revised cutting plane

16. Now switch to the *Elevation view*: **South Temp**.

Your elevation should look similar to Figure 8-1.6. If required, click on the cropping region and resize it to match Figure 8-1.6.

The atrium curtain wall and roof are now displayed in section because of the location of the "cutting plane" line in the plan.

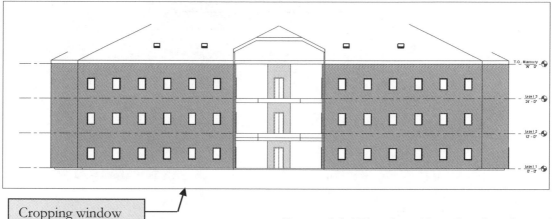

Cropping window

FIGURE 8-1.6 Elevation with cutting plane thru atrium

Notice that the roof is not fully visible. This is not related to the cropping window shown in Figure 8-1.6. Rather, it is related to the "Far Clip Plane" set in the plan view.

17. Adjust the "Far Clip Plane" in the **Level 1** plan view so that the entire roof shows in the **South Temp** view.

Next you will adjust the elevation tag to set up a detail elevation for the atrium curtain wall.

18. In **Level 1** plan view, adjust the elevation tag to show only the atrium curtain wall (Figure 8-1.7).

19. Switch to **South Temp** view to see the "detail" elevation (Figure 8-1.8).

20. Adjust the South Temp view's **Properties** to turn off the crop region's visibility; this can also be done via the *View Control Bar*.

21. **Save** your project as **ex8-1.rvt**.

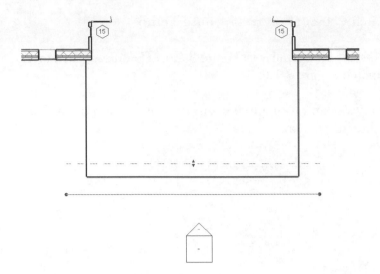

FIGURE 8-1.7 Atrium curtain wall detail elevation

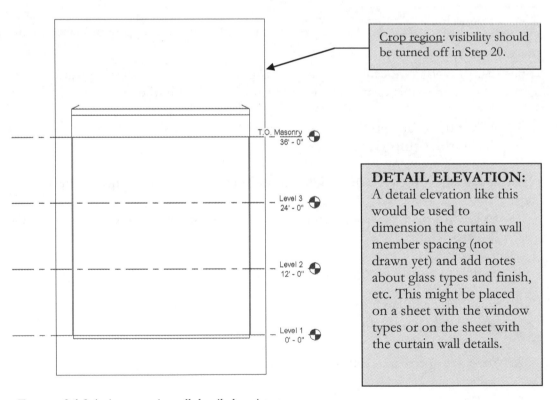

Crop region: visibility should be turned off in Step 20.

T.O._Masonry
36' - 0"

Level 3
24' - 0"

Level 2
12' - 0"

Level 1
0' - 0"

DETAIL ELEVATION:
A detail elevation like this would be used to dimension the curtain wall member spacing (not drawn yet) and add notes about glass types and finish, etc. This might be placed on a sheet with the window types or on the sheet with the curtain wall details.

FIGURE 8-1.8 Atrium curtain wall detail elevation

Exercise 8-2:
Modifying the Project Model: Exterior Elevations

The purpose of this exercise is to demonstrate that changes can be made anywhere, and all other drawings are automatically updated.

Modify an Exterior Elevation

1. Open ex8-1.rvt and **Save As ex8-2**.

2. Open the **East** exterior elevation view.

3. Use the **Window** tool and select _Fixed:_ **32″ x 48″** in the _Type Selector._

You will insert a window in elevation. This will demonstrate, first, that you can actually add a window in elevation not just plan view, and second, that the other views are automatically updated.

Notice, with the window selected for placement, you have the usual dimensions helping you accurately place the window. As you move the window around you should see a dashed horizontal cyan colored line indicating the default sill height.

4. Place a window as shown in **Figure 8-2.1**; make sure the bottom of the window "snaps" to the cyan sill line.

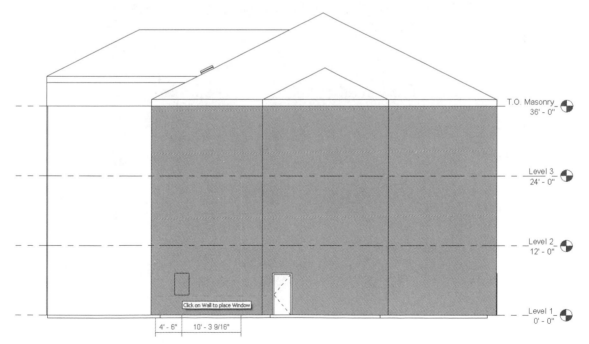

FIGURE 8-2.1 Placing a window

5. Switch to **Level 1** plan view; notice the window is added (Figure 8-2.2).

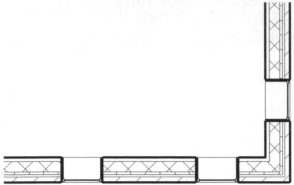

TIP: If the window is towards the inside, use the control arrows to flip the window within the wall. It should look like the window in Figure 8-2.2.

FIGURE 8-2.2 Level 1 – south-east corner

6. Switch back to the **East** elevation view.

7. Add windows as shown in Figure 8-2.3.

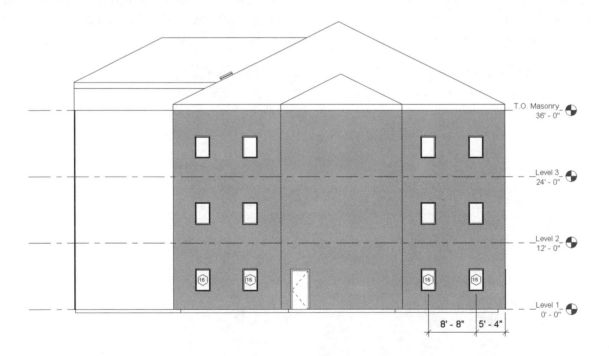

FIGURE 8-2.3 East elevation – windows added

If you laid out the interior walls as described in Lesson 3, you should get a warning message when inserting the windows on Level 1, towards the North side of the building. This is because the interior wall for the room in the north-east corner conflicts with the exterior window. Revit is smart enough to see that conflict and bring it to your attention. In this case you probably want the windows to be uniformly spaced, so you will ignore the conflict and move the wall in the plan view.

8. Click the red X (in the upper right) to ignore the wall/window conflict warning (Figure 8-2.4). *If you did not get this warning, skip this step.*

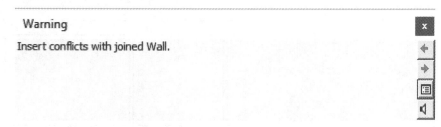

FIGURE 8-2.4 Conflict warning

9. Switch to **Level 1** plan view and revise the wall as shown in **Figure 8-2.5**.

    TIP: You will need to use the Split tool to break the wall where it offsets. You can then select the wall (just the wall; the doors will automatically move with the wall) and use the Move tool to move it north (check Disjoin on the Options Bar). Also notice that the windows on the east wall need to be flipped.

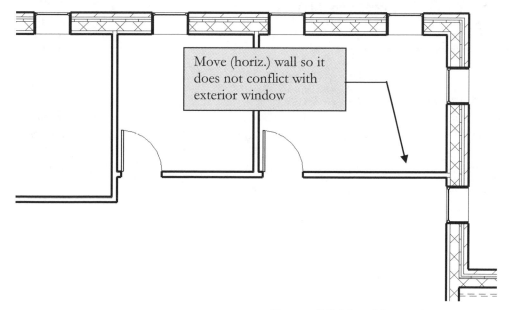

FIGURE 8-2.5 Level 1 – north-east corner

10. Switch to the **Level 1** RCP view. Fix the ceiling.

> *NOTE: The ceiling grids are likely not aligned with the revised walls. If this is the case, see the explanation below (Figure 8-2.6).*

Most of the time, when you move a wall, Revit will automatically update the ceiling grid to fit the new room. However, occasionally the definition of the room boundary is lost while making modifications. In this case, you will have to delete the grid and reinsert it - or select it and "edit sketch."

Deleting a Ceiling Grid:

When selecting a ceiling grid, Revit only selects one line. This does not allow you to delete the ceiling grid. To delete, hover cursor over a ceiling grid line and tap the TAB key until you see the ceiling perimeter highlight, then click the mouse. The entire ceiling will be selected. Press Delete.

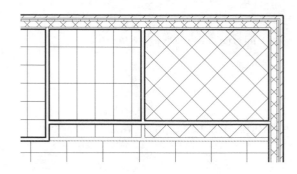

FIGURE 8-2.6 Level 1 RCP – north-east corner

11. Add the same layout of windows (Figure 8-2.2) to the West elevation.

> *TIP: Mirror the windows in plan view, each floor.*

12. **Save** your Project as **ex8-2.rvt**.

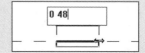

TIP: ENTERING DIMENSIONS IN REVIT
*As your experience with Revit grows, you will want to learn some of the shortcuts to using the program. One of those shortcuts is how you enter dimensions when drawing. You probably already know, maybe by accident, that if you enter only one number (e.g., 48) and press enter, Revit interprets that number to be feet (e.g., 48′-0″). So, if you want to enter 48″, you may be typing 0′-48″ or 48″. Both work but having to press the **Shift** key to get the inch symbol takes a little longer.*

Here are some options for entering dimensions:

0 48	*Revit reads this as 48″ (zero space forty-eight)*
48	*Revit reads this as 48′-0″*
5.5	*Revit reads this as 5′-6″*
0 5.5	*Revit reads this as 5 ½″*
2 0 1/4	*Revit reads this as 2′-0¼″ (two space zero space fraction)*

Exercise 8-3:
Creating and Viewing Interior Elevations

Creating interior elevations is very much like exterior elevations. In fact, you use the same tool. The main difference is that you are placing the elevation tag inside the building, rather than on the exterior.

Adding Interior Elevation Tag

1. Open project ex8-2.rvt and **Save As ex8-3.rvt**.

2. Switch to **Level 1** floor plan view, if necessary.

3. Select the **Elevation** tool.

4. Select _Elevation:_ **Interior Elevation** from the _Type Selector_ and then place an elevation tag, looking East, in the atrium area (Figure 8-3.1); place as shown in the center of the room.

**REMEMBER:** _The first thing you should do after adding a new view is to give it an appropriate name in the Project Browser list._

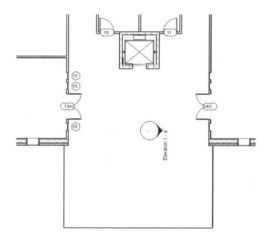

5. Change the name of the elevation to **East Atrium**.

6. Switch to the East Atrium view. _Try double-clicking on the pointer part of the elevation tag to open the view._

Initially, your elevation should look something like Figure 8-3.2. You will adjust this view next. _Notice how Revit automatically controls the line weights of things in section vs. things in elevations._

FIGURE 8-3.1 Level 1 - Atrium

**FYI:** _The elevation tags are used to reference the sheet and drawing number so the client or contractor can find the desired elevation quickly while looking at the floor plans. This will be covered in a later lesson. It is interesting to know, however, that Revit automatically does this (fills in the elevation tag) when the elevation is placed on a sheet and will update it if the elevation is moved._

7. Switch back to the **Level 1** view.

8. Pick the "pointing" portion of the elevation tag, so you see the view options (Figure 8-3.3).

You should compare the two drawings on this page (Figures 8-3.2 and 8-3.3) to see how the control lines in the plan view dictate what is generated/visible in the elevation view for both width and depth.

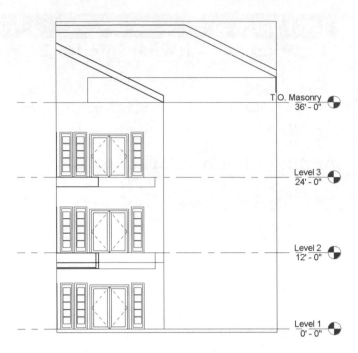

FIGURE 8-3.2 East Atrium – initial view

The goal is to set up an interior elevation of the entire east atrium wall, with the floor structure and roof shown in section.

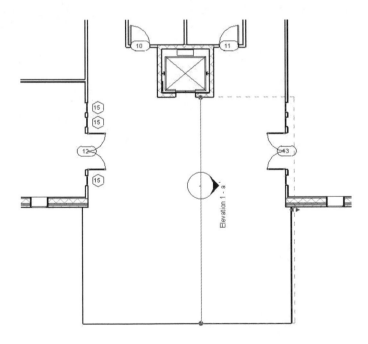

FIGURE 8-3.3 Elevation tag selected

9. Adjust the control lines for the elevation tag as shown in Figure 8-3.4. Drag the "cutting plane/extent of view" line to the location shown. Make sure the "far clip plane" extends past the door alcove, otherwise it will not show up.

10. Switch back to the **East Atrium** view.

Other than adjusting the height of the view, you have the view ready.

11. Select the cropping region and drag the top middle grip upward, to increase the view size vertically (Figure 8-3.5).

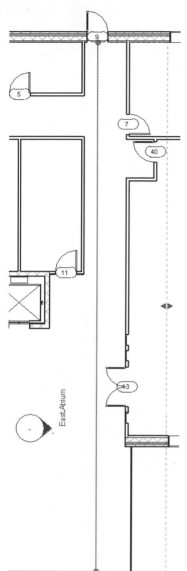

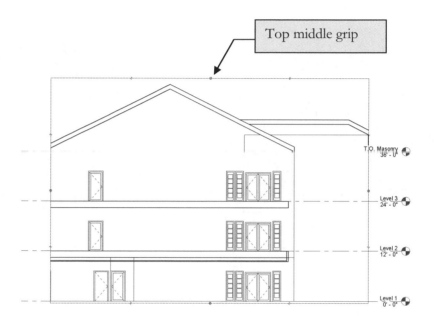

Top middle grip

FIGURE 8-3.5 East Atrium
Elevation – crop region selected

FIGURE 8-3.4 Elevation tag
adjustments

If the ceiling were drawn for the third floor (your instructor may have assigned this), you would probably stretch the crop region down to it. Interior elevations don't normally show walls, roofs and floors in section. An atrium elevation like this could be an exception for the floors.

12. Now stretch the top of the crop region down to approximately 9'-6" above Level 3. (Go to the ceiling if you have drawn one for Level 3.)

13. Stretch the bottom of the crop region up to align with the top of the Level 1 floor slab.

14. On the **View Control Bar**, set the scale to ¼"=1'-0".

Your elevation should look like Figure 8-3.6.

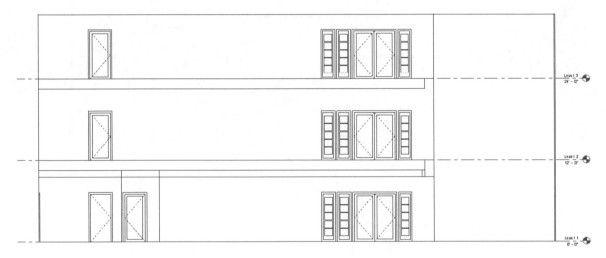

FIGURE 8-3.6 East Atrium Elevation

You can leave the crop region on to help define the perimeter of the elevation. You can also turn it off. However, some lines that are directly under the crop region might disappear. You could use the *Detail Line* tool to manually define the perimeter.

Also, notice the level datum automatically resized to match the new scale. When space permits, most interior elevations are ¼" = 1'-0".

15. Save your project as **ex8-3.rvt**.

Exercise 8-4:
Modifying the Project Model: Interior Elevations

This short exercise, similar to Exercise 8-2, will look at an example of Revit's ability to change anything anywhere. All drawings are generated from one 3D model.

Modify the Interior Elevations

1. Open ex8-3.rvt and **Save As ex8-4.rvt**.

2. Open the **East Atrium** elevation view.

You will move two doors and add one.

3. Select both of the single doors on Levels 2 and 3; use the Ctrl key to select multiple objects at one time.

4. Use the **Move** tool to move the door 6'-0" to the right (South). (Figure 8-4.1)

5. In the East Atrium elevation view, use the **Door** tool to place a _Sgl Flush:_ **36" x 84"** door on Level 2 to the far left (North). (See Figure 8-4.1.)

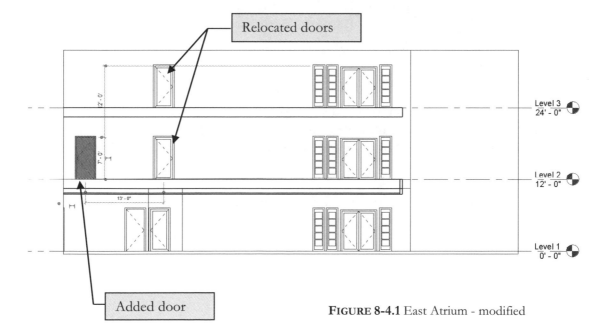

FIGURE 8-4.1 East Atrium - modified

Now it's time to see the effects to the plan views.

6. Switch to **Level 2** floor plan view (Figure 8-4.2).
    You can also see a similar change on Level 3.

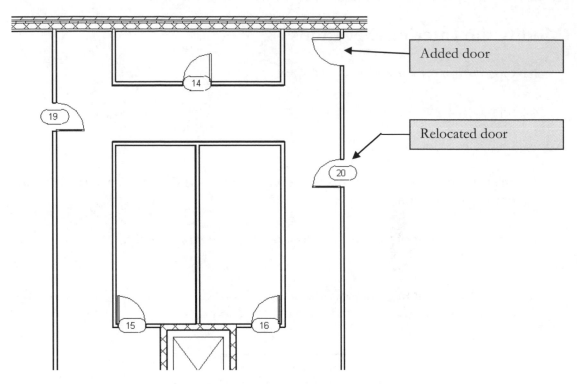

FIGURE 8-4.2 Level 2 plan with changes

When placing a door in elevation, you may have to switch to plan view to verify the door swing is the way you want it; you cannot control the door swing in elevation (well, actually, you can via a right-click on the door in elevation, but you would first need to know which way it is swinging).

In elevation, you can adjust many things this way. Some examples are ceiling height, interior and exterior windows, wall locations (perpendicular to the current view), etc.

7. Save your project as **ex8-4**.

Exercise 8-5:
Adding Mullions to a Curtainwall

This exercise will cover the steps involved in designing a curtainwall system (only from an aesthetic viewpoint, not structurally). This is surprisingly simple to do.

Adding Curtain Grid

First, you draw a grid on your curtainwall. This sets up the location for your mullions, which you will add later.

1. Open ex8-4.rvt and **Save As ex8-5.rvt**.

2. Switch to your **South Temp** view.

3. Select **Architecture → Build → Curtain Grid**.   Curtain Grid

4. Draw the grid as shown in **Figure 8-5.1**.
   Be sure to add horizontal grids at Levels 2 and 3.

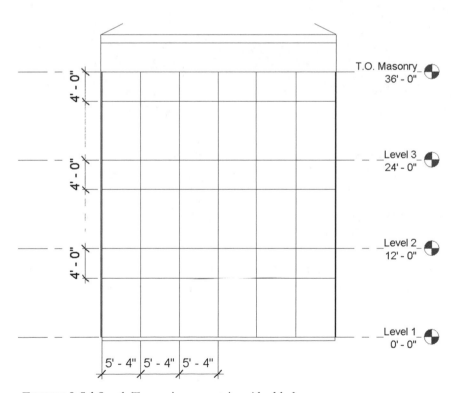

FIGURE 8-5.1 South Temp view – curtain grid added

If a curtain grid line did not land in the correct place, you can select it and adjust the dimensions that will appear on the screen. To select the grid you need to place your cursor over the grid line and press the *Tab* key until the curtain grid is highlighted, and then click to select.

5. Select the *Curtain Grid* tool and then click on the **One Segment** option on the *Ribbon* (Figure 8-5.2).

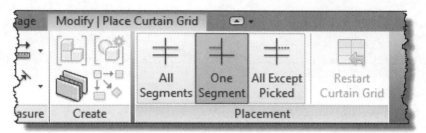

FIGURE 8-5.2 Ribbon options for Curtain Grid tool

6. Draw two vertical lines to set up the main entry door location (Figure 8-5.3). *You do not need to draw the dimensions shown.*

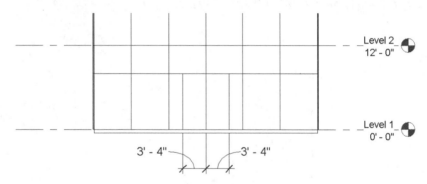

FIGURE 8-5.3 Curtain grid for door location

Notice that the *One Segment* option limits the grid to the "cell" you clicked in, rather than extending from top to bottom as the others did.

Next, you will set up the curtain grid lines around the corners. This can be done from the East or West views, similar to the previous steps. However, this can also be accomplished in a 3D view.

7. Click on the **3D** icon on the *Quick Access Toolbar*.

8. Using the **Curtain Grid** tool, add the grid lines shown in **Figure 8-5.4**. Starting at the outside corner, space the grids 5'-4" (the last space will be smaller).

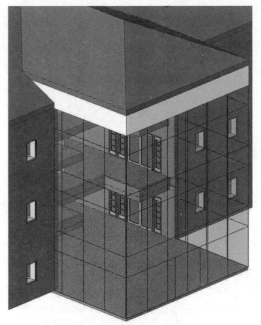

As you move the cursor while placing the horizontal grids, you should see the grid "snap" to the grid around the corner; that is when you click the mouse.

Drag your cursor on the ViewCube to see your project from different views. Clicking the "Home" icon (visible when your cursor is over the ViewCube) will reset the view.

FIGURE 8-5.4 Curtain grid – 3D view

Adding Doors

9. Switch back to **South Temp** view and select one of the 3′-4″ wide cells; place your cursor over the cell [edge] and tap the *Tab* key until that cell is highlighted and then click to select.

10. With the cell selected, pick **Door-Curtain-Wall-Single-Glass** from the *Type Selector* on the *Properties Palette* (Figure 8-5.5).

    TIP: *Per the steps outlined in chapter 3, load the door family into your project from the Door folder.*

11. Repeat the previous two steps for the other 3′-4″ wide cell (Figure 8-5.6).

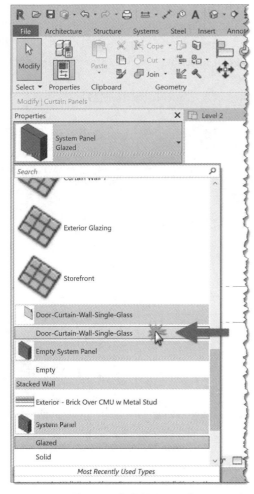

FIGURE 8-5.5 Type selector with curtain wall cell selected

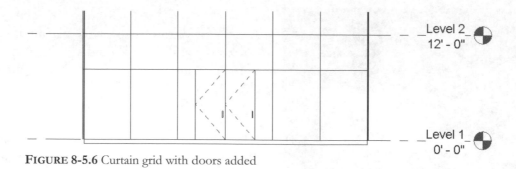

FIGURE 8-5.6 Curtain grid with doors added

Adding Curtain Wall Mullions

Thus far you have simply set up the spacing for the curtain wall mullions. Next you need to place the curtain wall mullions. This involves selecting a size for the mullion, as they typically come in many shapes and sizes. (The depth is usually related to the height of the curtain wall, as the mullion acts as the structure for the glass wall.)

12. Switch to the **3D** view.

13. Select **Architecture → Build → Mullion**.

14. Select *Rectangular Mullion:* **2.5″ x 5″ rectangular** from the *Type Selector* (Figure 8-5.7B).

15. Select all the grid lines you previously placed and the perimeter (excluding the outside corners and the bottom horizontal member on the south face).

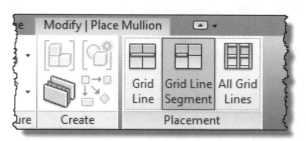

FIGURE 8-5.7A
Ribbon for Mullion tool; changed in step 16

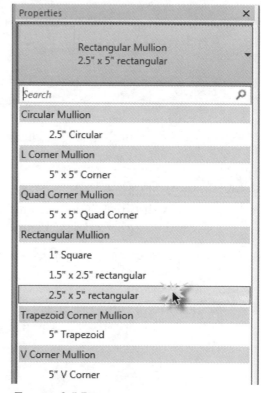

FIGURE 8-5.7B
Type selector for Mullion tool

Next, you will add the horizontal mullion at the bottom, on the south side. You need to place this mullion so it does not extend through the door openings.

16. With the Mullion tool selected, click **Grid Line Segment** from the *Ribbon* (Figure 8-5.7A).

17. Click on the bottom edge of the six cells (skipping the two door openings), to place the horizontal mullion (Figure 8-5.8).

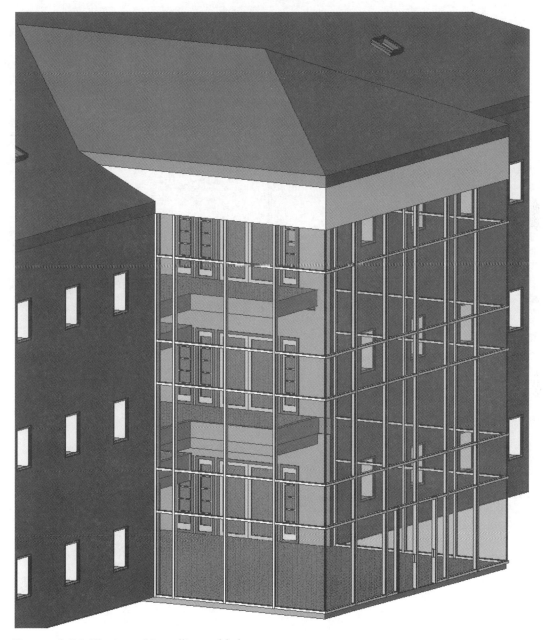

FIGURE 8-5.8 3D view with mullions added

18. Click on the two verticals next to each door to place a mullion (Figure 8-5.8).

In this example, you will not place a corner mullion. This will be a butt-joint condition where the two panes of glass are held together with silicon in the corner. However, figure 8-5.7B shows corner mullions that can be added.

All views will now be updated to show the curtain wall mullions.

19. Switch to the **Level 1** plan view to see the added curtain wall mullions and doors (Figure 8-5.9).

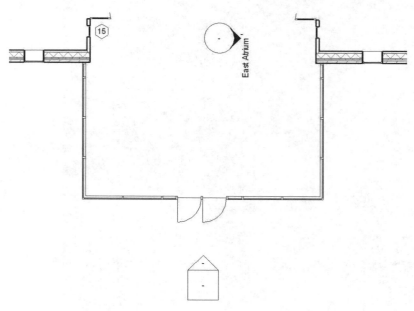

FIGURE 8-5.9 Level 1 plan view with curtain wall mullions

20. **Save** your project as **ex8-5.rvt**.

Exercise 8-6:
Design Options

This exercise will explore a feature called *Design Options*. This feature allows you to present two or more options for a portion of your design without having to save a copy of your project and end up having to maintain more than one project until the preferred design option is selected.

This feature can be used during the early design phase of a project or to manage bidding alternates all the way through construction documents. Design Options are different from phasing, which manages changes on the model over time, in that it manages changes to the model within the same time framc.

Design Options Overview

A Revit project can have several design option studies at any given time. You might have an (A) entry canopy options study, (B) an executive board room options study, and a (C) toilet room layout options study in a project. Each of these studies can have several design options associated with them. For example, the entry canopy study might have three options: 1. flat roof, 2. gable roof, and 3. sloped glass roof.

A particular study of an area within your project is called a *Design Option Set*, and the different designs associated with a *Design Options Set* are called *Options*. Both the *Design Options Set* and the *Options* can be named.

One of the *Options* in a *Design Option Set* is specified as the *Primary* option (the others are called *Secondary* options); this is the option that is shown by default in all new and existing views. However, you can adjust the *Visibility* of a *View* to show a different option. Typically you would duplicate a *View*, adjust the *Visibility*, and *Rename* it to have each option at your fingertips.

When the preferred design is selected, by you or the owner, you set that *Option* to *Primary*. Finally, you select a tool called *"Accept Primary"* which deletes the *Secondary Options* and the *Design Option Set*, leaving the *Primary Option* as part of the main building model.

Design Option Set

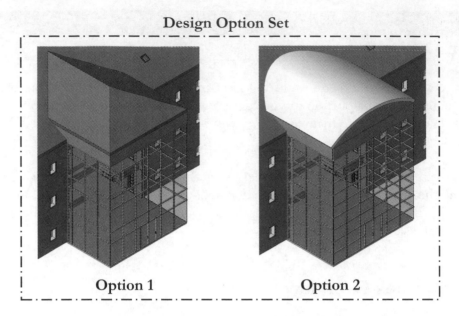

Option 1   **Option 2**

The following image (Figure 8-6.1) is an overview of the Design Options Dialog:

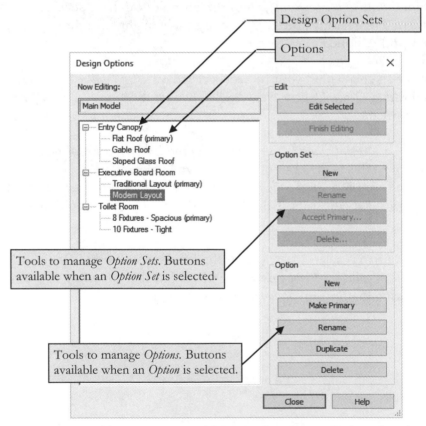

FIGURE 8-6.1 Design Options Dialog; Modern Layout Option selected

Notes about the *Design Options* dialog box:

Edit buttons:
You can edit a *Design Option* by selecting an *Option* (in the window area on the left) and then clicking the *"Edit Selected"* button. Next you add, move and delete elements in that Design Option.

When finished editing a *Design Option*, you reopen the *Design Options* dialog and click the *"Finish Editing"* button.

If you are currently in an Option Editing mode, the *Now Editing* area in the *Design Options* dialog displays the *Option* name being modified, otherwise it displays "Main Model."

Option Set buttons:
The *New* button is always available. You can quickly set up several *Option Sets*. Each time you create a new *Option Set*, Revit automatically creates a *Primary Option* named "Option 1."

The other buttons are only available when an Option Set label is highlighted (i.e., selected) in the window list on the left.

The *Accept Primary* button causes the Primary option of the selected Set to become a normal part of the building model and deletes the Set and Secondary Options. This is a way of "cleaning house" by getting rid of unnecessary information which helps to better manage the project and keeps the file size down.

Option buttons:
These buttons are only available when an *Option* (primary or secondary) is selected within an *Option Set*. You can quickly set up several *Options* without having to immediately add any content (i.e., walls, components, etc.) to them.

The *Make Primary* button allows you to change the status of a *Secondary Option* to *Primary*. As previously mentioned, the *Primary Option* is the *Option* that is shown by default in existing and new views.

The *Duplicate* button will copy all the elements in the selected Option into a new Option (this makes the file larger because you are technically adding additional content to the project). You can then use the copied elements (e.g., walls, furniture, etc.) as a starting point for the next design option. This is handy if the various options are similar.

Now you will put this knowledge to use!

Setting Up a Design Option Set

In this exercise you will create two *Design Option Sets*: one for the curtain wall and another for the roof area above the curtain wall. You will create an alternate roof and curtain wall design for the office building project.

You could just create one *Design Option Set* and have two design options total. However, by placing the curtain wall options in one *Option Set* and the roof in another, you actually get a total of four design options. You can mix and match the curtain wall and roof options.

Setting Up Design Options in Your Project

First you will set up the *Option Sets* and *Options*.

1. Open ex8-5.rvt and **Save As ex8-6.rvt**.

2. Select the **Design Options** icon on the *Status Bar* (see image below).

You are now in the *Design Options* dialog box (Figure 8-6.2). Unless you have modified your file to have *Option Sets*, your dialog will look like this one.

3. In the *Option Set* area click **New**.

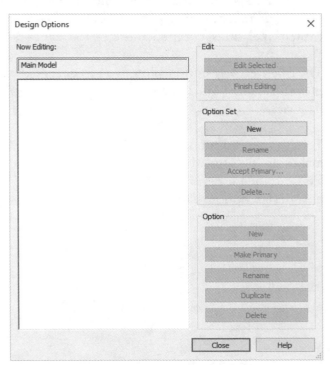

FIGURE 8-6.2 Design Options Dialog; initial view

Notice an *Option Set* named <u>Option Set 1</u> has been created. Revit also automatically created the *Primary Option* named <u>Option 1</u> (Figure 8-6.3). Next you will rename the *Option Set* to something that is easier to recognize.

FIGURE 8-6.3 Design Options Dialog; new option set created

4. (See warning below.) Select the *Option Set* currently named <u>Option Set 1</u> and then click the **Rename** button in the *Option Set* area.

> ***WARNING!*** *Be sure you are not renaming the Option but, rather, the Option Set.*

5. In the *Rename* dialog type **Curtainwall** (Figure 8-6.4).

6. Click **OK** to rename.

Giving the *Option Set* a name that is easy to recognize helps in managing the various options later, especially if you have several.

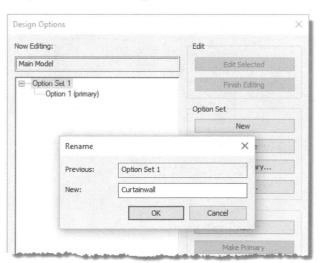

FIGURE 8-6.4 Rename Option Set Dialog; enter Curtainwall

Next you will create a *Secondary Option* for the <u>Curtainwall</u> *Option Set*.

7. With the <u>Curtainwall</u> *Option Set* selected (or any option in that set), click **New** in the *Option* area.

Notice a secondary option was created and automatically named Option 2. If you have descriptive names for the options in a set, you should apply them. In this example you can leave them as they are.

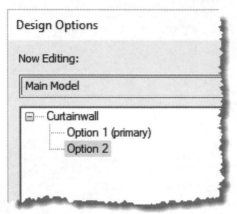

FIGURE 8-6.5 Design Options Dialog;
secondary option created

8. Create an *Option Set* for the roof (Figure 8-6.6):

    a. Name the set **Atrium Roof**.

    b. Create one secondary option.

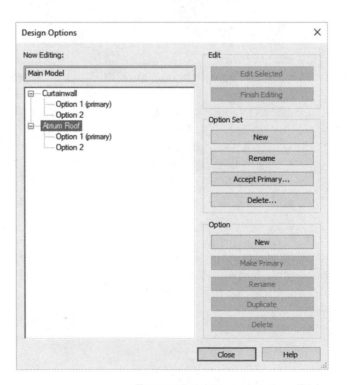

FIGURE 8-6.6 Design Options Dialog

The basic thinking with the *Design Options* feature is that you set up the *Option Set* and *Option*s and then start drawing the elements related to the current *Option*. So, you select the option you wish to edit via the *Design Options* drop-down list on the *Status Bar*, make the additions and modifications relative to that *Design Option*, and finally set the drop-down list back to "Main Model."

However, in your case, you want to move content already drawn to *Option 1*. Revit has a feature that allows you to move content to a *Design Option Set,* which means the content gets copied to each *Option* in the *Set* you select. This option will work for the curtain wall because the second option will be similar to the first one.

Curtain Wall Design Option

You are now ready to set up the different design options.

9. Switch to the **Default 3D view**.

10. Select the three curtain wall sections (three major areas around the atrium) and the three short walls above the curtain walls.

    TIP: *Make sure you click when the entire curtain wall area is selected, not just an individual mullion or cell. You may need to use the Tab key to cycle through the available options below your cursor.*

11. Select the **Add to Set** icon on the *Status Bar*.

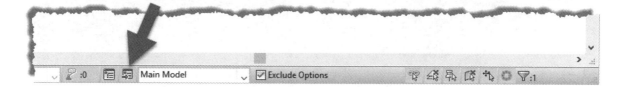

12. Select *Curtainwall* from the dialog and then click **OK** (Figure 8-6.7).

The selected items are now in both *Option 1* and *Option 2* under the *Option Set: Curtainwall*.

From this point forward you can only modify the curtain wall when in *Option 1* or *Option 2* "edit mode" (in which case the tables are turned and you cannot edit the main building model; this is because "exclude options" is selected on the *Status Bar*).

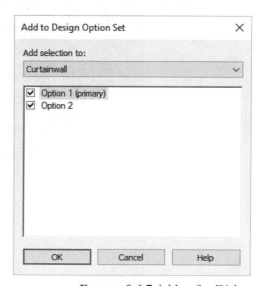

FIGURE 8-6.7 Add to Set Dialog

13. Select **Curtainwall: Option 2** (i.e., Option Set: *Curtainwall*; Option: *Option 2*) from the *Design Options* list on the *Status Bar*. See image to right.

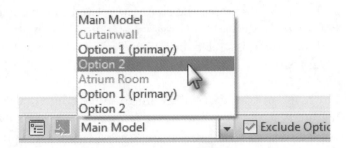

Now you should notice that the main building model is slightly grey and not editable. (It is not editable because "active option only" is selected on the *Status Bar*).

14. Zoom in on the curtain wall area in your 3D view.

15. Using the **Curtain Grid** tool on the *Modeling Tab*, add horizontal lines equally spaced between the larger vertical spaces as shown in Figure 8-6.8 (you should be able to place all the grid lines from this one view angle).

> *TIP:* Use the "All except picked" setting on the Ribbon when placing the grid on either side of the doors on Level 1.

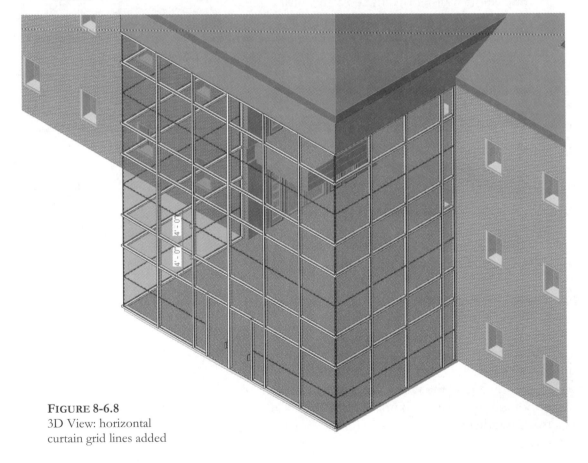

FIGURE 8-6.8
3D View: horizontal curtain grid lines added

16. Select the **Mullion** tool from the *Architecture* tab.

17. Select _Rectangular Mullion:_ **2.5″ x 5″ rectangular** from the _Type Selector_ (Figure 8-6.8).

18. Select each one of the grid lines to place the horizontal mullions (Figure 8-6.9).

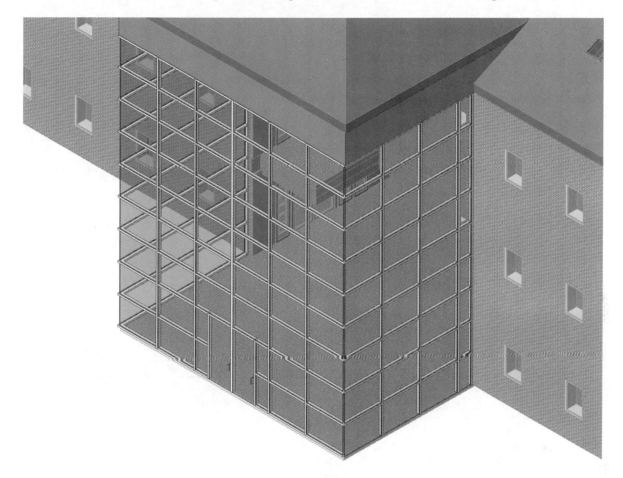

FIGURE 8-6.9 3D View: horizontal mullions added

19. To finish editing the design option select **Main Model** from the _Design Options_ list on the _Status Bar._

It now appears like all your changes disappeared, right? Well, if you recall from the introduction to this exercise, the _Primary Option_ is displayed by default for all new and existing views. So when you finished editing _Curtainwall:Option 2_ the _Default 3D View_ switched back to the primary option (which is currently set to Option 1).

Next you will create a new view and adjust its _Visibility_ to display _Option 2_ of the _Curtainwall Options Set._

First you will create a duplicate copy of the 3D view.

20. In the *Project Browser*, under *3D Views*, right-click on the **{3D}** label.

21. Select **Duplicate View → Duplicate** from the pop-up menu.

You now have a copy of the 3D view named *Copy of {3D}*.

22. **Rename** the new view to **Curtainwall Option 2**.

23. Switch to your new view (if required).

24. Type **VV** on the keyboard to open the *Visibility/Graphics Overrides* for the current view.

25. Click on the **Design Options** tab at the top of the dialog. Note that this tab did not exist until you created *Design Options* in the project.

26. Change the *Design Option* parameter for *Curtainwall* to *Option 2* (Figure 8-6.10).

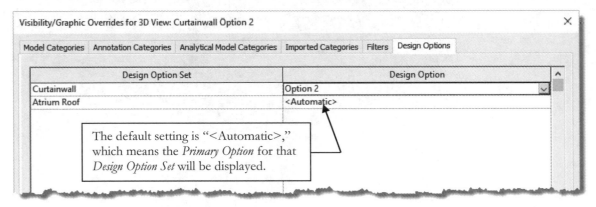

FIGURE 8-6.10 Visibility/Graphic Overrides dialog: modified Curtainwall design option visibility

27. Click **OK** to close the dialog.

Now, with the *{3D}* view and the *Curtainwall:Option 2* view, you can quickly switch between design options. Both views could be placed on the same sheet and printed out for a design critique. *FYI: When a design option is "hard wired" in this way you cannot ever edit any other curtain wall options in this view as they are not visible in this view (this only works when the view is set to automatic).*

Atrium Roof Design Option

Similar to getting things ready to create the second option for the curtainwall, you will do the same for the atrium roof. However, the second roof option is totally different from the first option, so it does not make sense to move the current roof to each option in the *Atrium Roof Option Set*; you would end up completely deleting the roof from *Atrium Roof:Option 2* (deleting the roof would be no problem in this case, but another scenario might have hundreds of entities that need to be moved to Option 1, which would be more difficult to delete from Option 2). Next you will explore how to move content to just one *Option* in an *Option Set*.

28. Switch to the *Default* **3D view**.

29. Select the roof over the atrium and curtain wall area.

30. Again, select **Add to Set** from the *Status Bar*.

You are now able to specify the *Option Set* and the specific *Options* to copy the selected elements to (Figure 8-6.11).

31. Select **Atrium Roof** from the drop-down list (Figure 8-6.11).

32. Uncheck "Option 2" from the list.

33. Click **OK** to close the *Add to Set* dialog.

You have now copied the roof into Option 1 for the *Option Set* named Atrium Roof, and Option 2 is still empty.

You are now ready to create the roof for Option 2.

34. Select **Atrium Roof: Option 2** from the *Design Options* list on the *Status Bar*. See image to the right.

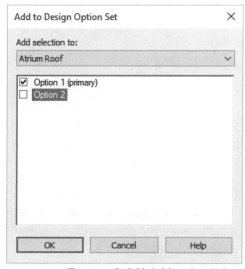

FIGURE 8-6.11 Add to Set dialog

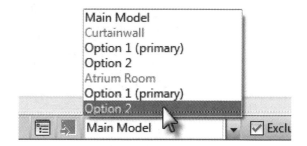

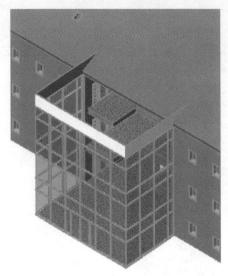

Your view should now look similar to the image on the left. The roof is gone because *Atrium Roof:Option 2* does not currently have a roof in it.

35. Switch to the **South** elevation view.

36. Zoom in to the area above the curtain wall.

Next, you will create an *In-place Family* to represent a curved roof option over the atrium area. Basically, you will create a solid by specifying a depth and then drawing a profile of the curved roof with lines.

37. Click **Architecture → Build → Component → Model In-Place**.

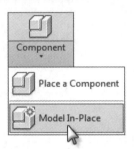

Immediately, you are prompted to select a *Family Category*. This allows Revit to understand how other elements should interact with the object(s) you are about to create and helps control visibility.

38. Select **Roofs** from the *Family Category* list (Figure 8-6.12).

39. Click **OK**.

Now you are prompted to provide a name for the new *Family*.

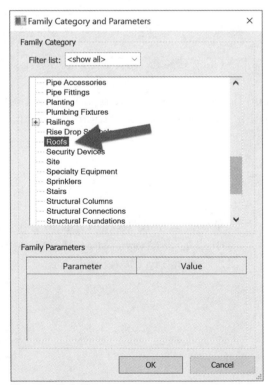

FIGURE 8-6.12
Family Category and Parameters dialog

For the *Family* name, type **Atrium Roof** (Figure 8-6.13).

FIGURE 8-6.13 Family name prompt

You are now in a mode where you draw the Atrium Roof. Notice that the *Architecture* tab on the *Ribbon* has changed (Figure 8-6.14) which has all the tools available to create a *Family*. You are continuously in the *Family* edit mode until you select *Green Check Mark* (to finish the roof) or the *Red X* (to cancel) from the *Ribbon*.

FIGURE 8-6.14
Model In-Place active; *Create* tab on the Ribbon

40. Select **Extrusion** from the *Ribbon* (Figure 8-6.14).

Finally, you are prompted to select a plane in which to start drawing the profile of the solid to be extruded. Even though the view is a 2D representation of a 3D model, Revit needs to know where you want the 3D Solid created. You will select the wall above the curtain wall as a reference surface to establish a working plane.

41. Select **Pick a Plane** and click **OK** (Figure 8-6.15).

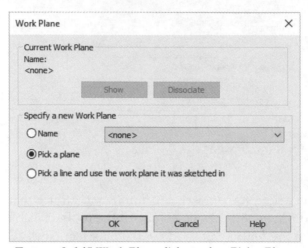

FIGURE 8-6.15 Work Plane dialog; select Pick a Plane

42. Move the cursor over the upper edge of the wall above the curtainwall and press the Tab key until a dashed line appears around the perimeter of the wall, and then click the mouse to select (Figure 8-6.16).

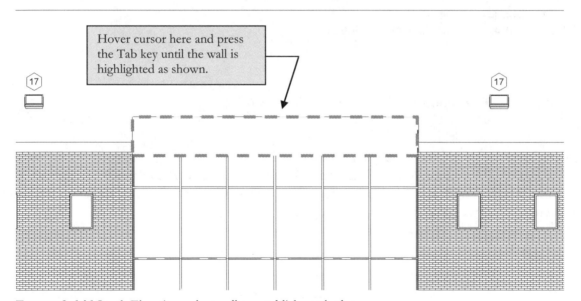

Hover cursor here and press the Tab key until the wall is highlighted as shown.

FIGURE 8-6.16 South Elevation; select wall to establish work plane

Next you will draw an arc to specify the bottom edge of the curved roof design option.

Notice the *Ribbon* changed again to show tools related to drawing an extruded solid (Figure 8-6.17).

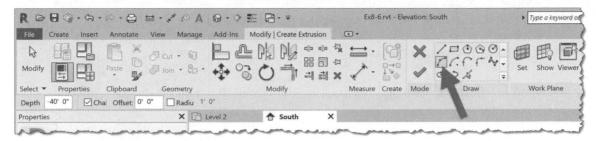

FIGURE 8-6.17 Create Extrusion contextual tab

43. On the *Options Bar*, enter **-40'-0"** for the **Depth** (Figure 8-6.18).

> *FYI:* *A positive number for the depth would cause the solid to project out from the curtain wall rather than back over the atrium.*

44. Click **Arc** (Start End Radius) from the *Draw* panel on the Ribbon (Figure 8-6.17).

45. Pick the three points shown in Figure 8-6.18 to draw the arc. The angle 71.847° is not critical; get as close as possible.

> *TIP:* *Zoom in on each arc endpoint to accurately select the corners.*

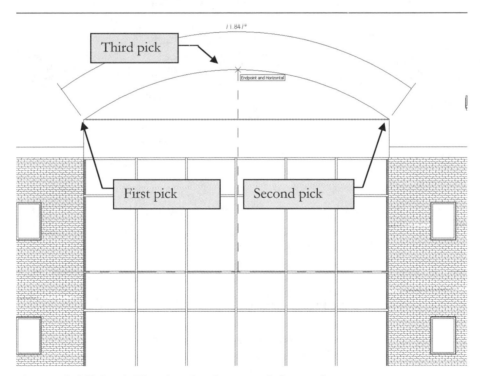

FIGURE 8-6.18 South Elevation; drawing arc to define roof

Now you will draw another arc 1'-0" above the previous one.

46. Select **Modify** on the *Ribbon*.

47. With the draw arc tool still selected, on the *Options Bar*, enter **1'-0"** for the *Offset*.

48. Pick the same three points shown in Figure 8-6.18.

Notice that an arc is drawn offset 1'-0" from the points you picked. If you pick the first two points in the other direction, the arc would be offset in the other direction (downward in this case).

Next, you will draw two short lines to connect the endpoints of the two arcs. This will create a closed area which is required before finishing the sketch. Think of it this way: you need to completely specify at least two dimensions before Revit can create the third.

49. Click the "straight" line icon on the *Draw* panel. This will switch you from drawing arcs back to drawing straight line segments.

50. Zoom in and draw a short line on each end of the arc as shown in Figure 8-6.19; try typing SE before picking to make it easier to snap to the arc's endpoints. (Make sure *Offset* is set back to zero.)

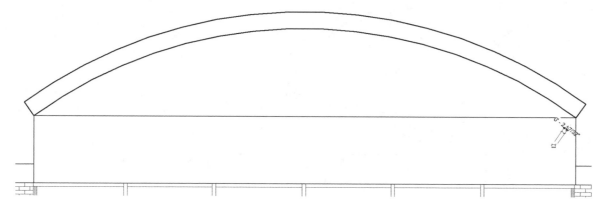

FIGURE 8-6.19 South Elevation; two arcs and two short lines define roof profile

51. Click the **green check mark** from the *Ribbon* to finish the solid extrusion.

> *TIP: If you get any warnings, it may be because one or more of the profile's corners do not create a perfect intersection. Zoom in to see and use the Trim tool to close the corners.*

You are still in the *Create Family* mode. Before you finish you will apply a material to the roof element.

52. Click the new roof to select it.

53. On the *Properties Palette*, click in the *Material* value field and then click the "**...**" icon that appears.

The Material dialog opens.

54. In the **AEC Material** library, under the **Metal** heading, double-click **Roofing, Metal** from the list of predefined *Materials* and then click **OK**.

    *FYI: Notice the rendering material is set to an Aluminum Anodized Bronze for **Roofing, Metal** on the Render Appearance tab.*

55. Click **Apply** to close the *Properties Palette*.

You are now ready to finish the Family.

56. From the *Ribbon* click **Finish Model** (Green Check Mark).

You are now also ready to finish editing the current design option for the time being.

57. Select **Main Model** from the *Design Options* list on the *Status Bar*.

As before, the option you were just working on was not the *Primary Option* in the *Atrium Roof Design Set*, so the current view reverted back to *Atrium Roof: Option 1* (which is the Primary view).

You will create a 3D view that has *Option 2* set to be visible for both the *Curtainwall* Option Set and the *Atrium Roof* Option Set.

58. Duplicate the *Default* **3D view**.

59. Rename the duplicated view to **Atrium – Option 2**.

60. Switch to the new view (Atrium – Option 2).

61. Type **VV** to access the *Visibility/Graphics Overrides*.

62. On the *Design Options* tab, set both *Option Sets* to **Option 2** in the *Design Option* column.

63. Click **OK** to close the dialog.

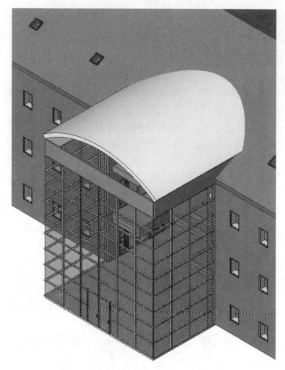

You can now see a 3D view of your new roof option. However, you realize that the walls above the curtain wall need to be modified based on the roof option, so it would make more sense to have the walls in the Atrium Roof Option Set rather than the Curtainwall Option Set.

Next, you will make this change and then modify the wall to conform to the curved roof option.

64. In the Default 3D View open each of the two *Options* (i.e., enter edit mode) in the *Curtainwall* Option Set, select the three walls above the curtain wall and **Cut to Clipboard**.

65. Now open each of the two *Options* for the *Atrium Roof* Option Set and **Paste** the three walls using *Paste Aligned\Same Place*.

FIGURE 8-6.20 Atrium – Option 2 view

The three walls should now exist in the *Atrium Roof:Option 1* set and the *Atrium Roof:Option 2* set. Next you will modify the *Option 2* walls.

66. Enter edit mode for *Atrium Roof:Option 2*.

67. In your newly created 3D view "Atrium Option 2," zoom in to the South wall above the curtain wall that needs to be extended up to the curved roof.

68. Click on the wall to select it.

69. Click the **Attach Top/Base** from the *Ribbon* (Figure 8-6.21).

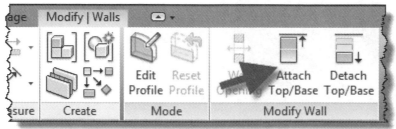

FIGURE 8-6.21 Ribbon with wall selected

This feature allows you to *Attach* a wall to another object. In this example you will pick the curved roof which will cause the wall to extend up and conform to the underside of the curved roof.

The *Edit Profile* button (also visible on the *Ribbon* when the wall is selected) would allow you to achieve the same results. Using that tool you sketch a new perimeter for the wall in an elevation or 3D view. This is particularly handy if you do not have another object to conform to; you simply want the top of the wall to do something unusual.

70. Hover the cursor over the curved roof until it highlights, and then click to select it.

Immediately the wall is modified; it should look similar to Figure 8-6.22.

71. Finish editing the current Design Option – switch to **Main Model**.

72. **Save** your project as **ex8-6.rvt**.

TIP: The Design Options feature can also be used to manage alternates, where both the base bid and the alternate(s) need to be drawn.

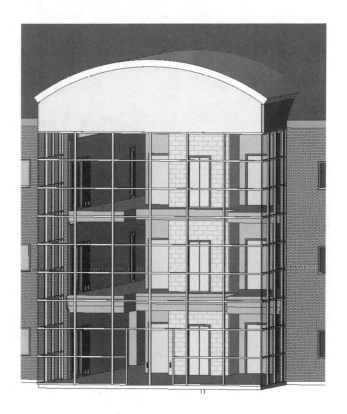

FIGURE 8-6.22 Atrium – Option 2 view; Wall attached to curved roof

Self-Exam:

The following questions can be used as a way to check your knowledge of this lesson. The answers can be found at the bottom of this page.

1. The plan is updated automatically when an elevation is modified but not the other way around. (T/F)

2. You can use the Elevation tool to place both interior and exterior elevations. (T/F)

3. You can rename elevation views to better manage them. (T/F)

4. You have to resize the Level datum symbols and annotations after changing a view's scale. (T/F)

5. How do you enter 5 ½″ without entering the foot or inch symbol?

   _____

Review Questions:

The following questions may be assigned by your instructor as a way to assess your knowledge of this section. Your instructor has the answers to the review questions.

1. The visibility of the crop region can be controlled. (T/F)

2. You have to manually adjust the line weights in the elevations. (T/F)

3. As you move the cursor around the building, during placement, the elevation tag turns to point at the building. (T/F)

4. There is only one part of the elevation tag that can be selected. (T/F)

5. You cannot adjust the "extent of view" using the crop region. (T/F)

6. What is the first thing you should do after placing an elevation tag? _____

7. In addition to the Window tool, if one window is already placed, you can use the _____ tool to place additional instances of that window.

8. With the elevation tag selected, you can use the _____ to adjust the tag orientation to look at an angled wall.

9. You need to adjust the Far Clip Plane to see elements, in elevations that are a distance back from the main elevation. (T/F)

10. What feature allows you to develop different ideas? _____

Lesson 9
Annotation:

This chapter covers annotation in Revit: text, dimensions, tags, shared parameters and keynotes. In printed form, the drawings are often not enough to convey the design intent. These tools provide for effective communication to those using the final printed drawings to both bid and build the building. Several of the chapters following this one will utilize many of these concepts.

All of the tools covered in this chapter are found on the Annotate tab. Three of the tools on this tab are also on the Quick Access Toolbar because they are used often; they are **Dimension**, **Tag by Category** and **Text**.

TIP: It is interesting to note that every tool on the Annotate tab is 2D and view specific.

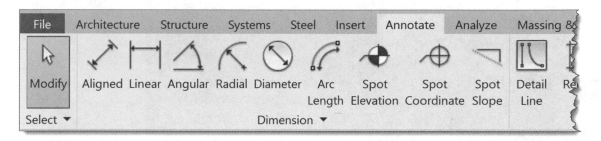

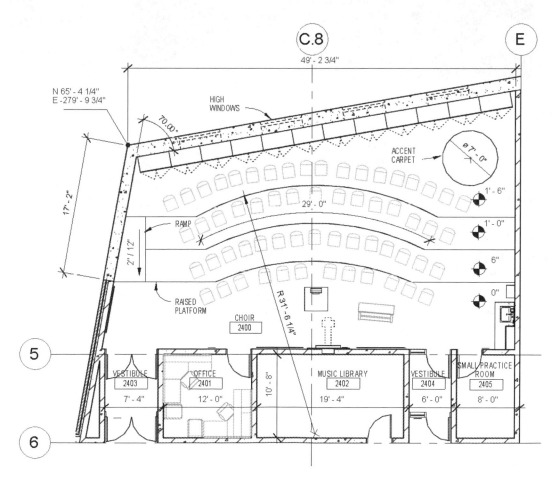

Exercise 9-1:
Text

This section covers the application of notes using the **Text** command in Revit. We will also look at Revit's **Spell Check** and **Find/Replace** tools, including their limitations.

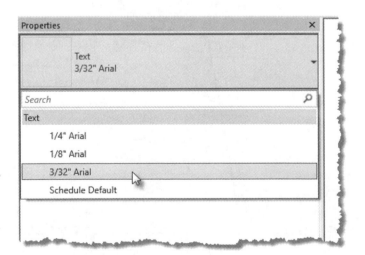

The first thing to know about the text tool is that it should be used as little as possible! Rather, live tags, keynotes and dimensions are preferred over static text to ensure the information is correct. Text will not update or move when something in the model changes—especially if the text is not visible or in the view where the model is being changed.

With that said, the text command is still necessary and used often within Revit.

Steps to add new Text to a view:
- Start the **Text** command
- Verify the **Text Style**
- Define the starting point; **click** or **click and drag**
- **Type** the desired text, typically without pressing *Enter*
- **Click away** from the text to finish the command

Steps to edit existing text:
- Select the text
- Click again to enter edit mode

Text

The Text tool can be started from the **Quick Access Toolbar**, the **Annotate** tab or by typing **TX** on the keyboard.

The first thing to do is consider the current text type in the Type Selector (Figure 9-1.1). The name of a text type should, at a minimum, contain the size and font. The size is the **actual size on the printed page** regardless of the drawing scale.

FIGURE 9-1.1 Type selector with text command active

There are additional steps that are optional, but at this point you could click, or click and drag, within the drawing window and start typing.

Here is the difference between **click** and **click and drag** options when starting text:

- **Click**: Defines the starting point for new text. No automatic return to next line while typing. An Enter must be pressed to start a new line; this is called a "hard return" in the word processing world.

- **Click and Drag**: This defines a windowed area where text will be entered. The height is not really important. The width determines when a line automatically returns to the next line while typing. Using this method allows paragraphs to be easily adjusted by dragging one of the corners of the selected textbox.

As long as "hard returns" are not used, the textbox width and number of rows can be adjusted at any time in the future. To do this, the text must be selected and then the round grip on the right (See Figure 9-1.1) can be repositioned via click and drag with left mouse button.

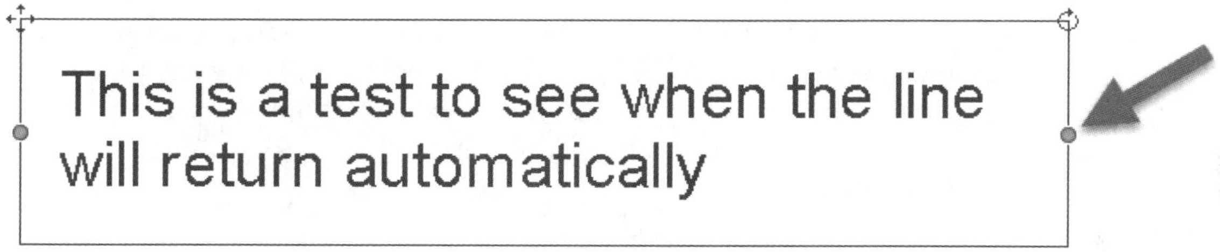

This is a test to see when the line will return automatically

FIGURE 9-1.2 Selecting text and adjusting width of the textbox

The image to the right, Figure 9-1.3, shows the result of adjusting the width of the textbox—the text element went from two lines to four.

If an Enter is pressed at the end of each line of text, when originally typed, the text will not automatically adjust as just described. An example of this can be seen in the following two images, Figure 9-1.4 & 5. Note that in some cases this is desirable to ensure the formatting of text is not changed.

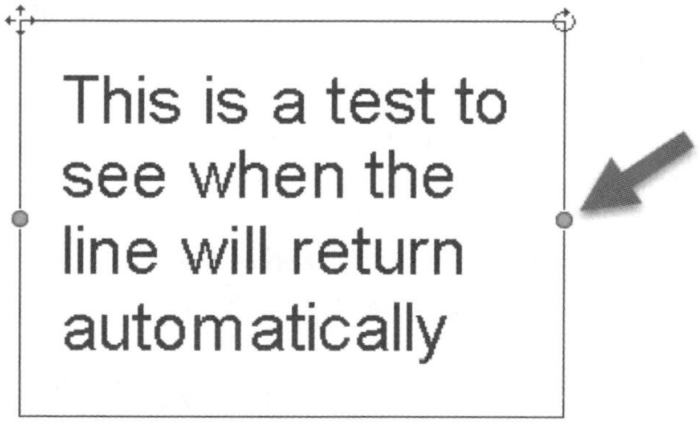

This is a test to see when the line will return automatically

FIGURE 9-1.3 Text width adjusted

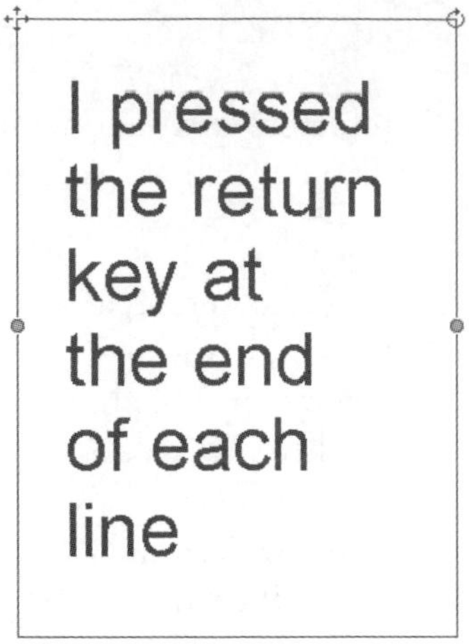

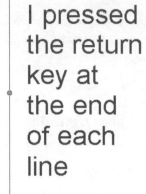

FIGURE 9-1.4
Text with hard returns

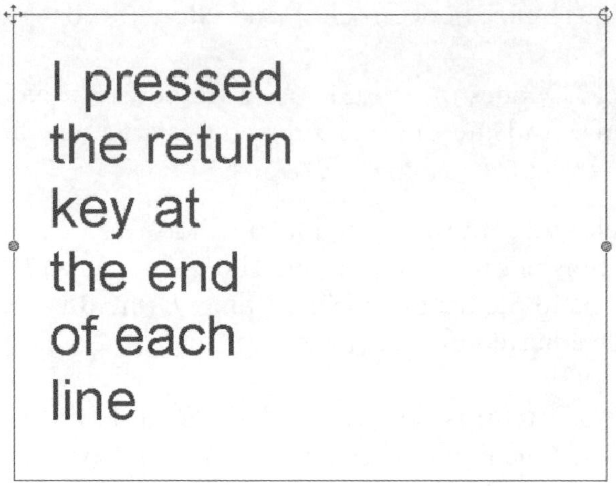

FIGURE 9-1.5
Text does not adjust when textbox width is modified

Keep in mind that this text tool is strictly 2D and view specific. If the same text is required in multiple views, the text either needs to be retyped or Copy/Pasted. However, for general notes that might appear on all floor plans sheets, for example, a Legend View can be utilized. Revit has a separate tool, on the Architecture tab, called **Model Text** for instances when 3D text is needed within the model.

Formatting Text

In addition to the basic topics just covered, there are a number of formatting options which can be applied to text. These adjustments can be applied while initially creating the text or at any time later.

The formatting options are mainly found on the Ribbon while in the Text command or when text is selected.

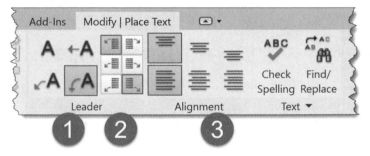

FIGURE 9-1.6 Formatting options on the Ribbon while creating text

The formatting options identified in the image are:
1. Leader options
2. Leader position options
3. Text Justification/Alignment

Leader Options

A leader is a line which extends from the text, with an arrow on the end, used to point at something in the drawing.

The default option, when using the text command, is no leader. This can be seen as the highlighted option in the upper left.

The remaining three options determine the graphical appearance of the leader: one segment, two segment or curved as seen in Figure 9-1.7. Often, a design firm will standardize on one of these three options for a consistent look.

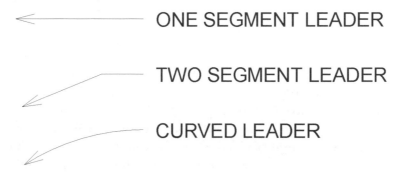

FIGURE 9-1.7 Leader formatting options

Here are the steps to include a leader with text:
- Start the **Text** command
- Select text type via **Type Selector**
- Select **Leader** option
- Specify **leader location**
- **Type** text
- Click **Close**, or click away from text, to finish

Leader Options

Once the text with leader is created, it can be selected and modified later if needed. In the next image, Figure 9-1.8, notice the two **circle grips** associated with the leader: at the arrow and the change in direction of the line (the other two circle grips are for the text box as previously described in this section). These two grips can be repositioned by clicking and dragging the left mouse button.

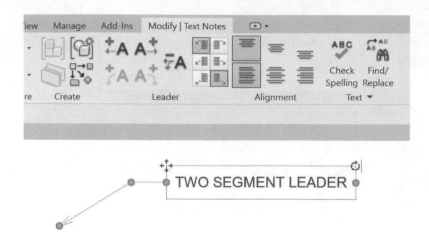

FIGURE 9-1.8 Text with leader selected; notice leader grips and Ribbon options

When text is selected the Ribbon displays slightly different options for leaders as seen in the image above. It is possible to have multiple leaders (i.e. arrows) coming off the text—denoted by the green "plus" symbol. In the next image, Figure 9-1.9, an additional leader was added to the left and one was also added to the right. It is not possible to have both curved and straight leader lines for the same text element. In this example, the curved leader options are grayed out as seen in Figure 9-1.8.

TWO SEGMENT LEADER

FIGURE 9-1.9 Multiple leaders added

Back in Figure 9-1.8, also notice that leaders can also be removed—even to the point where the text does not have any leaders. The one catch with the **remove leader** option is that they can only be removed in the reverse order added.

The **Leader Arrowhead** can be changed graphically (i.e. solid dot, loop leader, etc.). This will be covered in the section on Managing Text Types as this setting is in the Type Properties for the text itself.

> *Good to know…*
> Text can be placed in **any view** and on **sheets**. The only exception is the Text command does not work in schedule views.
>
> Text can also be in a **Group**. When the group only has elements from the Annotate tab, it is a Detail Group. When the group also has model elements, the text is in something called Attached Detail Group. When a Model Group is placed, selecting it gives the option adding the Attached Detail Group.

Leader Position Options

These six toggles control the position of the leader relative to the text as seen in the two images below (Figures 9-1.10 and 11). These options also appear in the Properties Palette when text is selected, called **Left Attachment** and **Right Attachment**.

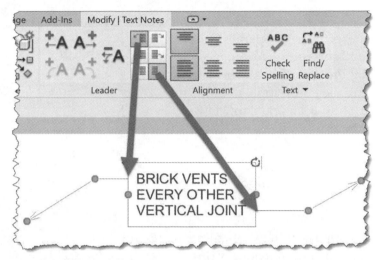

FIGURE 9-1.10 Leader position toggles – example A

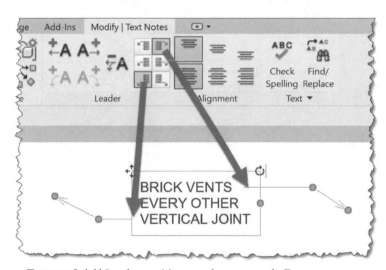

FIGURE 9-1.11 Leader position toggles – example B

Similar to leader type, a design firm will often select a standard that everyone is expected to follow so construction documents look consistent.

Text Justification

When text is selected there are three options for horizontal justification on the Ribbon: Left, Center and Right. The results can be seen in the three images for Figure 9-1.12.

This option also appears in the Properties Palette when text is selected, called **Horizontal Align**. Keep in mind that all options in the Property Palette are instance parameters—meaning they only apply to the instance(s) selected. Thus, each text entity in Revit can have different settings.

Text Formatting

The next section to cover is the options to make text Bold, Italic or be Underlined.

These options do not appear in the Properties Palette because they can be applied to individual words (or even individual fonts). In the example below, the word "Brick" is bold, "other" is italicized and "vertical" is underlined.

If all the text is selected and set to one of these three options, an edit made in the future will also have these settings.

FYI: Notice the formatting options are different when editing the text, compared to when the text element is just selected.

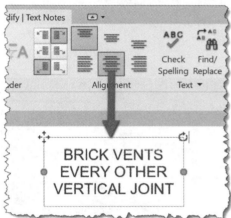

FIGURE 9-1.12 Text justification options

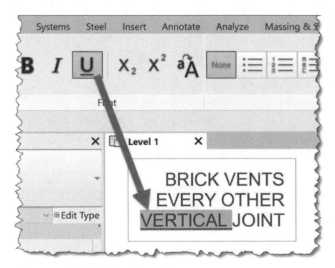

FIGURE 9-1.13 Text formatting

Text Formatting

The next section to cover is the **List** options on the Paragraph panel. This feature is only available while editing the contents of the text element; to do this, select the text and then click on the text to enter edit mode.

• 4" FACE BRICK	1. 4" FACE BRICK
• 1" AIR SPACE	2. 1" AIR SPACE
• 3" RIGID INSULATION	3. 3" RIGID INSULATION
• 8" CONCRETE MASONRY UNIT (CMU)	4. 8" CONCRETE MASONRY UNIT (CMU)
• 3 5/8" MTL STUDS AT 16" O.C.	5. 3 5/8" MTL STUDS AT 16" O.C.
• 5/8" GYP BD	6. 5/8" GYP BD
a. 4" FACE BRICK	A. 4" FACE BRICK
b. 1" AIR SPACE	B. 1" AIR SPACE
c. 3" RIGID INSULATION	C. 3" RIGID INSULATION
d. 8" CONCRETE MASONRY UNIT (CMU)	D. 8" CONCRETE MASONRY UNIT (CMU)
e. 3 5/8" MTL STUDS AT 16" O.C.	E. 3 5/8" MTL STUDS AT 16" O.C.
f. 5/8" GYP BD	F. 5/8" GYP BD

FIGURE 9-1.14 Four options to define a line within text

This is one case where you must press Enter to force the following text to a new line and automatically generate a list (i.e. bullet, letter or number).

Clicking the **Increase Indent** tool will indent the list as shown below (Figure 9-1.15). To undo this later, click in that row and select **Decrease Indent**. There does not appear to be a way to change what the indented listed value is. Using the backspace key and then indenting again allows an indent without a number/letter.

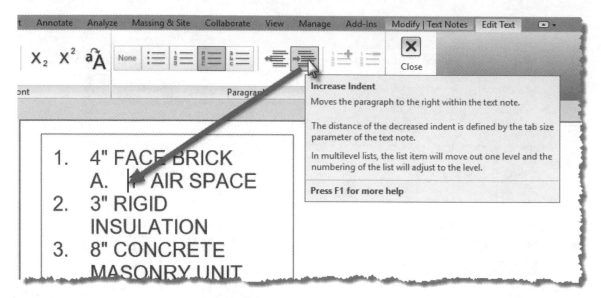

FIGURE 9-1.15 Indenting within a list

Clicking within the first row, clicking on the **plus** or **minus** icons will let you change the starting number/letter of the list.

The text formatting options also allow for **subscript** and **superscript** as shown in the example below (Figure 9-1.16).

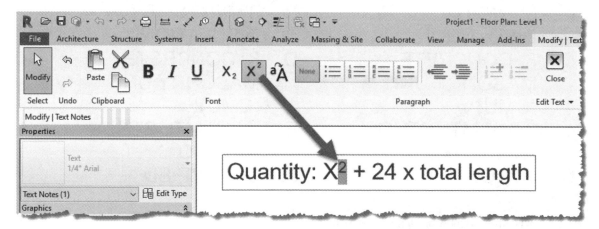

FIGURE 9-1.16 Superscript example

Also, notice the **All Caps** icon; clicking this icon will change selected text to all upper case. When this feature is used, Revit remembers the original formatting—thus, toggling off the All Caps feature later will restore the original formatting.

Managing Text Types

When using the text tool, the options listed in the Type Selector are the result of **Text Types** defined in the current project. Most design firms will have all the Text Types they need defined within their template.

There are two ways to access the text type properties. One is to start the Text command and then click **Edit Type** in the Properties Palette. The other is to click the arrow within the Text panel on the Annotate tab as pointed out in the image below.

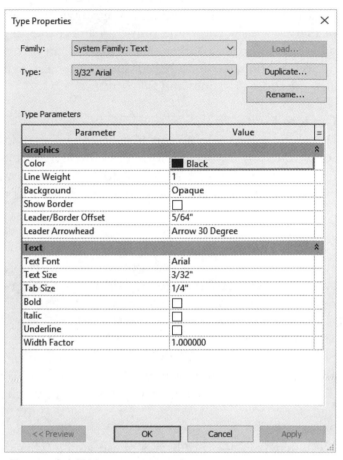

FIGURE 9-1.17 Text type properties

The **Type Properties**, as shown in the example to the right (Figure 9-1.17), are fairly self-explanatory. Below is a brief description of each.

Command	What it does…
Color	This can affect printing, so it is often set to black.
Line Weight	This is only for the leader.
Background	*Toggle:* Opaque or Transparent
Show Border	*Check box:* Show or Hide
Leader/Border Offset	Space between text and board and leader – helpful when Background is set to Opaque.
Leader Arrowhead	Select from a list of predefined arrow types
Text Font	Select from a list of installed fonts on your computer
Text Size	Size of text on the printed page
Tab Size	Size of space when Tab is pressed (size on printed paper)
Bold	Default setting – can be changed while in edit mode
Italic	Default setting – can be changed while in edit mode
Underline	Default setting – can be changed while in edit mode
Width Factor	Adjusts the overall width of a line of text

Colors

The color applies to the text and the leader. The color is often set to black. If any other color is used, this can affect printing. For example, any color becomes a shade of gray when printed to a black and white printer—similar to printing a document from Microsoft Word where green text is a darker shade of gray than yellow text. In the Print dialog there is an option to print all color as Black Lines which can make colored text black. However, this also overrides gray lines and fill patterns.

Custom Fonts

Be careful using custom fonts installed on your computer as others who do not have those fonts will likely not see the formatting the same as intended. Custom fonts can come from installing other software such as Adobe InDesign. In fact, Autodesk also installs several custom fonts which are supposed to match some of the special SHX fonts which come with AutoCAD.

Custom Fonts

It is not possible to create custom arrowheads. However, the list of arrowheads is based on styles defined here: **Manage → Additional Settings → Arrowheads**. This provides many options for how these items look.

Width Factor

Some firms will use a Width Factor like 0.75, 0.85 or something similar to squish the text to fit more information on a sheet. Getting any narrower than this makes the text hard to read. This option actually changes the proportions of each letter, not just the space between them.

Misc.

Note that Text does not have a phase setting. Thus the phase filters and overrides do not apply to text. It is sometimes desired to have text noting existing elements, such as ductwork, to be a shade of gray rather than solid—black being reserved for things that are new.

Check Spelling

Revit has a tool which allows the spelling to be checked. Keep in mind this only works on text created with the Text command and only for the current view. Revit cannot check the spelling of text in keynotes, tags or families. Neither can it check the entire project.

The Spell Check tool can be found on the Annotate tab or on the Ribbon when text is selected. When Spell Check is selected, the dialog to the right appears if there are any misspellings found (Figure 9-1.18).

When Revit finds a word not in the dictionary it will provide a list of possible correct words. Often the first suggestion is the right one. If not, select from the list.

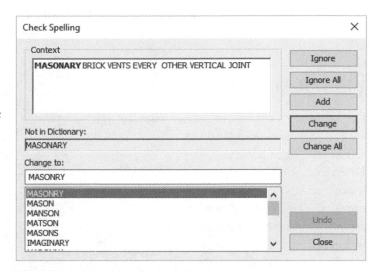

FIGURE 9-1.18 Check Spelling dialog

Clicking the **Change** button will correct the highlighted word.
Clicking **Change All** will change all of the words with this same misspelling in the current view.

Sometimes Revit will flag a word that is not misspelled. This might be a company name, your name, a product name or an industry abbreviation. In this case one might select the Add option to add the word to the custom dictionary so you don't have to deal with this every time you run spell check.

The image to the right shows the settings related to the Spell Check engine. The options are self-explanatory. In an office, consider placing the custom dictionary on the server and point all users to it.

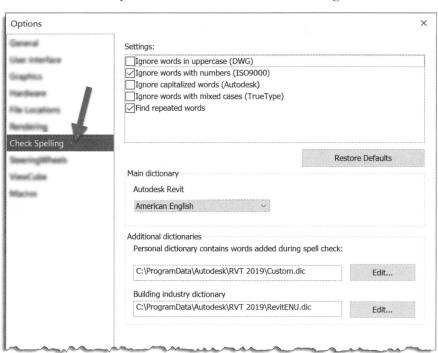

FIGURE 9-1.19 Check Spelling in Options dialog

Don't totally rely on Spell Check. A word may be spelled correctly but still be the wrong word. For example:

- Fill the **whole** with concrete and trowel level and smooth.
- File the **hole** with concrete and trowel level and smooth.

In this example the word "whole" is wrong but spelled correctly. Also keep in mind that Revit does not have a grammar check system like the popular word processing systems.

Find/Replace

Revit has a tool which allows words to be found or replaced within a view. Keep in mind this only works on text created with the Text command. Revit cannot find or replace text in keynotes, tags or families. Unlike the spelling tool, this tool can search the entire project.

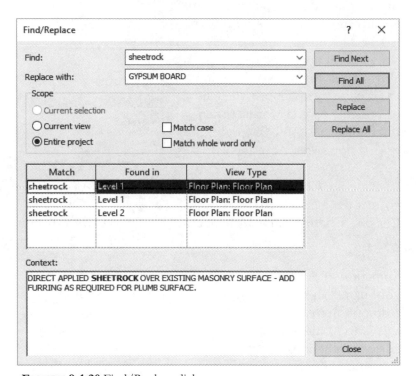

When this tool is selected the dialog to the right appears (Figure 9-1.20).

In this example, the current view is being searched for a brand name, **Sheetrock**, so it can be replaced with the generic industry standard term, **Gypsum Board**.

FIGURE 9-1.20 Find/Replace dialog

Selecting **Entire project** and then clicking **Find All** tells Revit to list all matches in the middle section of the dialog. For each row, you can click to select and see the context the found word(s) is used in.

When items are found, the **Replace** or **Replace All** buttons can be used to swap out the text in one location or all. When clicking Replace, only the selected row is replaced.

This tool can be used to just find something and not replace it. For example, on a large project with hundreds of views in the Project Browser, using the Find/Replace to search for the details with the word "roof drain" can significantly speed up the process of locating the desired drawing.

Replacing a Text Type

In addition to replacing content within a text element, there is a way to replace the text type as well. For example, some imported details use a different font and you want everything to match and be consistent. This is not really associated with the Find/Replace tool, but it is important to know how to accomplish this task.

> Replacing a Text Type within a view or project:
> - **Select** one text element within the project
> - **Right-click** (Fig. 9-1.19)
> - Pick **Select All Instances** →
>   - o Visible in View
>   - o In Entire Project
> - Select a different type from the **Type Selector**

This procedure will replace all the text in either the current view or the entire project. Even text which has been hidden with the "Hide in View" right-click option will be changed.

When a specific text type is selected, the selection count, in the lower right corner of the Revit window, will indicate the total number of elements selected (Figure 9-1.22). This can be used as a quick double check before replacing the text type. For example, if the intent was to just replace a few rogue text instances but the count was several hundred, this would be a clue that some other view uses this text type and perhaps should not be changed as it was created by someone else on the project. This can be especially true if multiple disciplines are working in the same project.

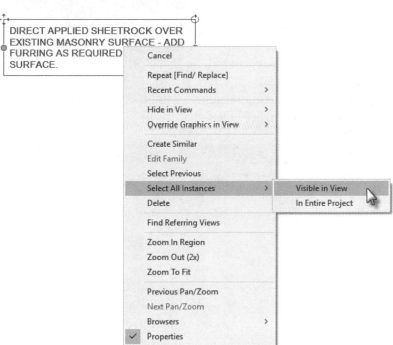

FIGURE 9-1.21 Select all instances via right-click

FIGURE 9-1.22 Total element selected count

Exercise 9-2:
Dimensions

This exercise will cover the ins and outs of dimensioning in Revit.

The first thing to understand about dimensions is that they are 2D and view specific—like all commands on the Annotate tab.

Dimension elements have an association with the thing(s) being dimensioned. If that thing(s) is deleted the dimension will also be deleted, even if the dimension is not visible in the current view. For example, if a wall is deleted from a 3D view, then any dimensions associated with that wall, in a floor plan view, which may not even be open, will be deleted.

Because dimensions have an association with specific elements in the model it is important to make sure the correct elements are selected while placing the dimension. In a floor plan, for example, the place to click to add a dimension may have several elements stacked on top of each other: grid line, wall edge, floor edge, window edge, wall sweep. It may be necessary to tap the Tab key to cycle through the options to select the correct item (while tabbing, the highlighted element will be listed on the status bar across the bottom of the screen).

Each dimension command will be covered in the order they appear on the Annotate tab as seen in the image below.

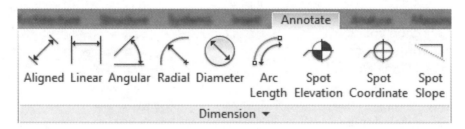

Command	What it does...
1. Aligned	Most used dimension tool; dimension between parallel references (e.g. walls or ducts) or multiple points
2. Linear	Dimensions between points and always horizontal or vertical
3. Angular	Measures angle between two references
4. Radial	Indicates the radius of a curved line or element
5. Diameter	Indicates the diameter of an arc or circle
6. Arc Length	Measures the length of a line/element along a curve
7. Spot Elevation	Lists the elevation at a selected point, on an element (e.g. floor, ceiling, toposurface)
8. Spot Coordinate	Indicates the N/S and E/W position of a selected point
9. Spot Slope	Used to indicate the slope of a ramp in plan or the pitch of a roof in elevation

The following example floor plan below shows each of the dimension types used. This is a middle school choir room with various conditions to dimension, such as angled walls, curved lines and multiple floor elevations.

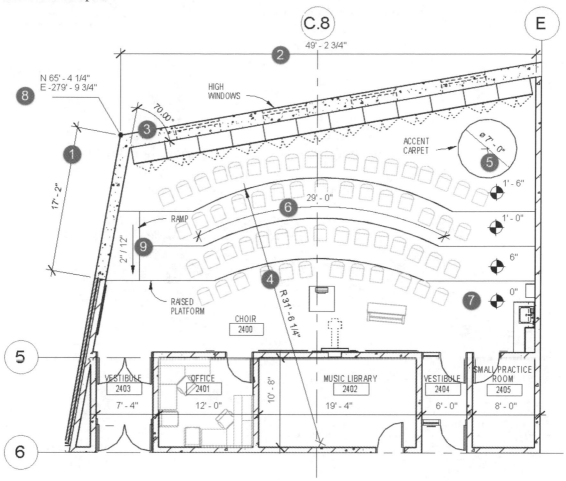

FIGURE 9-2.1 Example floor plan used for dimensioning study

Aligned - Dimension

This dimension tool is the most used of the dimension tools, and for this reason it is also located on the Quick Access Toolbar (QAT).

Steps to add Aligned dimensions:
- Review **Type Selector** and **Options Bar** selections
- **Select first reference**
  - Click on element *or*
  - Press tab to select specific reference or intersection
- **Select second reference**
- Optional: Select additional references (creates a dimension string)
- **Click away** from anything dimensionable to finish command

The **Align** dimension tool is able to create angled, horizontal and vertical dimensions. When the tool is first started, the default option is to select a **Wall Centerline**, per the selection on the Options Bar, just by clicking on a wall. This in turn starts a dimension line perpendicular to that wall. The second point could then be another wall at the same angle or the endpoint or intersection of lines.

In the example to the right, if the vertical wall is rotated the dimension will also rotate. If the dimension needs to remain horizontal, then the Linear dimension should be used.

The image below shows the User Interface while the Aligned dimension tool is active.

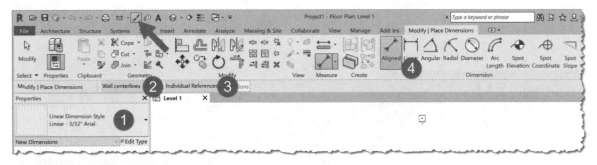

FIGURE 9-2.2 Using the Aligned dimension tool

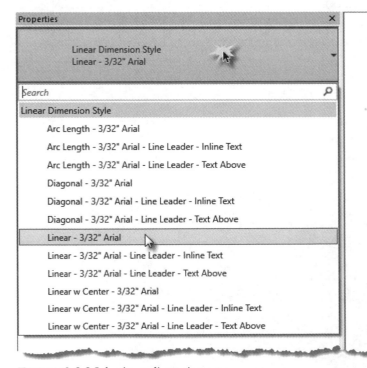

FIGURE 9-2.3 Selecting a dimension type

Here is a description of the numbered items in the image above:

1. Type Selector
2. Selection Preference
3. Individual or Automatic Dimension
4. Switch to another dimension tool

Type Selector:
Once the dimension tool is active, select the desired **Type** from the *Type Selector* (Figure 9-2.3). This list will vary depending on the template the project was started with and any modifications made.

Selection Preference:

The default is **Wall Centerlines** which means just clicking on a wall, even when zoomed way out, the dimension will reference the centerline of the wall. The other options listed (Figure 9-2.4) are self-explanatory. Another option, rather than changing this drop-down list, is to hover the cursor over the desired face, e.g. **Wall faces**, and then tap the Tab key until that face is highlighted and then click. This can save the time it takes to keep moving the cursor all the way up to the Options Bar and changing the formal setting.

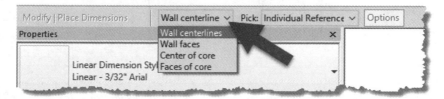

FIGURE 9-2.4 Specify the selection preference for dimensions

Individual versus Automatic Dimensions:

Revit defaults to **Individual References** option so the designer can deliberately select each reference to dimension to (Figure 9-2.5).

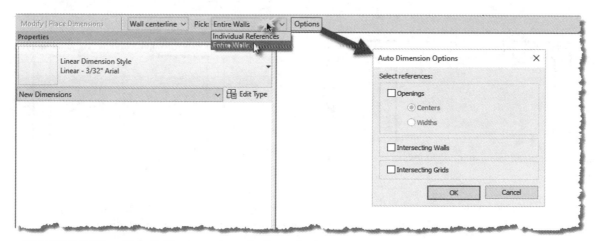

FIGURE 9-2.5 Specify individual or automatic dimensioning

Switching this option to **Entire Wall**, by default, will just add a dimension for the entire length of a selected wall. If **Wall Faces** is selected as well, the overall dimension is created as shown in the next image (Figure 9-2.6). Again, this dimension was created by simply clicking on the wall (with the door and windows); one click to specify the reference and another to position the line.

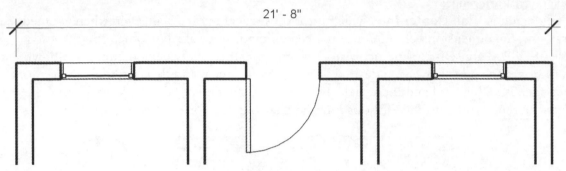

FIGURE 9-2.6 Automatic dimension created – example 1

When the Entire Wall option is selected, the Options button becomes active. Clicking this presents several options as shown in Figure 9-2.5. With the <u>Options Bar</u> set to **Wall Centerlines** and the <u>Options</u> dialog box set to **Openings\Centerlines** and **Intersecting Walls**, the string of dimensions shown below is automatically created (Figure 9-2.7).

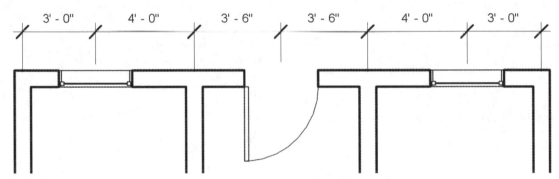

FIGURE 9-2.7 Automatic dimension created – example 2

In this next example, Figure 9-2.8, with the <u>Options Bar</u> set to **Wall Faces** and the <u>Options</u> dialog box set to **Openings\Width** and **Intersecting Walls**, the string of dimensions shown below is automatically created.

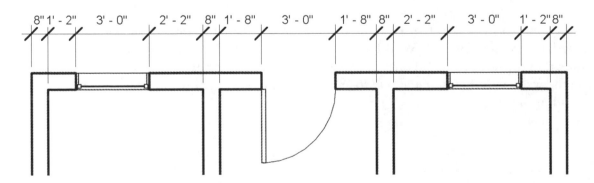

FIGURE 9-2.8 Automatic dimension created – example 3

Linear - Dimension

Use this tool to force Revit to maintain horizontal and vertical dimension segments or strings.

> Steps to add Linear dimensions:
> - Review **Type Selector** selection
> - **Select first reference**
>   - Intersection of walls and/or grids *or*
>   - Press tab to select specific reference or intersection
> - **Select second reference**
> - Optional: Select additional reference
> - **Click away** from anything dimensionable to finish command

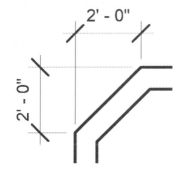

This tool will automatically pick points rather than the face of an element. In the example shown to the right, after picking the two points to be dimensioned, the direction the mouse is moved (in this case, up or to the left) will determine if the dimension is horizontal or vertical—pressing the Space Bar will also toggle between the two orientations.

If the model is adjusted, these dimensions will automatically update regardless of which view they are in. Similarly, if one of these elements is deleted, these dimensions will also be deleted.

Angular - Dimension

Use this tool to indicate the angle between two references.

> Steps to add Angular dimensions:
> - Review **Type Selector** and **Options Bar** selections
> - **Select first reference**
> - **Select second reference**
> - **Click** to place the location of the dimension line and text

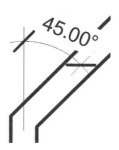

Using the angular tool provides a way to measure the angle between two elements. In the example to the right, the dimension could be in three different positions: one where it is, one in the lower right and the larger obtuse angle on the left.

If the two references are modified the angular dimension will update no matter which view it is in.

Radius - Dimension

Use this tool to indicate the radius of an arc or circular reference. This can be the edge of a floor, wall, duct or detail lines.

> Steps to add a Radius dimension:
> - Review **Type Selector** and **Options Bar** selections
> - **Select the reference**
> - **Click** to place the location of the dimension line and text

By default, the Radius dimension extends to the center point of the arc or circle. If a plan view is cropped, and Annotation Crop is active for the view, the center point must be visible within the cropped area.

Once the Radius is placed, the radius dimension can be selected and the grip at the center point can be repositioned closer to the arc. Additionally, it is possible to turn off the **Center Mark** symbol via the Type properties. However, in some cases the location of the center mark itself should be dimensioned so the contractor can accurately position the element within the building.

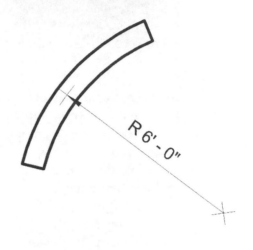

Diameter - Dimension

Use this tool to indicate the diameter of an arc or circular reference. This can be the edge of a floor, wall, duct or detail lines.

> Steps to add a Diameter dimension:
> - Review **Type Selector** and **Options Bar** selections
> - **Select the reference**
> - **Click** to place the location of the dimension line and text

Arc Length - Dimension

Use this tool to measure the length of a line along a curved reference. This can be the edge of a floor, wall, duct or detail lines.

> Steps to add an Arc Length dimension:
> - Review **Type Selector** and **Options Bar** selections
> - **Select the reference**
> - **Click** to place the location of the dimension line and text

Spot Elevation - Dimension

Use this tool to indicate the elevation of a surface.

Steps to add a Spot Elevation dimension:
- Review **Type Selector** and **Options Bar** selections
- **Select the reference**
- **Click** to place the location of the text and leader

The elevation listed is based on one of three options as listed below. This is a Type Property called Elevation Origin. Thus, it is possible to use all three options, even right next to each other.

Elevation Origin options for a Spot Elevation type:
- Project Base Point
- Survey Point
- Relative

Project Base Point:

This option is related to the **level datum** numbers in the project. For example, the default templates which come with Revit have Level 1 at elevation 0'-0". In this example, all Spot Elevations set to Project Base Point will be relative to 0'-0" within a project.

16' - 6"

Survey Point:

The Survey Base point is an **alternate coordinate system** used to align with the actual position on earth and elevation above sea level. Thus, any Spot Elevation set to Survey Point will display a value relative to the Survey Point settings. For example, where the Level 1 floor's Project Base Point is set to 0'-0" (or 100'-0") the Survey value for Level 1 might be 650'-0" to match the elevation numbers shown for the contours on the Civil Engineer's grading plan. Revit has the ability to track both coordinate systems within the context of the Spot Elevation tool.

116' - 6"

Relative:

When a Spot Elevation has the Elevation Origin set to Relative, the values listed are related to the plan view in which they are placed.

6"

Spot Coordinate

Use this tool to indicate the North/South, East/West position of a point within the model.

> Steps to add a Spot Coordinate dimension:
> - Review **Type Selector** and **Options Bar** selections
> - **Select the reference point**
> - **Click** to place the location of the text and leader

This feature can be used to indicate the position of one, or more, corners of the building on the site. This is usually relative to a predefined Survey Point which is relative to a bench mark or municipal coordinate system. On a typical commercial project this information is only provided on the civil drawings and therefore not required in the Revit model or documents.

N 65' - 4 1/4"
E -279' - 9 3/4"

Similar to the Spot Elevation tool, the Spot Coordinate can also be set to Relative or Project Base Point. This could be used to position items within the project but is not practiced in many cases as the contractor or installer would need to have a direct line of sight between the two points.

Spot Slope

Use this tool to indicate the **Slope** in a <u>floor plan view</u> or the roof **Pitch** in an <u>elevation view</u>.

> Steps to add a Spot Elevation dimension:
> - Review **Type Selector** and **Options Bar** selections
> - **Click** to both select a reference and place the location of the dimension element

<u>Plan views:</u>

In a plan view the Spot Slope can be used to indicate the slope of a floor, such as a ramp. However, as it currently works, the Ramp element cannot have a Spot Slope annotation applied to it. Thus, ramps should be modeled as a sloped floor, which can have a Spot Slope applied.

<u>Elevation View:</u>

In an elevation view the Spot Slope can be used to indicate the pitch of a roof as shown in the example to the right.

TIP: When placing the Spot Elevation on a hip roof, in elevation, use the Tab key to select the correct surface. By default the hip line will be selected, which is not the same slope as the roof face itself.

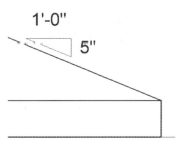

The Spot Slope element has an Instance Parameter, in the Properties Palette, which toggles between Arrow and Triangle, the two options shown above.

Modifying a Dimension

When dimensions need to be modified, there are a few things to know.

> Ways to Modify a Dimension:
> - Edit Witness Lines
> - Modify text
> - Reposition text
> - Lock a Dimension
> - Drive the location of geometry
> - Change type

Edit Witness Lines:
After a dimension is placed, a wall or opening may be added, and rather than deleting a dimension string and adding a new one, Revit provides the **Edit Witness Line** tool. Simply select a dimension and click the Edit Witness Line button, shown to the right, from the Ribbon. Once active, click new references to **add witness lines** and click existing references to **remove witness lines**.

Modify text:
When a dimension is selected, clicking on the text, the dimension value, the Dimension Text dialog appears (Figure 9-2.9).

FIGURE 9-2.9 Dimension Text dialog

Figure 9-2.10 shows the relative position of each of the "text fields."

When **Replace With Text** is selected, the dimension value can be replaced with text. For example, "PAINT WALL" or "EXISTING CORRIDOR WIDTH – VERIFY IN FIELD."

Revit will not allow a dimension value in the text replacement box.

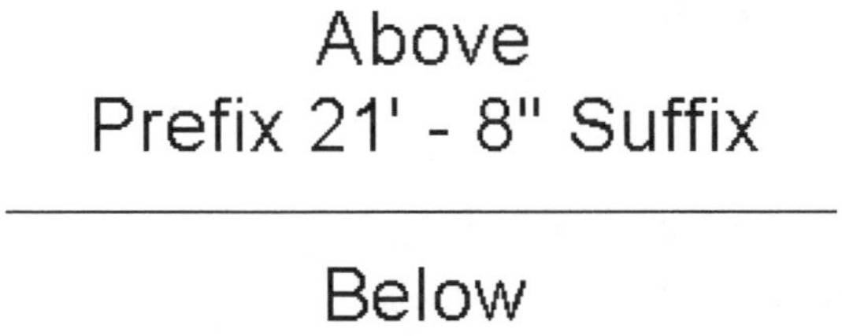

Above
Prefix 21' - 8" Suffix

Below

FIGURE 9-2.10 Dimension text field positions

Reposition text:
When a dimension is selected, clicking and dragging on the text grip allows the text to be repositioned.

When the text is moved past one of the witness lines, Revit will add a leader by default as shown to the right (Figure 9-2.11). Right-clicking on a dimension reveals related commands on the pop-up menu (Figure 9-2.12); selecting **Reset Dimension Text Position** will move the text back to the original location.

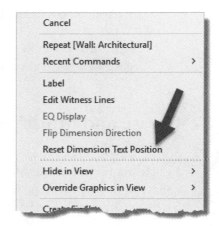

FIGURE 9-2.11 Dimension Text leader

Lock a Dimension:
When a dimension is selected, clicking the lock icon (Figure 9-2.13) will prevent that dimension value from changing. This will not prevent model elements from moving, but to maintain the dimension value, both reference elements will move. For example, if a dimension is locked between two walls which define a corridor, moving one wall will move the other wall to ensure the corridor width does not change.

When elements are selected which are in some way constrained, Revit will show the lock symbol. This is true even if the locked dimension is not visible in the current view (Figure 9-2.14). Clicking this icon will unlock the constraint.

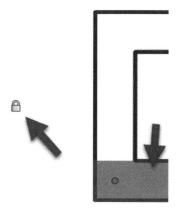

FIGURE 9-2.12 Right-click options

When a locked dimension is deleted, Revit asks if the constraint should also be removed. Thus, it is possible to delete the dimension but leave the constraint in place.

FIGURE 9-2.13 Dimension selected

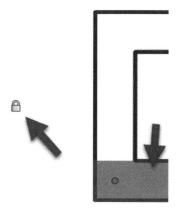

FIGURE 9-2.14 Locked element

Drive the location of geometry:

Like temporary dimensions, permanent dimensions can also be used to reposition geometry, such as walls, ducts and more. The key is to select the element to be repositioned first and then click on the dimension text. A common mistake is to select the dimension directly and then click the text. This only opens the Dimension Text dialog. Also, consider that Revit would not know which referenced element to move if just selecting the dimension and not the element: move the left one, move the right element or both equally?

TIP: A temporary dimension can be turned into a permanent dimension by clicking the "dimension icon" below the temporary dimension text.

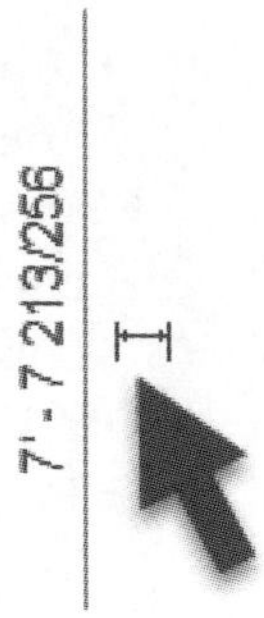

Change Dimension Type:

When a dimension is selected, click the Type Selector and click from the available options. Changing the type can affect the graphic appearance, the rounding and when alternate units appear (e.g. metric).

FIGURE 9-2.15
Temporary dimension selected

Create and Modify Dimension Types

Sometimes it is necessary to modify a dimension type to match a graphically firm standard or a client requirement. Revit allows dimension types to be created and modified. Use caution changing dimension type as all dimensions of that type will be updated in the current project. These changes will not have any effect on any other projects or templates.

To modify existing types, either start the dimension command and click Edit Type or select one of the "Types" options in the extended panel area of the dimension panel on the Annotate tab (Figure 9-2.16).

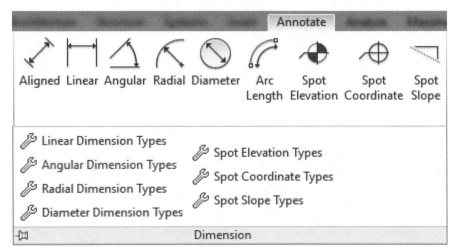

FIGURE 9-2.16 Editing Dimension types

To create new Types, click the **Duplicate** option in the Type Properties dialog.

Figure 9-2.17 shows the various options which can be changed.

One option is **Units Format**. Selecting this option opens the Format dialog shown in Figure 9-2.18. Notice a dimension style can be tied to the Project Units as set on the Manage tab. Un-checking this option allows this dimension style to round in a specific way and do a few other things like "Suppress 0 feet:" and control the symbol (e.g. monetary symbol).

Setting up all dimension styles typically needed in a template file will save a lot of time and help to enforce a firm's standard.

TIP: To see how many instances use a specific dimension style in a project, select one and then right-click and pick Select All Instances → In Entire Project. The total number selected will be listed in the lower right corner of the screen by the filter icon on the status bar.

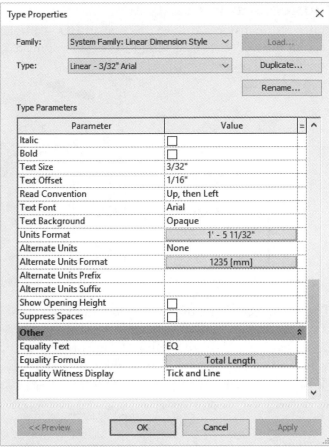

FIGURE 9-2.17 Dimension type properties

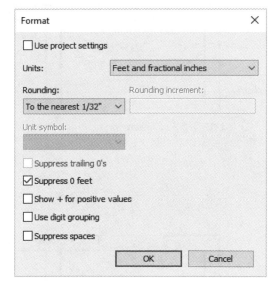

FIGURE 9-2.18 Dimension unit format

Dimension Equality

When a dimension string is selected, an **EQ** symbol appears with a red slash through it as seen in Figure 9-2.19. Clicking this toggle will make all dimension segments, in that string, equally spaced between the two ends.

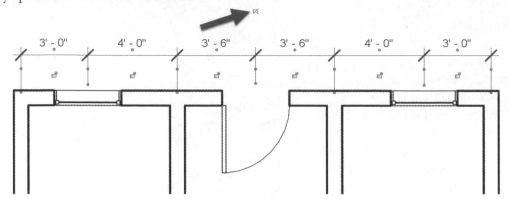

FIGURE 9-2.19 Dimension string selected and EQ symbol pointed

The image below (Figure 9-2.20) shows the result of clicking the EQ icon; the walls, windows and door all moved with the now equally spaced witness lines. Notice the EQ icon no longer has a slash through it. If one of the end walls are moved, all elements are moved to remain equally spaced. Setting the Dimension's **Equality Display** property to Value shows the dimension value rather than the "EQ" abbreviation.

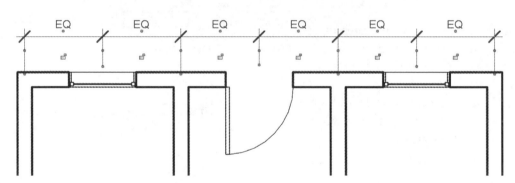

FIGURE 9-2.20 Dimension string with equality activated

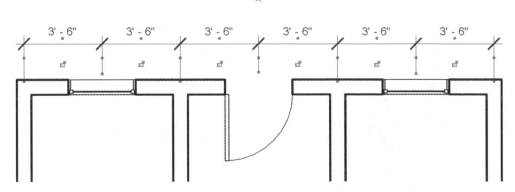

FIGURE 9-2.21 Dimension string with equality activated and dimension value shown

Exercise 9-3:
Tagging

This section will study how **tags** work in Revit. This is an important feature in Revit, one used extensively by designers. This feature is often used on casework and ductwork as well as many other elements within the Revit model.

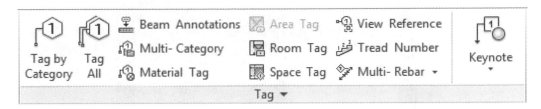

The basic premise of a *tag* is to be able to textually represent, in a drawing view, information found within an element, that is, an *Instance* or *Type Parameter*. One example found in nearly every set of construction documents is the door tag. This tag in Revit has been set up to list the contents of a specific door's **Mark** parameter, which is an *Instance Parameter*.

> The process to tag an element is simple:
> - Select the **Tag by Category** tool from the *Quick Access Toolbar*.
> - Adjust the settings on the *Options Bar* if needed.
> - Click the element(s) to be tagged.
>
>   *FYI: Different tags are placed based on the category of the element selected.*
>
> - Click the *Modify* button or press the Esc key when done.

A *tag* is view specific, meaning if you add the tag to the **Level 1 Finish Plan**, it will not automatically show up in the **Level 1 Furniture Plan**. If you want the tag to appear in two Level 1 plan views, you need to add it twice. *(TIP: Use Copy/Paste.)* The advantage is you can adjust the tag location independently in each view. One view may have furniture showing and the other floor finishes. Each view might require the tag to be in a different location to keep it readable.

Tags are dependent on the element selected during placement. You cannot simply move a tag near another similar element and expect Revit to recognize this change. Similarly, if an element is deleted, the tag will also be removed from the model, even if the tag is in another view in the project.

The image below has several tags added to a floor plan view. All the listed information is coming from the properties of the elements which have been tagged. For example, the "M1" tag within the diamond shape is listing the wall type (i.e., *Type Mark*). Because the wall tag(s) is/are listing a *Type Parameter*, all wall instances of that type will report the same value, i.e., "M1." The door number "1" is reporting the element's **Mark** value, which is an *Instance Parameter*. Therefore, each door instance may have a different number. Notice some tags have the *Leader* option turned on. This is especially helpful if the tag is outside the room. Any 3D element visible in a view may be tagged, even the floor. In this case the *Floor Tag* is actually reporting the *Type Name* listed in the *Type Selector*. The leader can be modified to have an arrow or a dot.

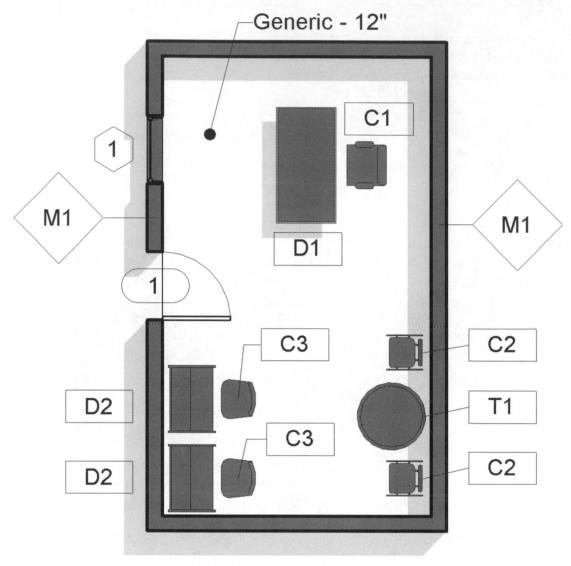

FIGURE 9-3.1 Floor plan with tags added

All of these tags were placed using the same tool: *Tag by Category*. Later you will learn how to specify which tag gets used for each category (e.g., walls, furniture, floors, windows, etc.).

The section / interior elevation below (Figure 9-3.2) shows many of the same elements tagged, as were tagged in the floor plan view on the previous page. If the wall's *Type Mark* is changed, the wall *Tag* will be instantly updated in all views: plans, elevations, sections, and schedules.

The tags had to be manually added to this view. They are not added here automatically just because you added one in the floor plan view, unlike when you add a section mark. If the view is deleted from the project, all these tags will automatically be deleted.

Some tags can actually report multiple parameter values found within an element. For example, the *Ceiling Tag* used in this example lists the *Type Name*, the ceiling height, and has fixed text which reads "A.F.F." Many tags can be selected and directly edited, which actually changes the values within the element. Changing the 8′-0″ ceiling height will actually cause the ceiling position to change vertically.

It is possible to tag an element multiple times. You could add the same tag more than once. In the example below, you could place a *Wall Tag* on each side of the door. You can also add different tags which report different information from the same element. The *Wall Tag* "M1" and the *Material Tag* "PT-1" are two different tags extracting information from the same wall.

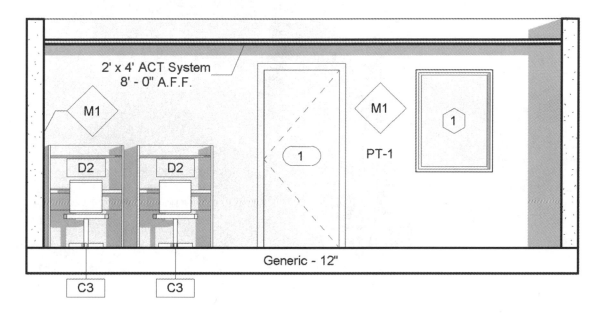

FIGURE 9-3.2 Section / interior elevation with tags added

Most design firms spend a little time adjusting the provided tags to make the text a little smaller and modify or delete the line work. This can easily be done by selecting a tag and then clicking *Edit Family* on the *Ribbon*. Revised tags can then be loaded back into the project and saved to a company template for future use.

With the more recent versions of Revit it is now possible to add tags in 3D views (Figure 9-3.3b). Before tagging a 3D view, you must **Lock** it (Figure 9-3.3a) to prevent changing the view and interfering with the position of the tags relative to the elements being tagged.

FIGURE 9-3.3A Locking a 3D view

When the *Leader* option is off, the tag is placed centered on the elected element. This initial position is not always desirable (e.g., note the D1 and C1 tags below) and you must move each tag so it is readable.

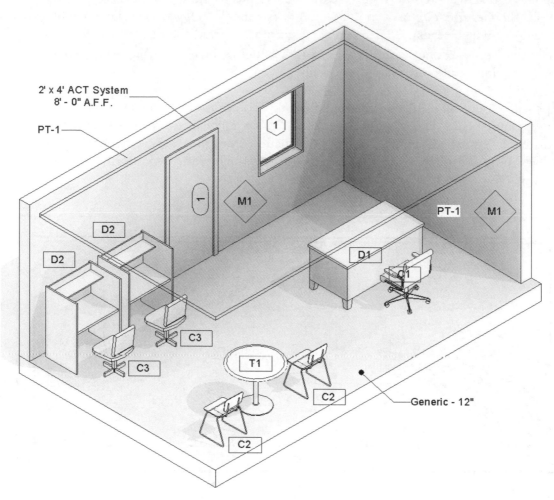

FIGURE 9-3.3B Locked 3D view with tags added

It is not possible to add *Room Tags* to 3D views yet.

It is helpful to understand where the information displayed in tags is stored. It is often necessary to change these values. In Figure 9-3.4 you can see that "PT-1" is the *Description* for the *Material* assigned to the wall type.

> ***TIP:*** *To find the* Material *assigned to a specific wall type: Select the wall* → *Edit Type* → *Edit Structure* → *observe the Material column.*

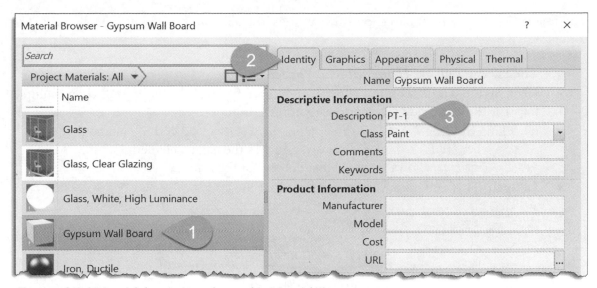

FIGURE 9-3.4 Material description value used in Material Tag

FYI: In the image above, the preview icons on the left with a gold triangle in the lower left corner are legacy materials, which are superseded by newer physically accurate materials.

A wall's *Type Mark* is found in the *Type Properties* dialog; select a wall → *Edit Type* from the *Properties Palette*. Notice in Figure 9-3.5 that the *Type Mark* is set to "M1." If you were to edit this value, all the tags for this wall will be updated instantly.

Seeing as all the information displayed in a tag is actually stored in the element being tagged, you will never lose any information by deleting a tag.

> *FYI: A tag will move with the element even if the element is moved in another view where the tag is not visible.*

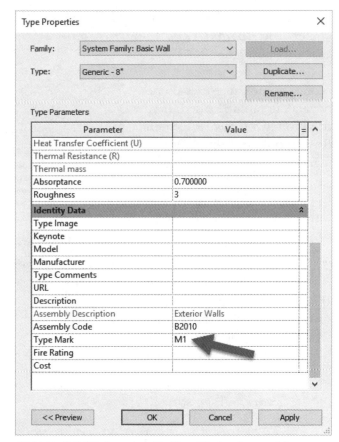

FIGURE 9-3.5 Selected wall's *Type Mark* value used in wall tag

The image below (Figure 9-3.6) shows the settings available on the *Options Bar* when the *Tag* tool is active.

- The tag can only be **Horizontal** or **Vertical**.
- Clicking the **Tags...** button allows you to specify which tags are used when multiple tags for the same category are loaded in the current project.
- Checking the **Leader** option draws a line between the tag and the element.
    - *Attached End:* the leader always touches the tagged element.
    - *Free End:* allows you to move the end of the leader.
- **Length**: This determines the initial length of the leader.

FIGURE 9-3.6 Tag options during placement

When clicking the **Tags...** button you get the **Loaded Tags** dialog as shown in Figure 9-3.7.

Notice some categories do not have a tag family loaded. If you tried to tag a *Casework* element, Revit would prompt you to load a tag.

Clicking on a tag name in the right-hand column will allow you to select from multiple tag types/families, if available in the current project.

If you use one tag type in plan views and another in elevation views, you would visit this dialog

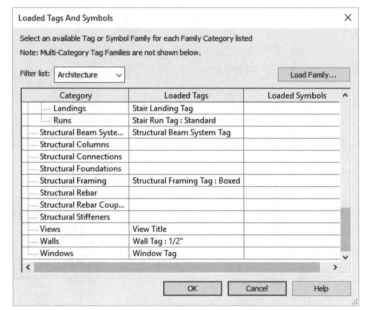

FIGURE 9-3.7 Loaded tags dialog

just before tagging in plan views and then again adjust before tagging in elevation views.

> **FYI:** *It is possible to just let Revit place the default tag and then select the placed tag and swap it out via the* Type Selector.

> **TIP:** *If you want to swap several tags with another type: Select one of the tags → right-click →* **Select All Instances** *and then pick either* **Visible in View** *or* **In Entire Project**. *You now have all the tags selected and can pick something else via the* Type Selector.

The reader should be aware that some tags may be placed automatically when content is being added to the model. For example, when placing a window, the *Ribbon* has a toggle called ***Tag on Placement*** (Figure 9-3.8). When this is selected, which it is NOT by default, a door tag is placed next to every door. Note that the tag is only added to the current view, which could be an elevation, plan, etc.

You typically want to turn this off in the early design stages as the tags will just get in the way. Also, existing elements are not usually tagged, thus you would toggle this option off so you do not have to come back and delete the tag later.

FIGURE 9-3.8 Tag on Placement option; **off** (shown on left) and **on** (on right)

If you did not add tags, or some were deleted along the way, you will need to add them at some point. The easiest way to do this is by using the ***Tag All*** tool; this tool used to be called *Tag All Not Tagged*. When you use this tool, you can add a door tag to every door in the current view that does not already have a door tag. This is great on large projects with hundreds of doors and you want to make sure you do not miss any. You may still have to go around and rotate and reposition a few tags, but overall this is much faster!

Before concluding this introduction to tags we will take a brief look at how tags are created. If you select a tag and then click *Edit Family* on the *Ribbon*, you will have opened the tag for editing in the *Family Editor*. Keep in mind that your project is still open in addition to the tag family.

If you select a furniture tag and edit it you will see text (sample value shown) and four lines (Figure 9-3.9). Here the lines can be selected and deleted if desired. Also, the text can be selected and then you can click *Edit Type* to change the text properties (e.g., height, font, width factor, etc.).

If you select the text and then click the ***Edit Label*** option on the *Ribbon*, you can change the parameter being reported, or report multiple parameters (Figure 9-3.9).

TIP: Open a tag family, edit it and then do a Save-As to make a new tag type.

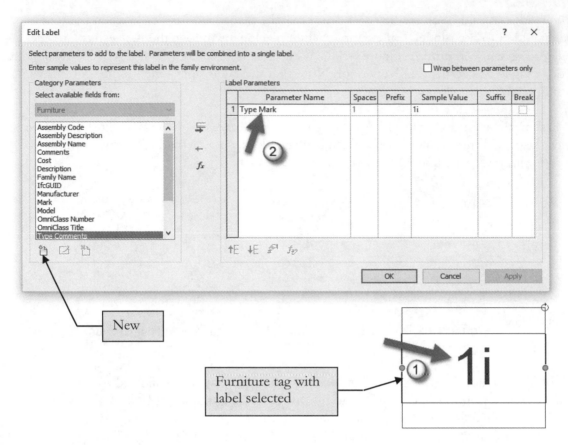

FIGURE 9-3.9 Editing a tag label in the family editor

In the *Edit Label* dialog above you can click the **New** icon in the lower left to load a *Shared Parameter* which allows you to create a tag which reports custom information. See the next section for more on this topic. Also, the last icon all the way to the right of the "New" icon allows you to override the formatting for numeric parameters (Figure 9-3.10).

This concludes the introduction to using tags in Revit. The next section continues this discussion with a more advanced concept called *Shared Parameters*.

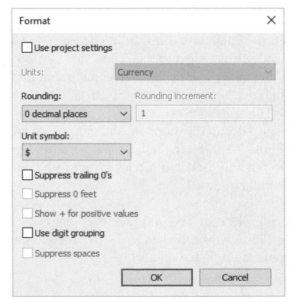

FIGURE 9-3.10 Editing a label's format

Exercise 9-4:
Shared Parameters

FYI: *You will not be using Shared Parameters in the tutorial but this information is essential for a designer to be proficient and properly leverage Revit's power.*

Introduction

Revit has many features which are unique when compared to other building design programs; one of these is *Shared Parameters.* The main idea with *Shared Parameters* is to be able to manage parameters across multiple projects, families and template files. This feature allows Revit to know that you are talking about the same piece of information in the context of multiple unconnected files. Here we will cover the basics of setting up *Shared Parameters,* some of the problems often encountered, and a few tricks.

The main reason for using **Shared Parameters** is to make custom information show up in tags; however, there are a few other uses which will be mentioned later. In contrast, a **Project Parameter** is slightly easier to make in a project but cannot appear in a tag; Revit has no way of knowing the parameter created in the family, info to be tagged, and the parameter created in the tag are the same bit of information. Both *Project* and *Shared Parameters* may appear in schedules.

The image below depicts the notion of a common storage container (i.e., the **Shared Parameter** file) from which uniquely coded parameters can be loaded into content, annotation and projects. This creates a connected common thread between several otherwise disconnected files.

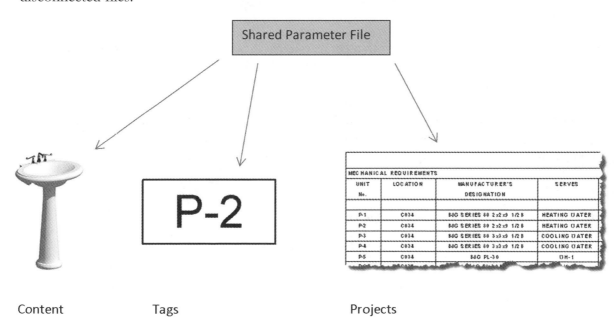

Content   Tags   Projects

Shared Parameter File

Creating a *Shared Parameter* is fairly straightforward, but afterwards, managing them can be troublesome. Select **Manage → Shared Parameters** from the *Ribbon* to open the *Shared Parameters* dialog box (Figure 9-4.1). This dialog basically modifies a simple **text file** which should not be edited manually. If a text file has not yet been created, you will need to select the *Create* button and provide a file name and location for your *Shared Parameter* file.

It is important to keep in mind that this file will be the main record of all your *Shared Parameters*. From this file you create *Shared Parameters* within Projects, Templates and Families. Therefore, it is ideal to only maintain one file for the entire firm, even if you have multiple locations. Of course, there is always an exception to the rule. This single file should be stored on the server and the software deployment should be set up to automatically point user computers to the *Shared Parameters* file. This can also be done manually by clicking the *Browse* button in the *Shared Parameters* dialog.

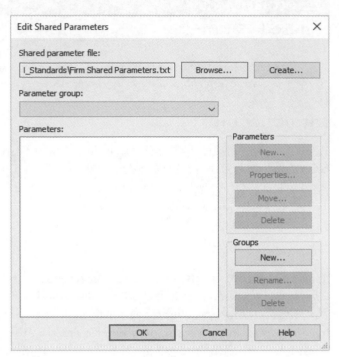

FIGURE 9-4.1 Edit Shared Parameters dialog

Once a *Shared Parameter* is loaded into a project, template or family, the text file is no longer referenced. So you technically do not need to send this file with your project file when transmitting to a consultant or contractor.

The name of the file should be simple and easy to find. You should create a new text file for each version of Revit you are using; include this in the file name. Some newer parameter types will cause older versions of the software to reject the text file.

Creating a Shared Parameter

Open the *Shared Parameters* dialog box. Here you can easily create *Groups* and *Parameters*. *Groups* are simply containers, or folders, used to organize the multitude of parameters you will likely create over time. When starting from scratch, you must first create one *Group* before creating your first *Shared Parameter*.

To create a new *Group*, click the **New** button under the *Groups* heading. In a multi-discipline firm you should have at least one group for each discipline. Parameters can be moved around later so do not worry too much about that at first.

Click the **New** button under the *Parameters* heading to create a new parameter in the current *Group*. You only need to provide three bits of information:

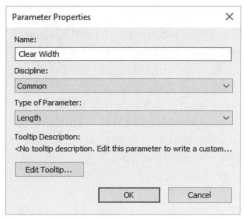

- Name
- Discipline
- Type of Parameter
- Tooltip

In this example, see image to the right (Figure 9-4.2); we will create a parameter called "Clear Width" which will be used in our door families (but can be used with other categories as this is not specified at this level). The other standard options, such as *Instance* versus *Type*, are assigned when the parameter is set up in the project or family.

FIGURE 9-4.2 Setting Parameter Properties

The **Discipline** option simply changes the options available in the *Type of Parameter* drop-down. The **Type of Parameter** drop-down lets Revit know what type of information will be stored in the parameter you are creating. Many programing languages require parameters to be declared before they are used and cannot later be altered. Revit is basically a graphical programming language in this sense.

The **Edit Tooltip** button allows a description of the parameter to be entered. Anyone working in the project will see the description when they hover their cursor over the parameter in the Properties Palette or Edit Type dialog.

When a new *Shared Parameter* is created, a unique code is assigned to it. The image below (Figure 9-4.3) shows the code created for the "Clear Width" parameter just created. For this reason, it is not possible to simply create another *Shared Parameters* file a few months from now and have it work the same.

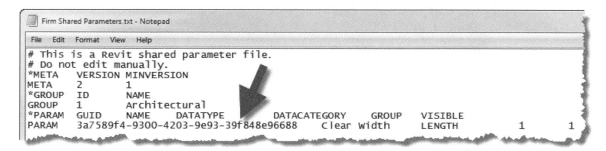

FIGURE 9-4.3 Shared Parameters text file

Creating a Shared Parameter in a Family

Now that you have created the framework for your *Shared Parameters*, i.e., the text file, you can now begin to create parameters within content; the next section covers creating *Shared Parameters* in project files.

Open a family file – in this example we will open the **Door - Single – Panel.rfa** file (C:\ProgramData\Autodesk\RVT 2021\Libraries\US Imperial\Doors). Select the *Family Types* icon. Click the *New Parameter* icon on the lower-left (Figure 9-4.5). Now select *Shared parameter* and then the *Select* button (Figure 9-4.4). This will open the shared parameter text file previously created. Select "Clear Width" and then **OK**.

Now you only have two bits of information left to provide. Is the parameter *Type* or *Instance*, and what is the "Group parameter under" option? This controls which section the parameter shows up under in the *Properties Palette*.

Notice how the bits of information specified in the *Shared Parameters* file are grayed out here.

FIGURE 9-4.4 Parameter properties

They cannot be changed as this would cause discrepancies between families and project files. This information is hard-wired.

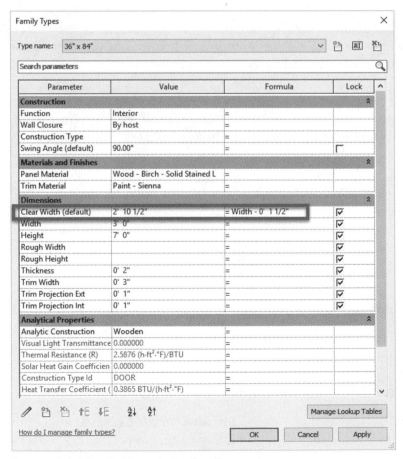

FIGURE 9-4.5 Family Types in family editor environment

Once the parameter has been created in the family it can be used just like a *Family Parameter*. In this example we created a formula to subtract the frame stops and the hinge/door imposition on the opening (Figure 9-4.5). Keep in mind that this value could now appear in a custom tag if desired, thus listing the clear width for each door in a floor plan. Maybe the code official has required this. Using *Shared Parameters* is the only way to achieve this short of using dumb text, that is, text manually typed which does not change automatically.

Next we will look at creating a door tag that lists the clear width. This is similar to the steps just covered for the door family as a tag is also a family. The only difference is this parameter will be associated to a *Label*.

Open the default **Door Tag.rfa** family (location: C:\ProgramData\Autodesk\RVT 2021\Libraries\US Imperial\Annotations\Architectural). Do a *Save As* and rename the file to **Door Tag – Clear Width.rfa**. Delete the linework, if desired. Select the text and click the *Edit Label* button on the *Ribbon*. In the *Edit Label* dialog you need to create a new parameter, by clicking the icon in the lower left; see image below. Your only option here is to select a *Shared Parameter*, as previously mentioned, only *Shared Parameters* can be tagged. Once the new parameter is created, click to move it to the right side of the dialog, in the *Label Parameters* column. Next, select the original *Mark* parameter and remove it. Finally, edit the *Sample Value* to something like 2'-10"; this is what appears in the family to give you an idea of what the tag will look like in the project.

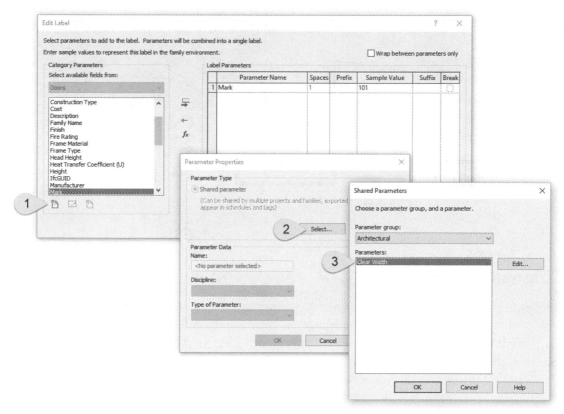

FYI: The Mark *parameter could be left in the tag if you wanted both the* Mark *and* Clear Width *to appear together. The other option is to have two separate tags which can be moved independently. Keep in mind it is possible to tag the same element multiple times. In this case you would have two door tags on the same door, each tag being a different type.*

Load your new door and door tag families into a new project file. In the next section you will see how these work in the project environment.

Using Shared Parameters in a Project

Now that you have your content and tags set up you can use them in the project. First, draw a wall and then place an instance of the *Single – Flush* door (turn off *Tag on Placement*). Select the door and notice the *Clear Width* parameter is showing up in the *Properties Palette*; it is an <u>Instance Parameter</u>. Add another door to the right of this one and change the width to 30″ via the *Type Selector*. Notice the *Clear Width* value has changed.

Next you will tag the two doors. Select the **Tag by Category** icon from the *Quick Access Toolbar*. Uncheck the *Leader* option and select each door. In the image below, the tag was also set to be *Vertical* via the *Options Bar*. If the tag placed is the door number, select the door tag and change it to the *Clear Width* option via the *Type Selector*.

TIP: If the text does not fit on one line you have to go back into the family and increase the width of the text box and reload the family into the project.

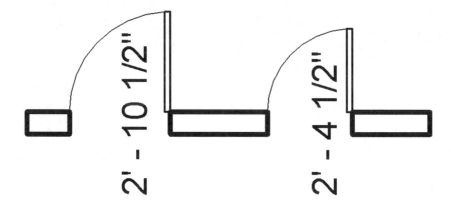

Things to Know

Only content which contains the *Shared Parameters* will display the parameter placeholder in a project. Try loading another door family (e.g., *Double – Flush*) into your test project without changing it in any way. Place an instance and then notice the *Clear Width* option does not appear in the *Properties Palette*. This means all content must have the *Shared Parameter* added to it. It is possible to create a *Project Parameter*, in the project/template file, but you have to specify *Instance* or *Type*. If this varies you cannot use a *Project Parameter*. The *Project Parameter* takes precedence and will change loaded content.

Selectively associating parameters with content is a great trick when it comes to certain categories having a variety of items, such as *Furniture* or *Mechanical Equipment*. Only loading

Shared Parameters into file cabinets or Air Handling Units (AHU) will make it so those parameters do not appear in your lounge chairs or VAV boxes.

> **FYI:** *This also works with* Family Parameters *but this information cannot be tagged or scheduled.*

Dealing with Problems

A number of problems can be created when *Shared Parameters* are not managed properly. The most common is when two separate *Shared Parameters* are created with the same name. This often happens when new users delete or otherwise lose a *Shared Parameter* file. They then try to recreate it manually, not knowing that the unique code, mentioned previously, is different and Revit will see this as a different *Shared Parameter*. The first place this problem typically shows up is when a schedule in the project has blank spaces even though the placed content has information in it. In this case, the schedule is using one specific version of the *Shared Parameter* and only the content using that same version of the *Shared Parameter* will appear in that schedule.

The fix for this problem is to open the bad content and re-associate the parameters with the correct *Shared Parameter*. It is not necessary to delete the bad parameter, which is good as this could cause problems with existing formulas.

If you get content from another firm you are working with and they have used *Shared Parameters* it is possible to export those parameters from the *Family Editor* into your *Shared Parameter* file. They will initially be located in the *Exported Parameters* group, but can then be moved to a more appropriate location. This allows that unique code to be recreated in your file.

If you have a *Shared Parameter* with the same name as another firm's *Shared Parameter,* and you are working on the same model, that is a problem. You will have to decide whose version to use. There are some tools one can use to add and modify *Shared Parameters* in batch groups of families. One is found in the new *Extensions for Revit 2021*, which is available via the subscription website.

Conclusion

Using *Shared Parameters* is a must in order to take full advantage of Revit's powerful features. Like any sophisticated tool, it takes a little effort to fully understand the feature and its nuances. Once you have harnessed the power of the feature and implemented it within your firm's content and templates you can be much more productive and have less potential for errors and omissions on your projects.

Exercise 9-5:
Keynoting

This exercise will present an overview of the keynoting system in Revit. There are no steps to be applied to the project in this section. At the end of the next chapter, Chapter 10 - Elevations, Sections and Details, you will apply some of these concepts.

Hybrid Method

A simple way to manage keyed notes is to use a custom Symbol family with the symbol and descriptive text, where the text visibility can be toggled off per family instance. Thus, the keyed notes are placed in plan with the text hidden and the same keyed notes are also placed in a legend (or drafting view) with the text visible.

Notice in the plan view below, the keyed note is selected and the custom instance parameter **Description Visible** is toggled off (Figure 9-5.1).

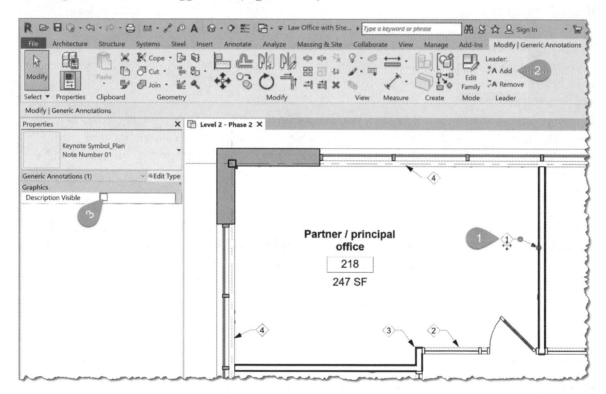

FIGURE 9-5.1 Keyed note in a floor plan view with 'Description Visible' toggled off

The next image shows the same family placed and then selected in a Legend view. In this view the **Description Visible** is checked (Figure 9-5.2).

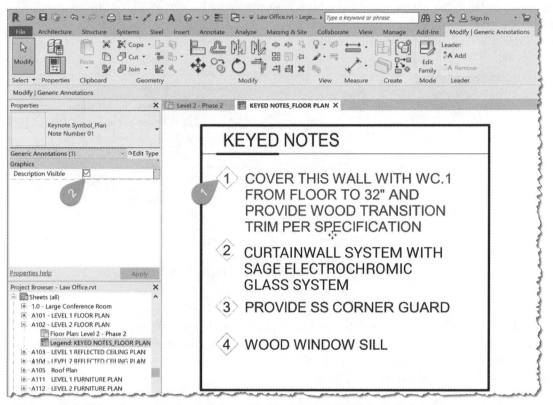

FIGURE 9-5.2 Keyed note in a legend view with 'Description Visible' toggled on

The image to the right shows the type properties for keyed note number 1. Notice the Keynote and Description are type parameters. This helps prevent someone from using the same number multiple times with a different description.

This method requires a different family for each plan type: e.g. Plan, Finish, Demo, Code, etc. Each family having one type for each note.

FYI: This example uses the **Multiline** parameter introduced in Revit 2016.

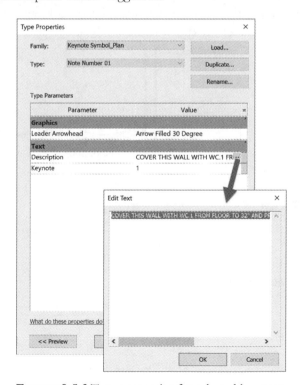

FIGURE 9-5.3 Type properties for selected key note

Material Keynotes

We often want to identify a material in the model with "smart" text. In this case, we are talking about a Revit material defined within a **System Family** (e.g. Wall, Ceiling, Floor), a **Loadable Family** (e.g. Furniture, Casework, Doors, etc.) or **Painted** on a surface (which overrides the material defined in the Family).

There are two ways to identify materials within a Revit model (Figure 9-5.4):

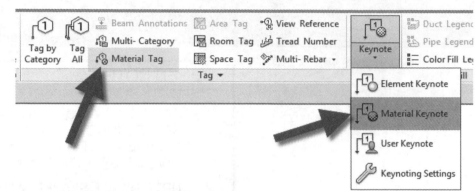

- **Material Keynote**

- **Material Tag**

FIGURE 9-5.4 Material keynote and tag tools on the Annotate tab

FYI: It should be pointed out, that even though we can define multiple materials for some elements, e.g. several layers within a wall (Brick, Insulation, Studs, Gypsum Board, etc.), we can only keynote/tag a material exposed in a view—i.e. we can click on it. Materials hidden within a wall or equipment family cannot be tagged.

The two examples, Material Keynote and Material Tag, are shown in the image below—referencing the wall and the wall/base cabinets (Figure 9-5.5). The boxed text is a Keynote and the adjacent text is a Material Tag.

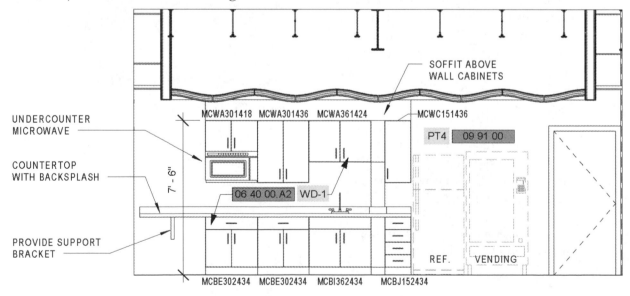

FIGURE 9-5.5 Material keynote applied to two elements (wall and cabinets)

Both values represented in the previous image are defined within the Material Browser as seen in the image below (Figure 9-5.6).

- Material Keynote → **Keynote**
    - o *The value must be selected from the Keynote file (more on this later)*
- Material Tag → **Mark**
    - o *The user can enter any value here*

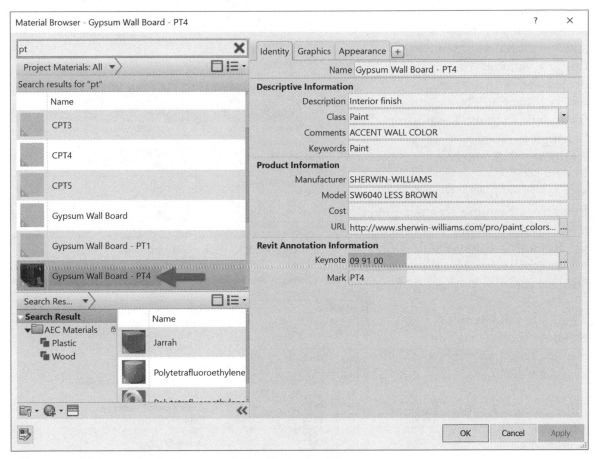

FIGURE 9-5.6 Annotation information within a Revit material

When either of these methods are used, the tags will update automatically if the family/type is changed (which also changes the material) or if a different material is painted on the elements being tagged.

When a Material is copied the Keynote value is also copied (i.e. same keynote value is listed for the new material); however, the Mark value is not. The latter is preferred if you want to ensure the same value is not used for multiple materials.

Element Keynotes

Keynotes can be used to nearly eliminate "dumb" text in Revit. Here we will quickly cover the basics of how keynotes work and then advance from there.

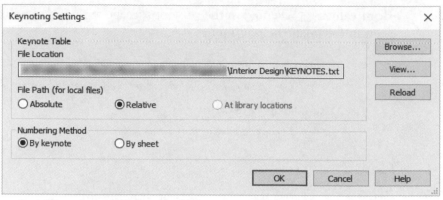

FIGURE 9-5.7 Keynoting settings dialog

<u>First</u>, the keynote information is stored in a simple text file (more on this in a minute). A Revit project can only reference one keynote file. The location to this file is defined in the **Keynoting Settings** file (Figure 9-5.7) via **Annotate → Keynotes**.

<u>Second</u>, every model-based element in Revit has a **Keynote** type parameter (Figure 9-5.8). Editing this field opens the **Keynotes** dialog (Figure 9-5.9), which is a list based on the aforementioned external text file.

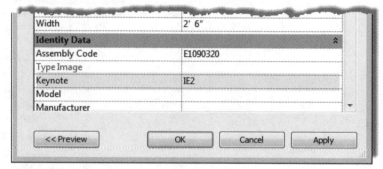

FIGURE 9-5.8 Keynote type parameter

<u>Third</u>, use the **Element Keynote** tool to "tag" an element based on the predefined keynote value.

<u>Fifth</u>, create a **Keynote Legend** and place it on any sheets with keynoted views.

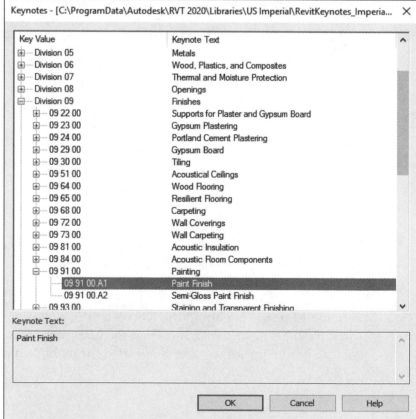

FIGURE 9-5.9 Keynotes dialog as viewed from within Revit

The result can be seen in the sheet view below (Figure 9-5.10). The default keynote tag shows the keynote number which corresponds to the same number and description in the legend. The result is "clean" looking drawings without excessive text.

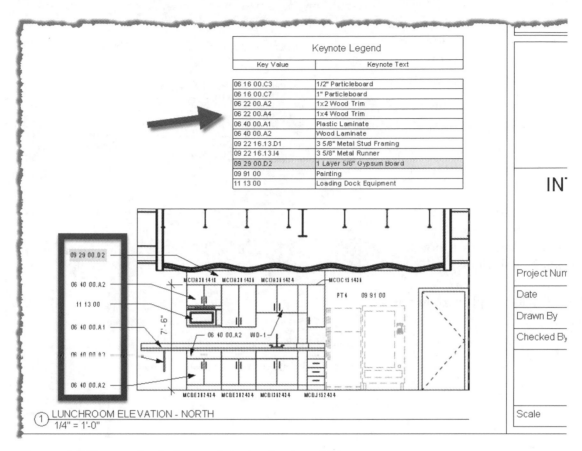

FIGURE 9-5.10 Example of a keynote legend and a view with keynotes

The keynote legend will only list keynotes that actually appear on a given sheet. Opening the keynote legend view directly will show all keynotes in the project. Keep in mind, the keynote legend only lists items which have been keynoted using the Keynote tool. It will not list a keynote just because an element's type property has a keynote value applied.

TIP: Use a Filter, in Schedule Properties, to limit the list if needed.

This represents a basic overview of how keynotes work in Revit – with an emphasis on the way it works out-of-the-box (OOTB). Next we will look at a variation on this system, using "normal" notes rather than MasterFormat section numbers (https://www.csiresources.org/practice/standards/masterformat).

For this alternate method we will start by creating a custom keynote file. Simply create an empty file using Microsoft Notepad and save it somewhere in your project folder (Figure 9-5.11).

To better understand the format of this file, note the following:

1. **Group definition**
   - unique number
   - descriptive text
2. **Keynote number**
3. **Keynote text**
4. **Group identifier**

The keynote text file must be stored on the server where all users have access to it (not necessarily edit rights).

Results below explained on the next page...

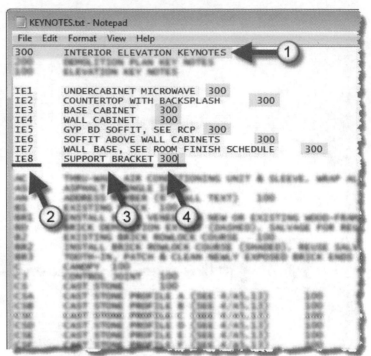

FIGURE 9-5.11 Anatomy of a keynote file

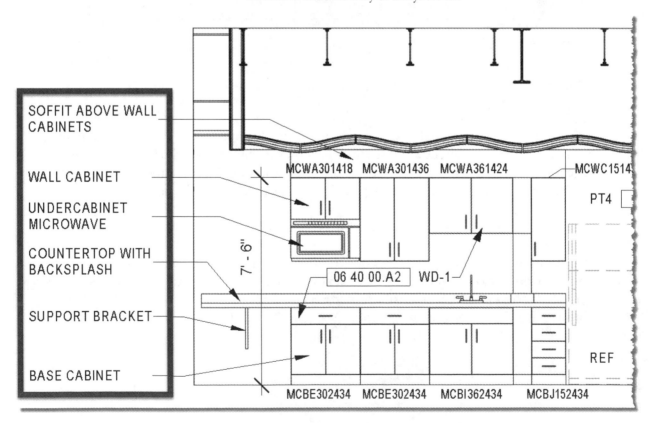

FIGURE 9-5.12 Keynotes added to elements in elevation - no "dumb" text

The elevation shown on the previous page has Element Keynotes added. The default Keynote tag has type properties to hide the keynote number & box and show the keynote number.

All the elements in the model have firm-approved standard notes applied to the Keynote type property. This information can be prepopulated within the content library to streamline production and produce consistency across the firm.

Keynote Manager + Add-in for Revit

One challenge with maintaining the keynoting system in Revit is the text file can only be edited by one person at a time. Also, after changes are made, the changes must be manually reloaded (note the **Reload** button on the Keynoting Settings dialog) to see the changes in the current session of Revit (Figure 9-5.12).

Using **Keynote Manager +**, a third-party add-in for Revit, can help. This tool has the following features:

- Multiple users
- Auto reload
- Spellcheck
- Indicates keynotes used in project (with house icon)
- Auto sequencing
- Add comments and URL links

FYI: To store the additional comments and external links, a same named XML file is created right next to the keynote file.

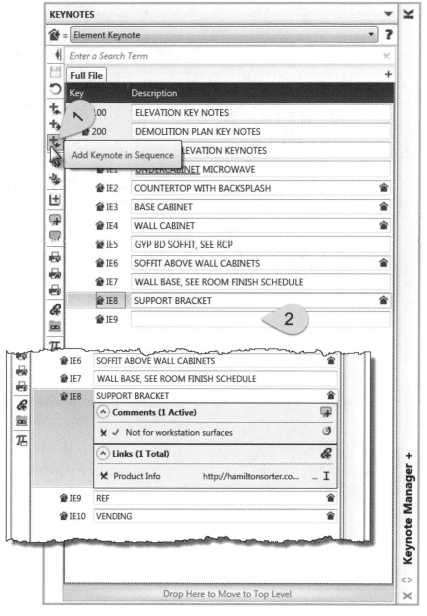

FIGURE 9-5.13 Keynote manager palette in Revit

Be sure to check out **Keynote Manager**, by Revolution Design, Inc., for a powerful add-in tool to help simplify the use of Keynotes in Revit. Multiple users can also work in the keynote file at the same time.
URL: http://revolutiondesign.biz/products/keynote-manager/features/

User Keynotes

User Keynotes work the same way as Element Keynotes with one exception…

When the **User Keynote** tool is selected, the user is prompted to select an element to keynote. Next, a selection must be made from the keynote list (i.e. the text file).

Although this adds some flexibility to the process, it provides an opportunity for selecting the wrong keynote. In the example below, Figure 9-5.14, the selected base cabinet is tagged with three keynotes (highlighted in yellow). One is an **Element Keynote** and two are **User keynotes**—both user keynotes indicate something completely different.

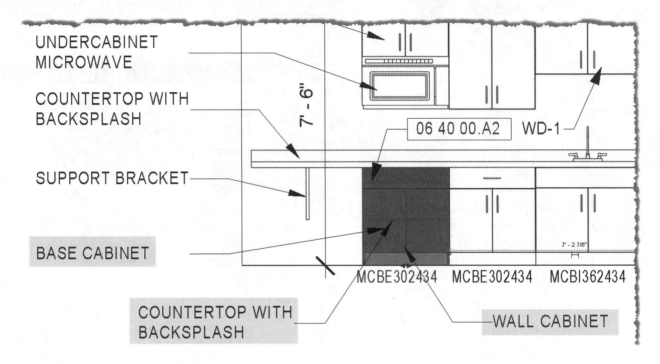

FIGURE 9-5.14 Multiple user keynotes applied to a single model element

Used correctly, the User Keynote tool can be helpful. However, those using it must be trained and pick carefully every time they use this tool.

Self-Exam:

The following questions can be used as a way to check your knowledge of this lesson. The answers can be found at the bottom of this page.

1. Deleting a dimension with equality toggled on will always also delete the constraint in the model. (T/F)

2. To move an element, simply click on the dimension and edit the text. (T/F)

3. Material Keynote and Material Tag are the two ways to annotate a material. (T/F)

4. Pressing Enter in a text element limits the ability to modify using the text box grips and have the number of rows adjust. (T/F)

5. To measure the length along a curved line, use this dimension tool:

   _____ .

Review Questions:

The following questions may be assigned by your instructor as a way to assess your knowledge of this section. Your instructor has the answers to the review questions.

1. A dimension string set to Equality can only display 'EQ' and not the actual dimension value. (T/F)

2. A single text element can have both curved and straight leaders. (T/F)

3. Spell check only works on text elements, not tags or keynotes. (T/F)

4. The use of basic text should be minimized. (T/F)

5. The same element can be tagged more than once, even in the same view. (T/F)

6. Use a Linear dimension to ensure a dimension remains horizontal. (T/F)

7. Revit's keynoting system requires an external text file. (T/F)

8. Text leaders can only be removed in the reverse order they were added. (T/F)

9. Revit can dimension all the openings in a wall at once. (T/F)

10. A 'user keynote' presents an opportunity for user error by selecting the wrong item from a list each time this tool is used. (T/F)

SELF-EXAM ANSWERS:
1 – F, **2** – F, **3** – T, **4** – T, **5** – Arc Length

Notes:

Lesson 10
SECTIONS and DETAILS:

Sections are one of the main communication tools in a set of architectural drawings. They help the builder understand vertical relationships. With traditional CAD software, architectural sections can occasionally contradict other drawings, such as mechanical or structural drawings. One example is a beam shown on the section is smaller than what the structural drawings call for; this creates a problem in the field when the duct does not fit in the ceiling space. The ceiling gets lowered or the duct gets smaller, ultimately compromising the design to a certain degree.

Revit takes great steps toward eliminating these types of conflicts. Sections, like plans and elevations, are generated from the 3D model. So, it is virtually impossible to have a conflict between the architectural drawings. As structural and mechanical engineers begin to use Revit, the coordination gets better as the various disciplines' models are linked together.

Exercise 10-1:
Specify Section Cutting Plane in Plan View

Similar to elevation tags, placing the reference tags in a plan view actually generates the section view. You will learn how to do this next.

Placing Section Tags

1. Open ex8-6.rvt and **Save As ex10-1.rvt**.

2. Switch to **Level 1** view.

Section

3. Select **View → Create → Section** button from the *Ribbon*.

4. Draw a Section tag as shown in Figure 10-1.1. Start on the left side in this case. Use the *Move* tool if needed to accurately adjust the section tag after insertion. The section should go through the doors in the stair shaft (Figure 10-1.1).

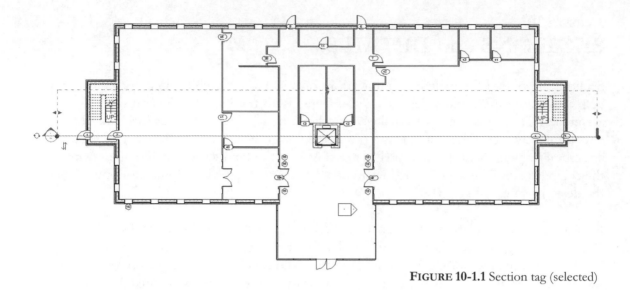

FIGURE 10-1.1 Section tag (selected)

Figure 10-1.1 shows the section tag selected. The section tag features are very similar to the elevation tags covered in the previous lesson. You can adjust the depth of view (*Far Clip Plane*) and the width of the section with the *Adjustment Grips*.

Section views are listed under that heading in the *Project Browser*. Similar to newly created elevation views, you should name section views as you create them.

5. Rename the new section view to **Longitudinal Section**.

6. Switch to the **Longitudinal Section** view (Figure 10-1.2).

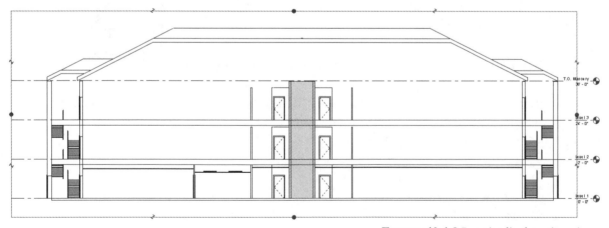

FIGURE 10-1.2 Longitudinal section view

You can see that the stairs are cut off on the back side because of the *Far Clip Plane* location in the plan view (Figure 10-1.1). Also, you can see the roof is shown in section exactly where the section line is shown in plan. Figure 10-1.2 also shows the *Crop Region*.

7. Adjust the *Far Clip Plane* in plan view so the entire stair shows and the *Crop Region* is not visible (Figure 10-1.3).

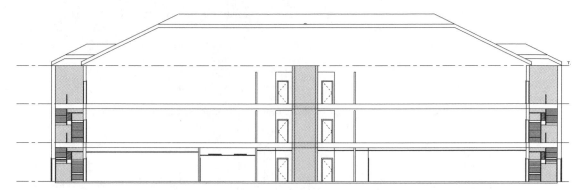

FIGURE 10-1.3 Longitudinal view - updated

8. On the *View Control Bar*, change the view *Properties* so the **Detail Level** is set to **Medium**. *(Notice how the walls change to show more detail.)*

9. Zoom in to the elevator shaft area as shown in **Figure 10-1.4**.

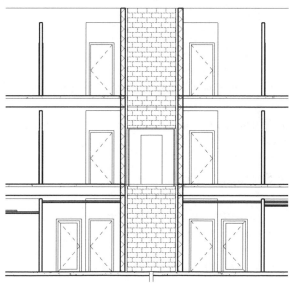

FIGURE 10-1.4 Section view – zoomed in

You should notice an added level of detail in the section view—for example, the concrete hatch in the floor and the CMU joint lines in the elevator shaft. This added detail helps the drawing read better.

Next you will add a cross sectional view.

10. Create a **Section** as shown in **Figure 10-1.5**.

> *TIP: You can use the control arrows to make the section look the other direction.*

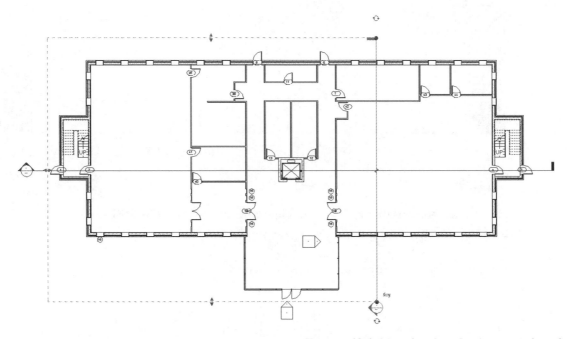

FIGURE 10-1.5 Level 1 view; Section tag (selected)

11. Rename the new section view to **Cross Section 1** in the *Project Browser*.

12. Adjust the **Far Clip Plane** (if required) so the entire atrium roof will be visible in the **Cross Section 1** view.

13. Switch to the **Cross Section 1** view.

14. Set the *Detail Level* to **Medium** and turn off the **Crop Region** visibility via the *View Control Bar* (Figure 10-1.6).

> *TIP: Both of these settings can be controlled via the Properties Palette as well.*

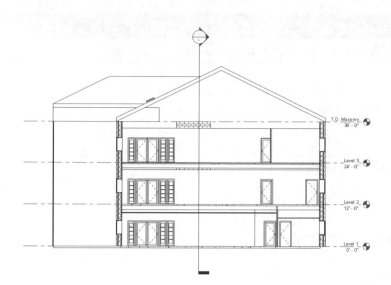

Revit automatically displays lines heavier for objects that are in section than for objects beyond the cutting plane (i.e., in projection) and shown in elevation.

Also, with the *Detail Level* set to medium, the walls and floors are hatched to represent the material in section.

FIGURE 10-1.6 Cross Section 1 view

Notice that the Longitudinal Section tag is automatically displayed in the Cross Section 1 view. If you switch to the Longitudinal Section view you will see the Cross Section 1 section tag. Keeping with Revit's philosophy of change anything anywhere, you can select the section tag in the other section view and adjust its various properties, like the Far Clip Plane.

   15. **Save** your project as **ex10-1.rvt**.

FYI: In any view that has a Section Tag in it, you can double-click on the round reference bubble to quickly switch to that section view.

Exercise 10-2:
Modifying the Project Model in Section View

Again, similar to elevation views, you can modify the project model in section view. This includes adjusting door locations and ceiling heights.

Modifying Doors in Section View

In this section you will move a door and delete a door in section view.

1. Open ex10-1.rvt and **Save As ex10-2.rvt**.

2. Open **Cross Section 1** view.

3. On **Level 2**, move the *Single Glass* door **5′-0″** to the north and **delete** the door added in a previous lesson; see modified section view Figure 10-2.1. (See Figure 10-1.6 for an unmodified view.)

4. Adjust the **ceiling height** in the lower right room to be **9′-0″** above Level 1 (Figure 10-2.1).

    TIP: Select the ceiling and simply change the temporary dimension that appears (or use the Properties Palette).

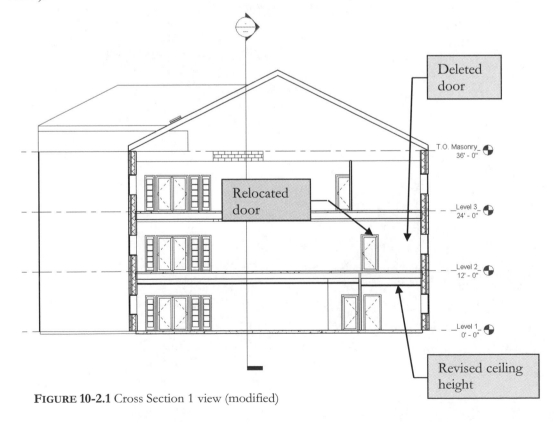

FIGURE 10-2.1 Cross Section 1 view (modified)

5. Switch to the **Level 2** view (Figure 10-2.2).

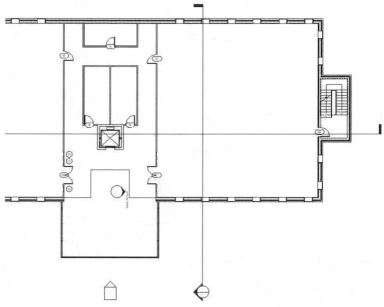

FIGURE 10-2.2 Level 2 view

You should see the door in its new location and the other door has been deleted.

6. Switch to the **East Atrium** view (Figure 10-2.3).

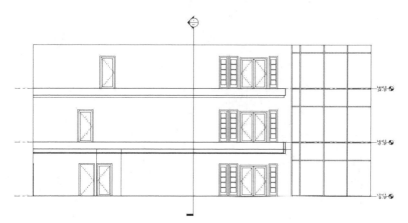

FIGURE 10-2.3 East Atrium view

You can see the changes here as well. Compare this elevation with Figure 7-4.1 from lesson 7. Also, notice that the section mark was automatically added to the elevation. Remember, you can double-click on the section bubble to switch to that view.

7. Save your project as **ex10-2.rvt**.

Exercise 10-3:
Wall Sections

So far in this lesson you have drawn building sections. Building sections are typically 1/16″ or 1/8″ scale and light on the details and notes. Wall sections are drawn at a larger scale and have much more detail. You will look at setting up wall sections next.

Setting Up the Wall Section View

1. Open ex10-2.rvt and **Save As ex10-3.rvt**.

2. Switch to the **Cross Section 1** view.

Callout

3. Select **View → Create → Callout**.

4. Place a **Callout** tag as shown in Figure 10-3.1.

   TIP: Pick in the upper left and then in the lower right (don't drag) to place the Callout tag.

   a. Select *Section:* **Wall Section** from the *Type Selector*.

5. Use the *Control Grips* for the *Callout* tag to move the reference bubble as shown in Figure 10-3.1.

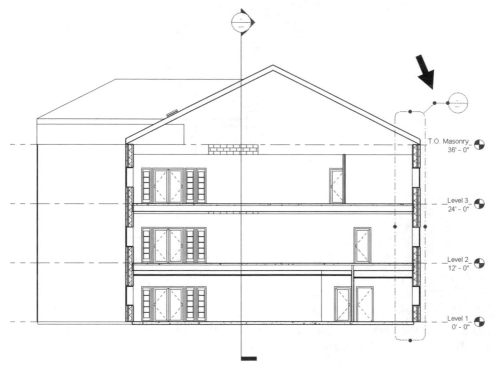

FIGURE 10-3.1 Cross Section 1
view with Callout added

Notice that a view was added in the *Sections* category of the *Project Browser*. Because *Callouts* are detail references off of a section view, it is a good idea to keep the section view name similar to the name of the callout.

Additionally, *Callouts* differ from section views in that the callout is not referenced in every related view. This example is typical, in that the building sections are referenced from the plans and wall sections are referenced from the building sections. The floor plans can get pretty messy if you try to add too much information to them.

6. Double-click on the reference bubble portion of the *Callout* tag to open the **Callout of Cross Section 1** view (Figure 10-3.2).

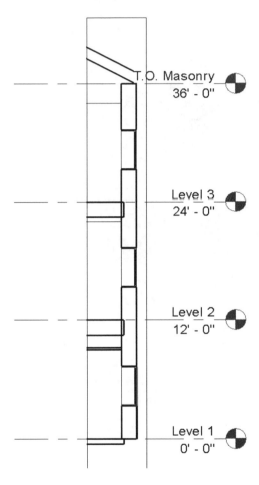

FIGURE 10-3.2 Callout of Cross Section 1

7. In the View properties, set the *View Scale* to **¾″ = 1′-0″** and the *Detail Level* to **Fine** (Figure 10-3.3).

    TIP: *This can all be done from the View Control Bar as well.*

    Notice the Level datum symbol size changed as well as the Detail Level (Figure 10-3.4).

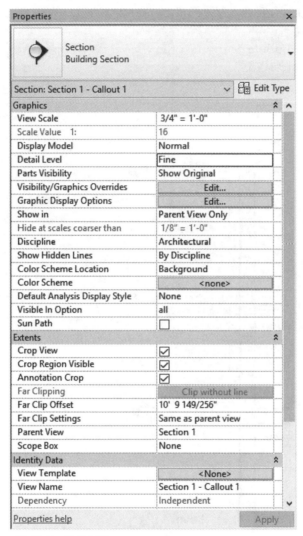

FIGURE 10-3.3 View properties

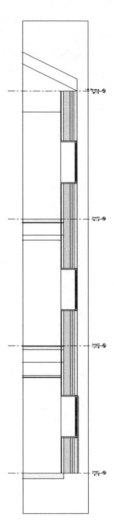

FIGURE 10-3.4 Revised Detail Level and view scale

If you zoom in on a portion of the *Callout* view, you can see the detail added to the view. The wall's interior lines (i.e., veneer lines) are added and the materials in section are hatched (Figure 10-3.5).

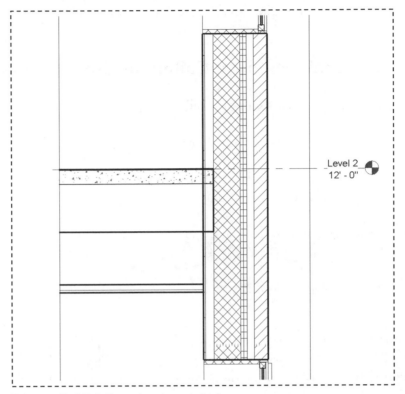

Level 2
12' - 0"

FIGURE 10-3.5 Callout view (zoomed in)

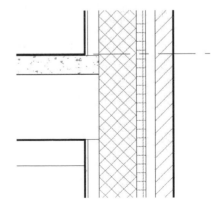

🔲 Join ▾

You can use the **Join** tool to clean up the floor to wall condition shown above. Simply click the icon and the two elements you want to join (this works on many types of elements). See Figure 10-3.6 for a "joined" condition; every view is updated!

You can use the **Detail Lines** tool to add more detailed information to the drawing. For example, you could show the masonry coursing, window trim, brick vents/weeps and flashing.

As before, you can turn off and adjust the crop region.

8. **Save** your project as **ex10-3.rvt**.

FIGURE 10-3.6 Joined wall/floor

Exercise 10-4:
Annotations

This exercise will explore adding notes and dimensions to your wall section.

Add Notes and Dimensions to Callout of Cross Section 1

1. Open ex10-3.rvt and **Save As ex10-4.rvt**.

2. Switch to **Callout of Cross Section 1** view.

3. Adjust the View Properties so the crop region is not visible (via the *Properties Palette* when nothing is selected in the model).

4. Add two dimensions and adjust the Level datum symbol location as shown in **Figure 10-4.1**.

   TIP: Dimension to the masonry opening.

These dimensions are primarily for the masons laying up the CMU and Brick. Typically, when an opening is dimensioned in masonry, the dimension has the suffix M.O. This stands for Masonry Opening, clearly representing that the dimension identifies an opening in the wall. You will add the suffix next.

5. Select the dimension at the window opening and pick the **Blue text** (i.e., 4'-0″).

6. Type **M.O.** in the Suffix field (Figure 10-4.2).

7. Click **OK**.

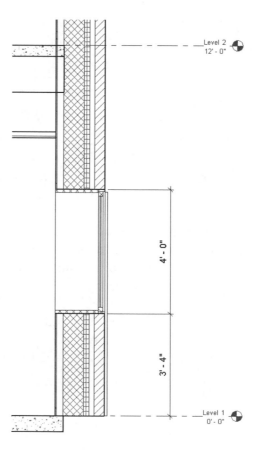

FIGURE 10-4.1 Added dimensions

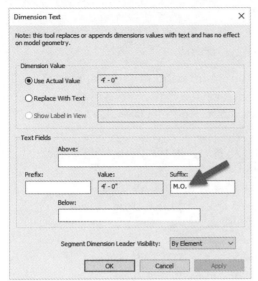

FIGURE 10-4.2 Editing dimension text

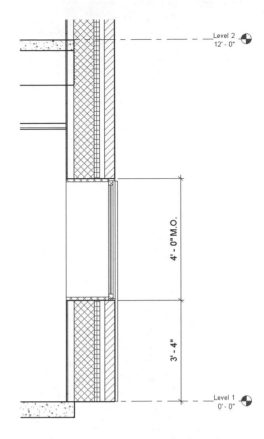

FIGURE 10-4.3 Dimension with suffix

Figure 10-4.3 shows the dimension with the added suffix.

8. Add the additional dimensions shown in **Figure 10-4.4**; be sure to add the suffixes.

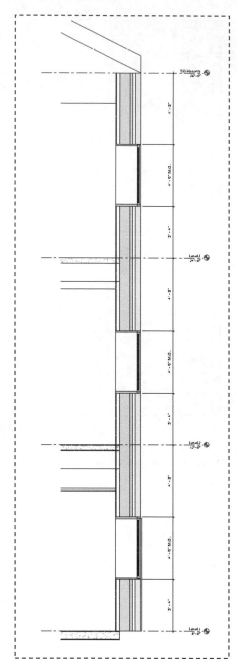

FIGURE 10-4.4 All dimensions added

9. Add the notes with leaders shown in Figure 10-4.5. *(See step 10.)*

    a. Aluminum window system, typical
    b. Brick cavity wall
    c. Concrete and metal deck and stl. bar joists
    d. Concrete slab on grade over vapor barrier

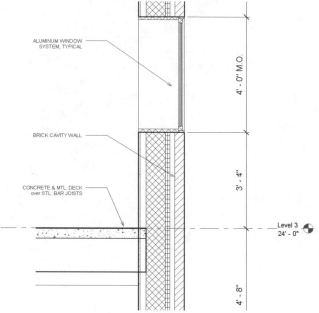

FIGURE 10-4.5 Notes added

10. The text style should be set to **3/32″ Arial**; it may still be set to the last text style you used (¼″ Outline Text, in this book).

11. Select the text and use the grips and the justification buttons to make the text look like **Figure 10-4.5**.

FYI: Architectural text is typically all uppercase.

12. **Save** your project.

Exercise 10-5:
Details

This exercise shows you how to develop 2D details which are not tied to the 3D model in any way. Why would you want to do this, you might ask? Many design firms have developed detail libraries over the years for typical conditions. These details contain a significant amount of embodied knowledge of the firm as a whole. Many notes and dimensions have been added which cover certain situations that have come up and created problems or cost the firm money. For example, a window head detail might show flashing, which directs any moisture in the wall out – rather than into – the window or inside the building. Well, a note might have been added to instruct the contractor to turn the flashing up at each side of the window to ensure the moisture does not just run off the end of the flashing and stay in the wall.

If every detail were a live cut through the model, the designer would have to spend the time adding all these notes and dimensions, and more importantly try not to forget any, even if typing them from a printed reference page – similar to what you are doing with this book. (Have you missed anything yet, and had to go back and make a correction?) Furthermore, if the part of the model changes, the detail could be messed up. Or, an item being detailed from the live model might change and not be the typical condition anymore.

So, as you can see, there are a number of reasons a design firm maintains and utilizes static 2D details. *NOTE: Sometimes these details are used as starting points for similar details. This saves time not having to start from scratch.*

It should also be pointed out that standard details should always be reviewed before "dumping" them into a project. If a note says, "Apply fireproofing to underside of metal roof deck" and your building has precast concrete plank, you need to change the note and the drawing. All other parts of the detail may perfectly match the project design you are working on.

Linking an AutoCAD Drawing

This first exercise will explore linking AutoCAD drawings into Revit when the need to use legacy details arises.

It is better to recreate these details in native Revit format rather than linking an AutoCAD drawing. Any external files linked in have the potential to slow your BIM experience and introduce corruption. This is especially true with site plans created in AutoCAD or AutoCAD Civil 3D. Site plans are often a great distance from the origin (i.e., 0,0,0 coordinate in an AutoCAD drawing) and this creates several issues.

In general, it is best to avoid AutoCAD DWG files within Revit. However, when it is required, they should always be *Linked* in and not *Imported*, and never *Exploded*. Importing DWG files makes them difficult to manage and exploding them creates lots of extra text styles, fill patterns and other items that clutter the BIM database.

1.   Open ex10-4.rvt and **Save As ex10-5.rvt**.

AutoCAD DWG files can be linked directly into a plan view and be used as an underlay to sketch walls and place doors and windows, when modeling an existing building in Revit that has been drawn in a traditional CAD program.

In our example, we have a DWG file which contains a detail we want to reference and place on a sheet for our office building project. To do this, you create a *Drafting View* and link the CAD file into the drafting view. A *Drafting View* is a 2D drawing within the BIM project that has no direct relationship to the 3D model.

2.   Click **View → Create → Drafting View** from the *Ribbon*.

Drafting
View

Now you are prompted for a name and scale for the new drafting view; this can be changed later if needed.

3.   Enter the following (Figure 10-5.1):

    a.   *Name:* **Typical Roof Drain Detail**

    b.   *Scale:*   **1½″ = 1′-0″**

4.   Click **OK** to create the new *Drafting View*.

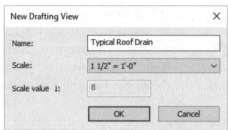

FIGURE 10-5.1 Creating a drafting view

You now have a new section, under *Views* in the *Project Browser*, called *Drafting Views* (Detail). Within this section is your new *Drafting View* – <u>Typical Roof Drain Detail</u>.

Within this new drafting view you could begin sketching a detail from scratch using the various tools on the *Annotate* and *Modify* tabs. Or, in this example, you may link in a DWG file.

This roof drain detail is a good example of why 2D details are still useful in Revit, either DWG or native Revit. As previously mentioned, many firms spend years developing standard details. These details have notes that have been added to and edited as building materials change and problems occur. It would take a lot of time to cut a section at a 3D roof drain in the model and then add all the notes, if one can even remember what all the notes are. Now, repeat this for 20 to 50 other items throughout the building project.

Now you will link in the DWG file using the provided online file.

5.   While in the newly created drafting view, select **Insert → Link → Link CAD** from the *Ribbon*.

Link
CAD

6.   Browse to the **DWG Files** folder within the files downloaded from the publisher's website.

7. Select the file **Typical Roof Drain Detail.DWG**.

8. Set the *Colors* options to **Black and White** (Figure 10-5.2).

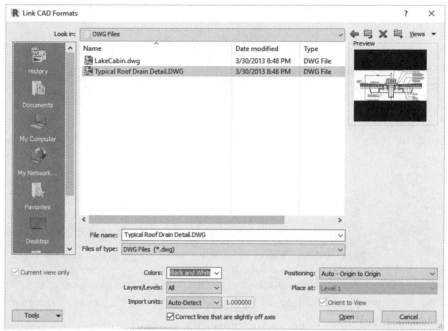

FIGURE 10-5.2 Linking an AutoCAD detail file

9. Click **Open** to place the linked AutoCAD DWG file.

10. Type **ZF** (for *Zoom Fit*) on the keyboard; do **not** press **Enter**.

You should now see the roof drain detail, with line weights.

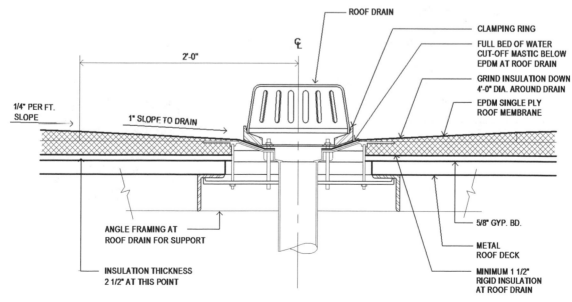

FIGURE 10-5.3 DWG file linked into drafting view

The drawing can be selected and moved around within the drafting view, but it cannot be edited. If you need to make changes to this drawing, you would have to do it in AutoCAD. Revit will automatically update any linked files when the project file is opened. It can also be done manually using the *Manage Links* tool.

When DWG files are linked into Revit, a specific set of line weights are used. These settings can be seen by clicking the small arrow (i.e., the dialog launcher) in the lower right corner of the *Import* panel on the *Insert* tab.

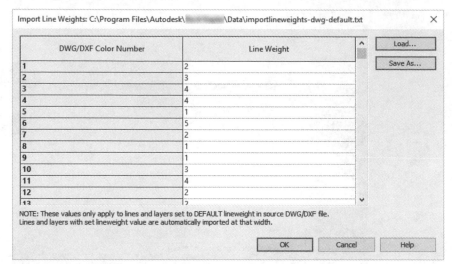

FIGURE 10-5.4 DWG color to Revit line weight conversion

Applying line weights is a onetime conversion process when the DWG file is linked in. Changing the line weight setting after a DWG has been linked in has no effect on it (only on new DWG files to be linked in).

> **TIP:** *Go to* **Manage → Settings → Additional Settings → Line Weights** *to see what each line weight number is equal to.*

If you type **VV** in the drafting view and then select the *Imported Categories* tab, you can see the AutoCAD *Layers* that exist in the imported DWG file (Figure 10-5.5). Unchecking a *Layer* will hide that information within the drafting view. You can also control the color and line weight of the lines on each *Layer*.

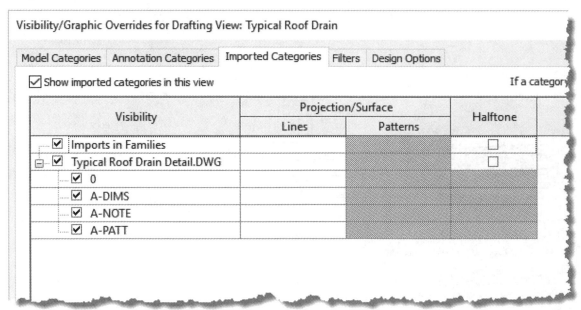

FIGURE 10-5.5 Controlling DWG layer visibility in Revit

Creating 2D Details

Autodesk Revit has a large array of 2D detail components that can be used to create 2D details. These components allow for efficient detail drafting and design. Not every detail in Revit is generated from the 3D model; the amount of modeling required to make this happen is restricted by time, file size and computing power. The following is an outline of the overall process; this will be followed by a few exercises for practice:

- To create a 2D detail one would create a **drafting view** via *View > Create > Drafting View*, providing a name and selecting a scale.

- Once the *drafting view* has been created, **Detail Lines** and **Filled Regions** (via the *Annotate* tab) can be added.

- In addition to *Detail Lines* and *Filled Regions*, one can insert pre-drawn items from the <u>Detail Library</u>.

  i. Select *Component > Detail Components* from the *Annotate* tab.

  ii. Select **Load Family** from the *Ribbon*.

  iii. Open the **Detail Items folder** in the **English-Imperial** library folder.

  iv. **Browse** to the specific "CSI organized" folder; for example, *Div 5-Metals→052100- Steel Joists Framing → K-Series Bar Joist-Side.rfa*.

  v. Click **Open** to place the component.

- Add notes and dimensions to complete the 2D detail.

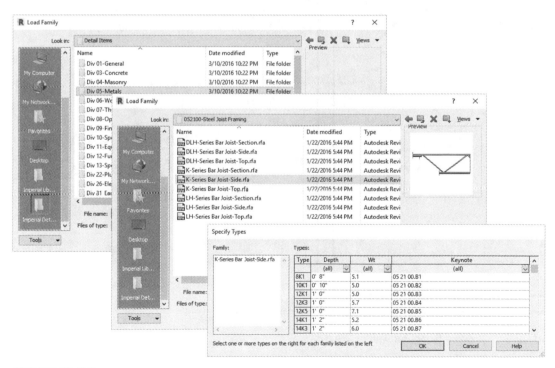

FIGURE 10-5.6
Loading detail components

Flooring Details

The first two details you will draw are simple details consisting of *Detail Lines*, *Filled Regions*, *Text* and *Dimensions*.

You will draw a high-end floor and wall base detail known as terrazzo. This finish is poured in a liquid state, allowed to dry and then polished to a smooth finish. The colors and aggregate options are virtually unlimited (for example, you could use a clear epoxy resin and place leaves within the flooring).

1. Per the steps previously covered in this section, create a new drafting view.

    a. *Name*: **Terrazzo Base Detail**
    b. *Scale*: **3″=1′-0″**

You will now draw the detail shown below. See the next page for specific steps.

Terrazzo floor example with brass inlay

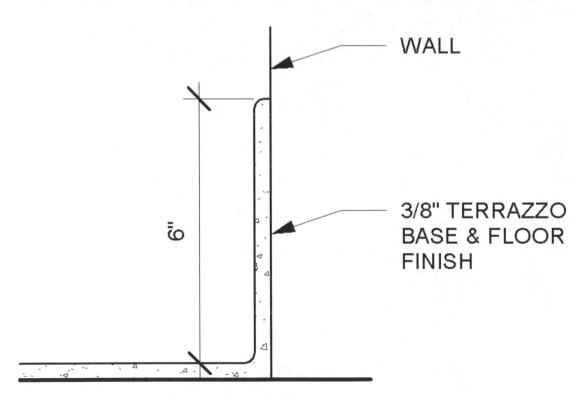

FIGURE 10-5.7 Terrazzo floor and wall base detail

2. Using the **Detail Lines** tool from the *Annotate* tab, draw the floor line **8″** long using **Wide Lines**.

> *FYI:* The eight inch dimension is random and does not represent anything other than a portion of the floor surface.

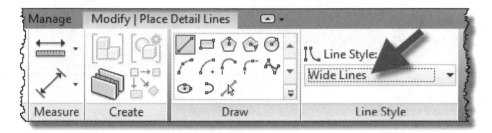

3. Draw a vertical line, also **8″** long, using the **Medium Lines** style. This line should be about 2″ from the right edge of the floor line (just so your detail is generally proportional to the one presented in the book).

Next, you will offset the two lines just drawn to quickly create the terrazzo floor and base.

4. Select **Modify → Offset** on the *Ribbon*.

5. On the *Options Bar*, enter an offset value of **3/8″**.

6. Pick the floor line when the preview line appears above the horizontal line.

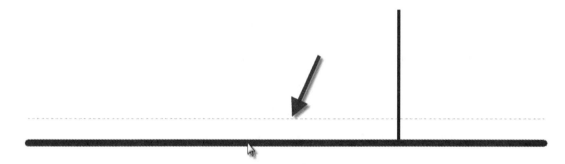

7. Now **Offset** the vertical line **3/8″** to the left.

Your drawing should now look like the image to the right (less the arrows). Next you will change the top horizontal line to a lighter line weight, offset it up 6″ to create the top of the wall base and then use the *Fillet Arc* feature to round off the corners.

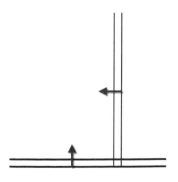

8. Select the top horizontal line and change the *Line Style* from *Wide Lines* to **Medium Lines** via the *Ribbon*.

9. **Offset** the top horizontal line upward **6″** inches.

10. Select the new horizontal line and then drag its right end grip over to the vertical line.

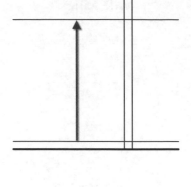

Next you will trim and round the corners in one step.

11. Select the **Detail Line** tool from the *Annotate* tab.

    a.   Select **Fillet Arc** from the *Draw* panel.
    b.   Set the *Line Style* to **Medium Lines**.
    c.   Check and set the *Radius* to ¼″.

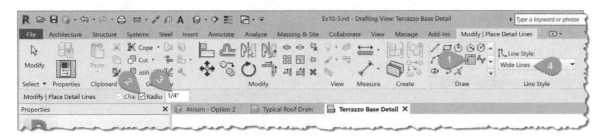

12. Click on the portion of the two lines you want to remain (see the two numbered clicks in the image to the right).

The line is now trimmed and an arc has been added.

13. Repeat these steps to round off the top edge of the wall base.

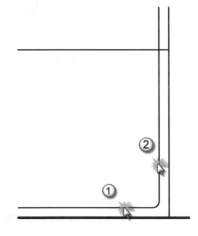

Next you will add a *Filled Region* to represent the terrazzo material with a pattern when viewed in a section. When creating the perimeter of a *Filled Region* you also specify a *Line Style* (similar to the *Detail Line* tool). In this case you will use *Thin Lines* for all but the left edge of the floor thickness. There you will change the *Line Style* to be an invisible line so as not to suggest a joint or the end of the flooring but rather that the flooring material continues.

14. Select **Annotate** → **Detail** → **Region** → **Filled Region** from the *Ribbon*.

    a.   Select the **Pick Lines** option from the *Draw* panel.

    b.   Select **Thin Lines** from the *Lines Style* panel.

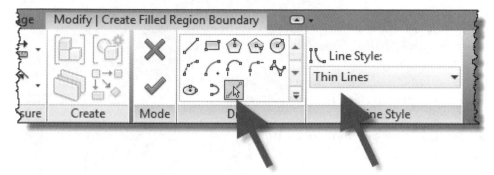

Selecting *Pick Lines* will allow you to quickly create the boundary of your *Filled Region* based on lines already drawn. So you simply click on a line rather than pick two points to define the start and endpoints of each edge. However, with the *Line* and *Arc* options you could also snap to the endpoints of the previously drawn linework. The one drawback to using the *Pick Lines* option is you will have to trim a few corners because the *Filled Region* tool requires that a clean perimeter be defined (similar to the floor and roof tools).

15. Pick the five lines and two arcs which define the edges of the floor and wall base.

16. Switch to the **Line** option in the *Draw* panel and then set the *Line Style* to **<Invisible Lines>**.

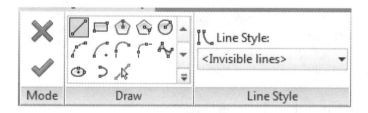

While in *Sketch Mode* (i.e., the green check mark and red X are visible) for the *Filled Region* tool, the invisible lines are not actually invisible. This allows you to select and modify them as needed.

17. Draw a line to close off the open edge of the flooring on the left hand side.

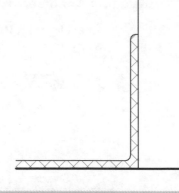

18. Use the **Trim** tool to clean up the two corners where the lines run past.

19. Click the **green check mark** to finish the *Filled Region*.

Your drawing should look like the one shown to the right. The default pattern is a cross hatch. You will change this next.

20. Select the *Filled Region*; you must click on one of the edges (and you may need to use *Tab*).

21. Expand the **Type Selector** to see the options currently available.

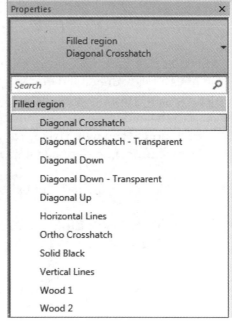

Looking at the list, we notice an option for concrete is not listed (which is what we decided we want). Next you will learn how to add this.

22. Press **Esc** to close the *Type Selector*.

23. Click **Edit Type**.

24. Click **Duplicate**.

25. Enter **Concrete** for the name.

26. Click in the *Foreground Fill Pattern* field and then **click the icon** that appears to the right.

27. Select **Concrete** from the list and click **OK**.

Notice the option to hatch the fill background opaque or transparent and the line weight setting.

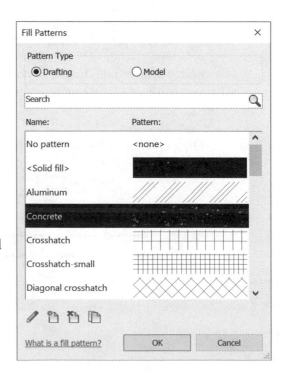

28. Click **OK** to close the *Type Properties*.

Your terrazzo now has a concrete pattern. You can import additional fill patterns (using any AutoCAD hatch pattern file) or create custom ones with specific line spacing. This *Filled Region* tool can be used in floor plans as well; maybe you want to highlight the corridors or private office areas. When a *Filled Region* is selected the square foot area is listed in the *Properties Pallet*. Thus, you could quickly create a *Filled Region* just to list the area and then delete the *Filled Region*.

The last thing you will do is add the notes and dimensions. These will be the correct scale based on the *View Scale* setting (which should be 3″ = 1′-0″). Once you place the dimension you will learn how to adjust the dimension style so the 0′ does not show up.

29. Add the dimension and two notes as shown.

30. Select the dimension.

31. Select **Edit Type**.

32. Click the button to the right of **Units Format**.

33. Uncheck **Use Project Settings**.

34. Check **Suppress 0 feet**.

35. Click **OK** twice to close both open dialog boxes.

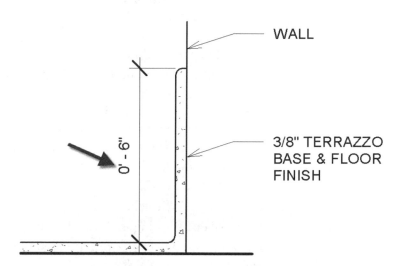

WALL

0′ - 6″

3/8″ TERRAZZO BASE & FLOOR FINISH

Your dimension should now only say 6″ rather than 0′-6″. Because you changed this in the *Type Properties*, all dimensions will have this change applied (both previously saved and new). If you want to have both options, you would first need to *Duplicate* the dimension type, which is similar to creating new wall, door and window types!

The last thing you will look at is adjusting the arrow style for the notes.

36. Select one of the notes.

37. Click **Edit Type**.

38. Change the *Leader Arrowhead* to **Arrow Filled 30 Degree**.

39. Click **OK**.

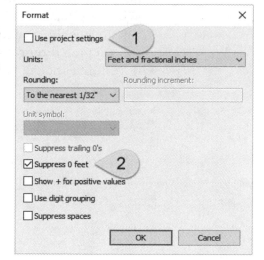

Your detail should now look similar to the one presented at the beginning of this exercise. This detail can now be placed on a sheet (covered in chapter 12). You can also export this detail and save it in a detail library that you and others in your firm can utilize. To export the detail, simply right-click on the view name in the *Project Browser* and select *Save to new File*. Place the file on a server so everyone can get at it. To load this file into another project select **Insert → Insert from File → Insert Views from File**.

Now you will draw another detail using the same tools and techniques. Keep in mind, every detail needs to go in its own *drafting view*. This is required for Revit to manage the reference bubbles on the sheets.

40. Create a new **drafting view**:
    a. *Name:* **Floor Transition Detail**
    b. *Scale:* **3″ = 1′-0″**

41. Draw the detail per the following guidelines:
    a. The detail will be plotted at 3″=1′-0″ (this determines the text and leader size).
    b. Tile pattern is **Diagonal Up–Small**; this requires a new *Fill Region* style (see the previous exercise for more information).
    c. The grout (i.e., area under tile) is to be hatched with **Sand – Dense**; this also requires a new *Fill Region* style.
    d. Draw the tile ¼″ thick and 4″ wide.
    e. The grout is ¼″ thick.
    f. The resilient flooring is shown ⅛″ thick.
    g. The solid surface (e.g., Corian) threshold is 1⅞″ wide; draw an arc between the two floor thicknesses.
    h. Hatch the threshold with the solid hatch:
        i. **Duplicate** the **Solid Black** *Filled Region* style.
        ii. Name the new style **Solid Gray**.
        iii. Set the **foreground pattern color** to a light gray (RGB color 192).
    i. The bottom concrete floor line is to be the heaviest line.

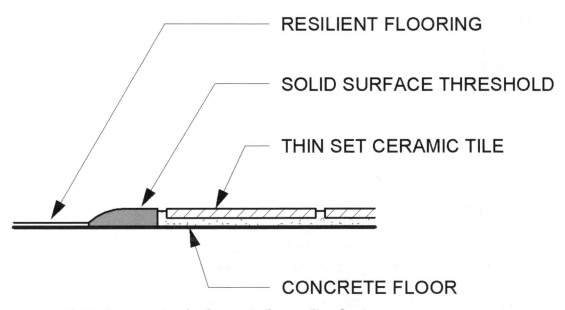

RESILIENT FLOORING

SOLID SURFACE THRESHOLD

THIN SET CERAMIC TILE

CONCRETE FLOOR

FIGURE 10-5.8 Floor transition detail: ceramic tile to resilient flooring

The previous drawing would typically occur in a door opening and the location of the door would also be shown in the detail. This lets the contractor know that the threshold is to occur directly below the door slab.

Base Cabinet with Drawers

This section will dive right into drawing cabinet details. These are often based on industry standard dimensions so many of the dimensions and material thicknesses can be omitted (assuming the project manual/specification covers this).

42. Create a **drafting view**:

    a. *Name:* **Base Cabinet Detail – Drawers**
    b. *Scale:* **1″ = 1′-0″**

Revit provides many *Detail Components* which aid in creating 2D details. Things such as side views of bar joists, section views of steel beams and angles, and more are available in the *Detail Component* library. The detail below takes advantage of three *Detail Components* which ship with Revit: particle board, lumber and the countertop. The only things drawn with the *Detail Line* tool are the tops of the drawers, the drawer pulls (i.e., handles) and the heavy wall/floor lines.

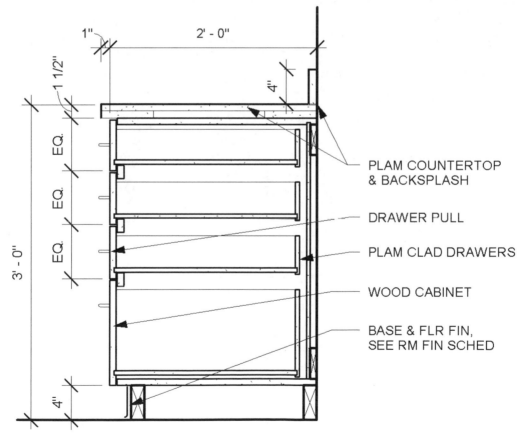

FIGURE 10-5.9 Cabinet section (with drawers)

43. Using the *Detail Line* tool, draw the floor and wall line, shown in the image to the right, using the **Wide Lines** style. (Do not add the dimensions.)

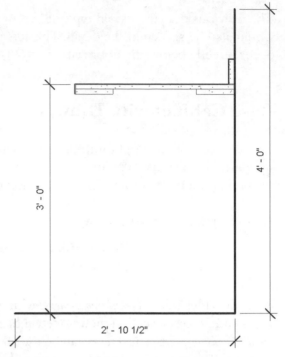

Next, you will load the countertop *Family* from the *Detail Component* library.

44. Select **Annotate → Detail → Component → Detail Component** from the *Ribbon*.

45. Click **Load Family** from the *Ribbon*.

46. Click the **Detail Library** shortcut on the left (or just open the *Detail Items* folder).

47. Now, browse to *Div 12-Furnishings → 123000-Casework →123600-Countertops*.

48. Select **Countertop-Section.rfa** and then click **Open**.

Countertop-Section with type *24″ Depth* is current, in the *Type Selector*, and ready to be placed.

49. Place the countertop as shown:
    a. Aligned with wall
    b. 3′-0″ above the floor (dimension to the line second from the top; the open heavy line is an exaggeration to highlight the added plastic laminate surface).
    c. With the countertop selected, adjust the values in the *Properties Palette*:
        i. *Backsplash Depth:* **0′ 1″**
        ii. *Counter Depth:*   **2′-1″**
        iii. *Thickness:*   **0′ 1⅝″**
    d. Do not add dimensions yet.

Be careful not to click on the grips when the countertop is selected as this will adjust its dimensions; this is because the values are associated with an *instance parameter* rather than a *type parameter*.

Next, you will load and place the 2x lumber. The two on the floor are 4″ high, which are cut down from a 2x6. So you will load a 2x4 *Family* and then create a duplicate and adjust the height from 3½″ to 4″.

50. Per the step just covered, load **Nominal Cut Lumber-Section.RFA** from the following location: *Detail Item → Div 06-Wood and Plastic → 061100-Wood Framing*.

To minimize the number of *Types* for the lumber *Family*, you are presented with the *Specify Types* dialog. This lets you pick just the sizes you want – more can be added later.

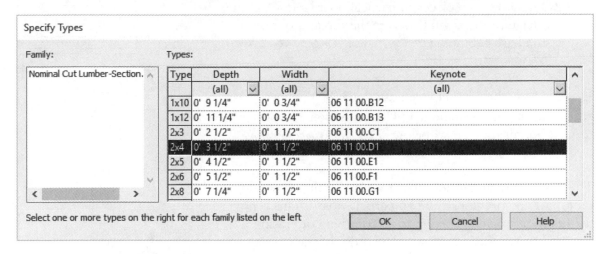

51. Hold the **Ctrl** key and select **1x4** and **2x4** and then pick **OK**.

52. Select **2x4** in the *Type Selector*.

53. Click **Edit Type**.

54. Click **Duplicate**.

55. Enter **2x4 Base Cabinets** for the name.

56. Change the *Height* from 3½″ to **4″**.

57. Click **OK** to close the *Type Properties*.

58. Place the two **2x4 Base Cabinet** detail components on the floor as shown in the image to the right. (Do not add the dimension.)

59. Place the two **1x4** components as shown.

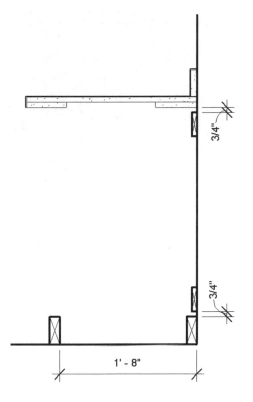

Next, you will load the detail component used to draw particle board. This *Family* is somewhat like the *Wall* tool. You pick two points and two lines and a fill pattern is generated. The *Type Selector* also has a number of standard thicknesses ready to use.

60. Per the step just covered, load **Particle board-Section.rfa** from the following location: *Detail Items* → *Div 06-Wood and Plastic* → *061600-Sheathing*.

61. Select the **3/4″** type from the *Type Selector*.

62. Draw the two pieces of particle board shown in the image to the right; these are the top and bottom of the base cabinet. *TIP: Use the space bar to flip the thickness while drawing, if needed.*

63. Draw the cabinet back; use the **3/8″** thickness option and extend it **1/4″** into the top and bottom boards. See image below. *TIP: Draw temporary detail lines so you have a place to pick if needed. Delete them when done.*

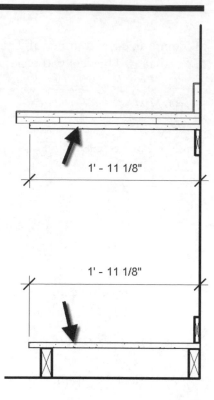

The newest drawn particle board automatically shows up on top of any previously drawn particle board. This is how the notch is created. There was no trimming or erasing required. If you need to change the order of the overlap, simply select the component and use the *Arrange* options on the *Ribbon*.

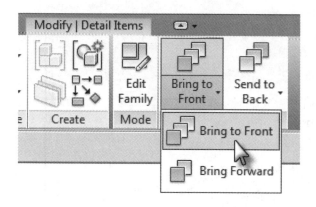

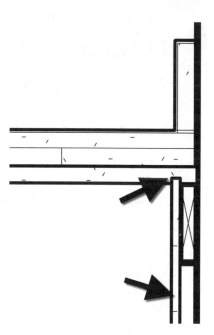

64. Draw the rest of the base cabinet using the techniques previously covered and the following information:

    a. The "EQ" dimensions are 6″.

    b. The drawer bottoms and backs are all ½″ particle board; everything else is ¾″.

    c. Use **Thin Lines** for the pulls and the top edge of the side drawer panel (seen in elevation in **Figure 10-5.9**). Use **Medium Lines** for the rubber base.

    d. Add dimensions not provided which can be approximated; make it look like the image in the book as much as possible.

    e. Add all the notes and dimensions shown on the first image only.

Some things you should know about detailing

In the previous steps you drew a typical detail showing a standard base cabinet with drawers. Interior designers occasionally draw these details, but more often, they simply review them for finishes.

The countertop material needs to match that which is specified in the **Project Manual** and intended for the project. For example, a PLAM (i.e., plastic laminate) countertop would not be appropriate in a laboratory where chemicals would be used.

The base of the cabinet typically has the same wall base as the adjacent walls. For example, if the walls have a rubber base (also referred to as a resilient base), the toe-kick area of the base cabinet would also receive a rubber base; this is the type of base shown in the previous detail. If the project only had ceramic tile wall base, you would show that.

The **notes for details** (or any drawing) should be simple, generic and to the point. Notice in previously drawn detail that the note for the base does not indicate whether the base is rubber, tile or wood. This helps avoid contradictions with the room finish schedule. The note simply says the cabinet and floor are to receive a finish and instructs the contractor to go to the *Room Finish Schedule* to see what the finish is. This is particularly important in buildings that have several variations of floor and base finishes.

Notes should not have any **proprietary or manufacturers' names** in them either. For example, you should not say "Sheetrock" in a note because this is the brand name; rather, you should use the generic term "gypsum board." Similarly, you would use the term "solid plastic" rather than "Corian" when referring to countertops or toilet partitions. In any event, whatever term you use on the drawings should be the same term used in the Project Manual!

One last comment: the *Construction Documents* set should never have **abbreviations** within the drawings that are not covered in the *Abbreviations* list, usually located on the title sheet. *Construction Documents* are legal, binding documents, which the contractor must follow to a "T." They should not have to guess as to what the designers meant in various notes all over the set of drawings. It is better to spell out every word, if possible, only abbreviating when space does not permit. You would not want a bunch of abbreviations in your bank loan or mortgage papers you were about to sign! Plus, non-documented abbreviations would probably not have much merit before a judge or arbitrator in the case of a legal dispute!

65. Using the same steps just covered, draw the base cabinet shown below. Name the drafting view **Base Cabinet Detail**. The *Scale* is also **1″ = 1′-0″**.

> ***TIP:*** *Duplicate the Base Cabinet Detail – Drawer view and modify it to be this detail; much of this detail is exactly the same.*

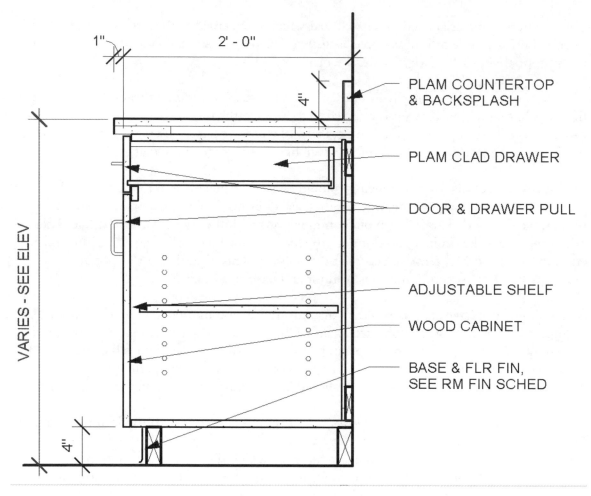

FIGURE 10-5.10 Cabinet section (door + drawer)

Cabinet details do not need to have every nook-and-cranny dimensioned because they are very much a standard item in the construction industry. Furthermore, the Project Manual usually references an industry standard that the contractor can refer to for typical dimensions, thicknesses and grades of wood.

The vertical dimension shown in the cabinet detail above says "VARIES - SEE ELEV." This notation, rather than an actual number, allows the detail to represent more than one condition. The interior elevations are required to have these dimensions, which may be the standard 36″ or the lower handicap-accessible height.

Ceiling Detail

Next, you will draw a typical recessed light trough detail at the ceiling. This is used often in commercial/public toilet rooms.

66. Create a new drafting view:

    a. *Name*: Toilet Room Ceiling Detail

    b. *Scale:* 1½″ = 1′-0″

    c. *See the next page for additional notes.*

67. Load the following *Detail Components* into your project:

    a. Div 09-Finishes → 092000-Plaster and Gypsum Board → 092200-Supports → 092216-Non-Structural Metal Framing

        i. **Interior Metal Runner Channels-Section.rfa**

        ii. **Interior Metal Studs-Side.rfa**

    b. Div 09-Finishes → 092000-Plaster and Gypsum Board → 092900-Gypsum Board → **Gypsum Wallboard-Section.rfa**

    c. Div 09-Finishes → 095000-Ceilings → 095100-Acoustical Ceilings

        i. **Suspension Wall Angle-Section.rfa**

        ii. **Suspended Acoustic Ceiling-Square Edge-Section.rfa**

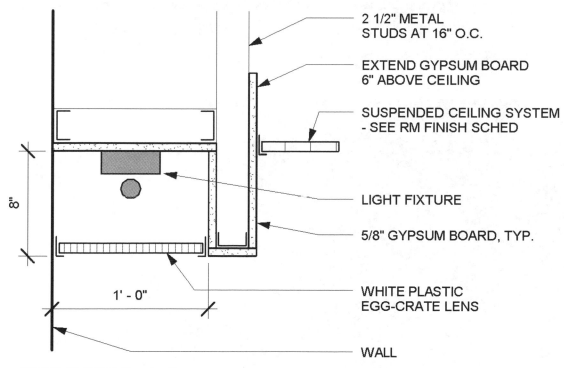

FIGURE 10-5.11 Ceiling detail

TIP: The notes in details should align on one edge as shown. Additionally, the leaders should not cross dimensions or other leaders unless it is totally unavoidable.

When using the *Detail Components* you will occasionally run into a few challenges getting things to look exactly the way you want them. For example, many firms have traditionally shown gypsum board as if it were continuous – not indicating the joints. Most contractors understand this and the designer usually does not want to imply the gypsum board be installed in a specific way or order. However, the *Detail Component* feature forces these edge lines to appear. Unfortunately we **cannot** use the *Linework* tool to set some of the lines in a *Detail Component* to be invisible.

Another problem you will run into, when using *Detail Components*, is the fact that everything is drawn true to life size. You may think, "Isn't that a good thing?" Usually it is, but some things often need to be exaggerated so they are legible on the printed page. Take the metal stud runner for example; when placed next to the *Gypsum Board* detail component it is totally hidden because the gypsum line weight is heavier than the runner stud. The best thing to do is to edit the *Family* so the runner stud has a heavier line weight and its thickness is exaggerated inward (you don't want to change the overall width). However, this type of change is outside the scope of this exercise, so you will do the following:

68. Erase the runner stud *Detail Components* and draw **Detail Lines** using **Medium Lines** – in bound from the gypsum board.

69. Make the light fixture **4½″ x 1¾″** with a **1″** circle for the light. Use the **Solid Gray** *Fill Region* previously created.

70. Sketch the egg-crate lens using **Detail Lines**. Create a new *Fill Region* using the **Vertical-Small** fill pattern.

71. Add the notes and dimensions shown.

Fixed Student Desk at Raised Seating Classroom

This is a continuous student desk in a classroom setting.

72. Create a new **Drafting View**:
    a. *Name:* **Fixed Student Desk**
    b. *Scale:* **1½" = 1'-0"**
    c. *See the next page for additional comments*

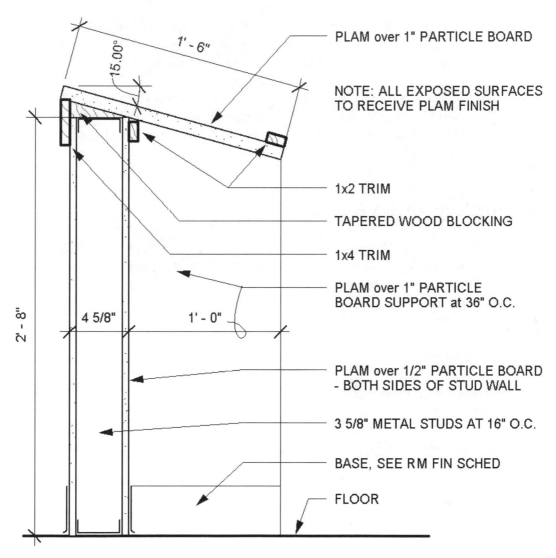

FIGURE 10-5.12 Fixed desk detail

73. Load the following detail component:
    a. Div 06-Wood and Plastic\062200-Millwork\Standard Millwork-Section.rfa
        i. Load the **1x2** and **1x4** sizes.

74. Develop the fixed student desk following these guidelines:
    a. All particle board and trim to be Detail Components.
    b. Tapered wood blocking to have **Wood 2** *Fill Pattern*.
    c. The curvy line pointing to the 1'-0" dimension should be drawn like this:
        i. First add the "PLAM" note using a regular leader;
        ii. Use the **Detail Line** tool;
        iii. Select the **Spline** *Draw* option;
        iv. Set the *Line Style* to **Thin**;
        v. Sketch the curvy line starting at the corner of the leader.
    d. Add a horizontal Detail Line at the top of the sloped work surface so you have something to pick when adding the angle dimension.

As you can see, some of the *Detail Components* have odd line weights when placed side-by-side. Both the wood trim and particle board are in section so they should be the same line weight. You would have to edit the family to make this change, which will not be covered at this time.

Using the Keynotes Feature

The content that ships with Revit, both 2D and 3D, has a default keynote value assigned to it. Keynotes are used to save room and make details look neater; it is a reference number rather than a full note, and then an adjacent legend lists what each number means. This legend is for all the details on a sheet. You will learn how this works next. You will make a copy of the *Fixed Student Desk*, add keynotes and then create a keynote legend.

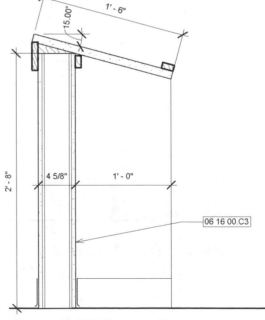

75. Right-click on the **Fixed Student Desk** item in the *Project Browser*.

76. Select **Duplicate → Duplicated with Detailing**.

77. Rename the new view **Fixed Student Desk – Keynotes**.

78. **Erase all the notes**, but leave the dimensions.

79. Select **Annotate → Tag → Keynote** from the *Ribbon*.

FIGURE 10-5.13 Keynote added

80. With the *Keynote* tool active, click the ½″ particle board shown in Figure 10-5.13; you will see it highlight just before selecting it.

81. Click two additional points to define a leader and text location, just like placing text with a leader.

You now have a keyed note placed in your drawing. This only works on *Detail Components* (in drafting views) and not *Detail Lines* as they are too generic. Next, you will see where this keynote notation is coming from.

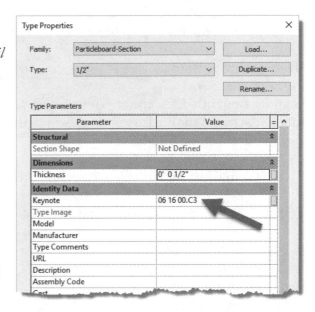

82. Select the ½″ particle board; go to its *Type Properties*.

Notice the *Keynote* value listed. This was defined in the *Family* you loaded.

Now you will view the *keynote text file* so you can see how a family could be changed to "mean" something else. You might change a family directly or create a duplicate *Type* first.

83. Click in the cell listing the *Keynote*.

84. Click the **small icon** that appears to the right (not in the "equals" column).

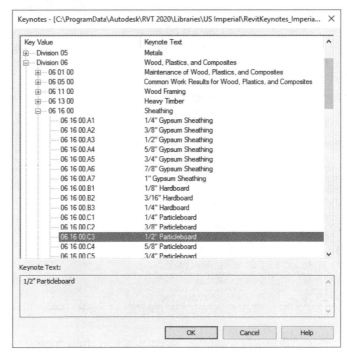

You now see a rather extensive listing of keyed notes. Take a minute to explore the various sections and descriptions for the keynote references.

Notice the path listed at the top. This is the location of the text file being used for the keynotes. The path to this file is set via the *Keynoting Settings* icon located in the *Tag* panel expanded area on the *Annotate* tab.

85. After reviewing the keynote text file click **Cancel**.

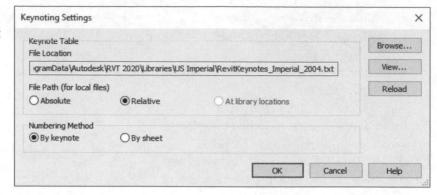

86. Click **OK** to close the *Type Properties* dialog.

The last thing to learn is how to create the **Keynote Legend**. This legend can be added to any sheet with keynotes. If the "By sheet" option is selected in the dialog box shown to the right, only keynotes actually found on that specific sheet will be listed.

87. From the *View* tab, select **Legend → Keynote Legend**.

88. Click **OK** to accept the name **Keynote Legend**.

89. Click **OK** to accept the default properties and to create the legend.

You should now see the *Keynote Legend* shown below.

FYI: *Make sure the* numbering method *is still set to* By keynote *in order to see all the keynotes at this time.*

Later, in Chapter 12, you will learn how to place views on sheets.

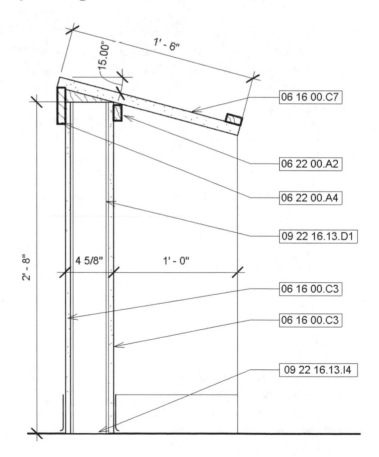

Keynote Legend	
Key Value	Keynote Text
06 16 00.	1/2" Particleboard
06 16 00.	1" Particleboard
06 22 00.	1x2 Wood Trim
06 22 00.	1x4 Wood Trim
09 22 16.	3 5/8" Metal Stud Framing
09 22 16.	3 5/8" Metal Runner
09 29 00.	5/8" Gypsum Wallboard

90. Add the remaining keynotes and then **Save**.

Finally, you can create elevations and sections that reference a drafting view rather than a true view of the model. This is another one of those places where you are breaking the intelligence of Revit's drawing sheet and number coordination.

To do this, you select the *Elevation* or *Section* tool, and rather than picking points in the drawing right away, you select a view that already exists in the project from the *Options Bar*. Even though you have not placed any roof drains, you could switch to the **Roof** plan view and add a section mark that references the roof drain detail.

The following steps do not need to be performed at this time:

91. Switch to the **Roof** floor plan view.

92. Select the **Section** tool on the *View* tab.

93. On the *Ribbon/Options Bar* settings (Figure 10-5.14):

    a. *Type Selector:*   **Detail View: Detail**

    b. *Reference other view:*   **check**

    c. *Reference other view drop-down list:* **Typical Roof Drain Detail**

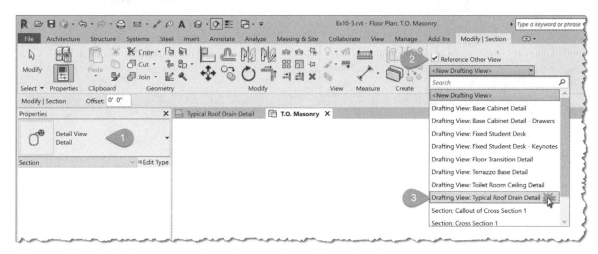

FIGURE 10-5.14 Placing a section that references the roof drain detail

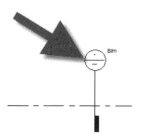

94. Pick two points, roughly as shown in the image to the right.

Try double-clicking on the blue bubble head; it should bring you to the roof drain detail. Once the detail is placed on a sheet, the bubble head will be automatically filled out!

95. Save as **10-5.rvt**.

Self-Exam:

The following questions can be used as a way to check your knowledge of this lesson. The answers can be found at the bottom of this page.

1. The controls for the section tag (when selected) are similar to the controls for the elevation tag. (T/F)

2. In large-scale <u>elevations</u> (and areas elevated within a section), Revit displays the masonry coursing. (T/F)

3. In large-scale <u>sections</u> (i.e., wall sections), Revit displays the masonry coursing in addition to the material hatching. (T/F)

4. The "Crop Region" is represented by a black rectangle in the section view. (T/F)

5. Use the _____ tool to reference a larger section off a building section.

Review Questions:

The following questions may be assigned by your instructor as a way to assess your knowledge of this section. Your instructor has the answers to the review questions.

1. The visibility of the crop region can be controlled. (T/F)

2. It's not possible to draw a leader (line with arrow) without placing text. (T/F)

3. When a section tag is added to a view, all the other related views automatically get a section tag added to it. (T/F)

4. It is possible to modify objects (like doors, windows and ceilings) in section views. (T/F)

5. You cannot adjust the "depth of view" using the crop region. (T/F)

6. What is the first thing you should do after placing a section tag?
   _____

7. If the text appears to be excessively large in a section view, the view's
   _____ is probably set incorrectly.

8. The abbreviation M.O. stands for _____.

9. Describe what happens when you double-click on the section bubble:
   _____ .

10. Revit provides _____ different leader options within the text command.

Lesson 11
INTERIOR DESIGN:

This lesson explores the various "features," if you will, of a floor plan, such as toilet room layouts (i.e., fixtures and partitions), cabinets and casework (e.g., reception counters and custom cabinets). Additionally, you will look at placing furniture into your project.

Exercise 11-1:
Toilet Room Layouts

Toilet room layouts involve placing water closets (toilets), toilet partitions and sinks. These rooms have many code issues related primarily to handicapped accessibility. These codes vary from state to state (and even city to city).

You will start this exercise by loading several components to be placed into your project.

1. Open ex10-5.rvt and **Save As ex11-1.rvt**.

2. Select **Load Family** from the *Insert* tab and load the following items into the current project:

    Local Files *(i.e., on your hard drive)*
    a. Plumbing\Architectural\Fixtures\Water Closet\
        Toilet-Comercial-Wall-3D.rfa
    b. Plumbing Fixtures\Architectural\Fixtures\Urinals**Urinal-Wall-3D**

    Online Files *(i.e., SDC Publications, see inside front cover)*
    c. **Sink-Wall-Rectangular**
    d. **Grab Bar-3D**
    e. **Toilet Stall-Accessible-Front-Braced-3D**
    f. **Toilet Stall-Braced-3D**
    g. **Urinal Screen-3D**

These files represent various families that will be used to design the toilet room. It is possible to create custom families for non-typical conditions.

3. Switch to **Level 1** view.

The next step will be to place the toilet stalls.

4. Using the **Component** tool, pick **Toilet Stall-Accessible-Front-Braced-3D: 60″ x 60″ Clear** from the *Type Selector*.

5. Zoom in to the toilet rooms (North of the elevator).

6. **Place** the toilet stall as shown and then move into place using the **Move** tool and your snaps or the *Align* tool (Figure 11-1.1).

Once you move the toilet stall North, you will have your first stall in place.

Next, you will place two standard size toilet stalls.

7. Place two toilet stalls: **Toilet Stall-Braced-3D: 36″ x 60″ Clear** as shown in Figure 11-1.2.

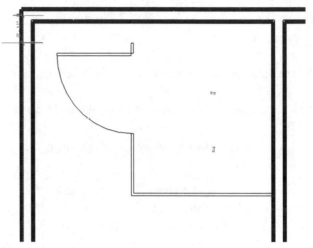

FIGURE 11-1.1 Accessible Toilet stall

As with most projects, you will need to modify the model as you develop the design. In this case we notice that toilets that are back-to-back and stacked on each floor will require a thicker wall to accommodate the fixture brackets (W.C.'s are not hung on the wall by light gauge metal studs) and larger piping. You will make this adjustment next.

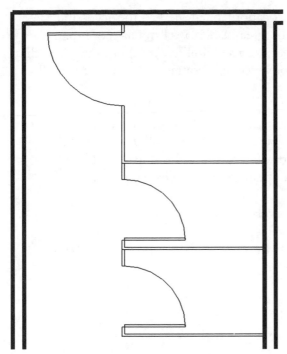

FIGURE 11-1.2 Toilet stalls placed

8. Select the middle wall and the West wall and **Move** them **6″** to the West.

9. **Move** the far East wall (of the two toilet rooms) **6″** to the east.

10. Add an additional **4⅞″** gyp. bd. wall as shown in **Figure 11-1.3**.

TIP: Make sure wall height and base offset are correct.

11. Modify the North wall of the elevator shaft to have furring and gyp. bd. on the toilet room side.

TIP: Use the wall type you created for the stair shafts.

12. Add the additional components as shown in **Figure 11-1.3**.

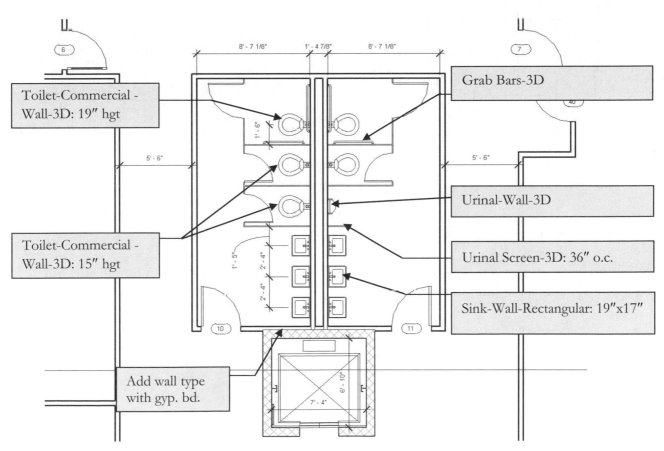

FIGURE 11-1.3 Toilet room layout

As mentioned previously, building codes vary by location. The toilets in the accessible stall area are usually mounted higher than the typical fixtures. When a room has more than one urinal, one is usually required to be mounted lower for accessibility. Another example is that Minnesota requires a separate vertical grab bar above the horizontal grab bar on the wall next to the toilet.

You will now copy the revised walls and toilet room layout to the other levels. The elevator shaft extends through each floor, so you will not have to copy that wall. Looking at the upper levels you can see the revised elevator shaft wall and the old wall layout (Figure 11-1.4). It will be easier to delete the stud walls rather than modify the existing walls.

13. Delete the walls, per **Figure 11-1.5**, for Levels 2 and 3.

14. **Copy** the walls and toilet room layout (for Level 1) to the clipboard and **Paste Aligned** to Levels 2 and 3.

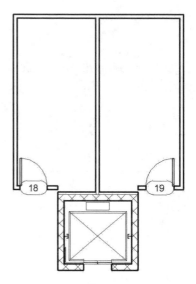

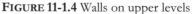

FIGURE 11-1.4 Walls on upper levels   **FIGURE 11-1.5** Walls deleted

Interior Elevation View

Next, you will set up an interior elevation view for the Men's Toilet Room. You will also add a mirror above the sinks in elevation view.

15. Switch to **Level 1** and place an interior **Elevation** tag looking towards the west wall (wall with fixtures on it); see Figure 11-1.6.

16. Rename the new view to **Men's Toilet – Typical** in the *Project Browser*.

17. Switch to the new view. Adjust the *Crop Region* so the concrete slab is not visible. Your view should look like Figure 11-1.7.

You may see the building section reference as shown in Figure 11-1.7. This would not typically be shown in an interior elevation view, especially because it does not intersect the elevation view. You will remove the reference in the next step. You cannot simply delete it because that will remove it from all views and delete the section.

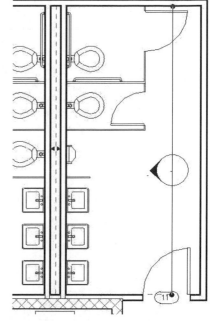

FIGURE 11-1.6 Elevation tag added

18. If you see the building section tag, click on the section reference to select it and then right-click and pick **Hide in View** → **Elements** from the pop-up menu (Figure 11-1.8).

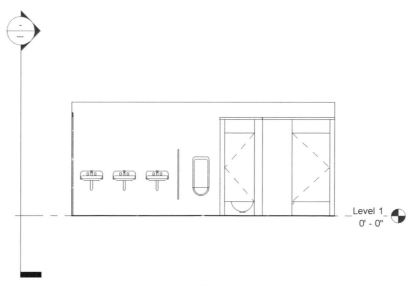

FIGURE 11-1.7 Men's Toilet – Typical view

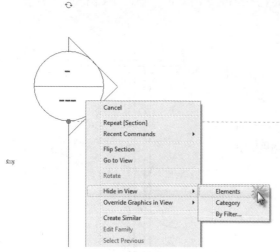

FIGURE 11-1.8 Hide annotation

19. Load component *Specialty Equipment\ Toilet Room Specialties* **Mirror.rfa** from the Online Revit library.

20. While in the interior elevation view, place a **72″ x 48″ Mirror** on the wall above the sinks. Use the *Align* tool to align the mirror with the middle sink (Figure 11-1.9).

21. Add the notes and dimensions shown in Figure 11-1.9. Adjust the heights and locations of the fixtures/components as required.

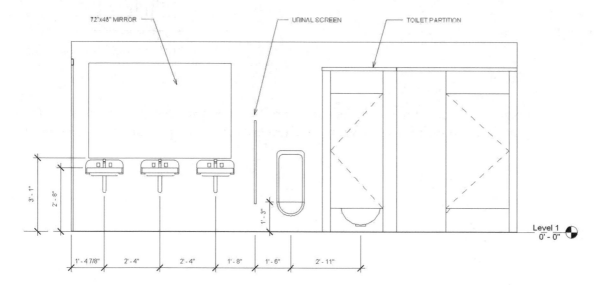

FIGURE 11-1.9 Updated interior elevation

FYI: *Keep in mind that many of the symbols that come with Revit (or any program for that matter) are not necessarily drawn or reviewed by an architect. The point is that the default values, such as mounting heights, may not meet ADA, national, state or local codes. Items such as the mirror have a maximum height off the floor to the reflective surface that Revit's standard components may not comply with. However, as you apply local codes to these families, you can reuse them in the future.*

Adjusting the Reflected Ceiling Plan

Because you added a wall in the East toilet room, the definition of the room that the reflected ceiling plan uses is incorrect. You will adjust that next. You will have similar problems on Levels 2 and 3 if you added ceilings there because you deleted walls and then pasted new walls.

22. Switch to **Level 1 RCP** (Figure 11-1.10).

23. Hide the Interior Elevation tag from this view (via right-click and Hide in View).

24. Delete the ceiling in the Men's Toilet room.

25. Place a new ceiling to fit within the room.

26. Select the ceiling grid in the atrium, and then pick **Edit Boundary** to adjust the reference lines for the perimeter of the ceiling grid in the atrium area (Figure 11-1.11).

> *FYI: You could have used this method for Steps 24 / 25 as well.*

27. Correct the ceilings on Levels 2 and 3.

28. **Save** your project as **ex11-1.rvt**.

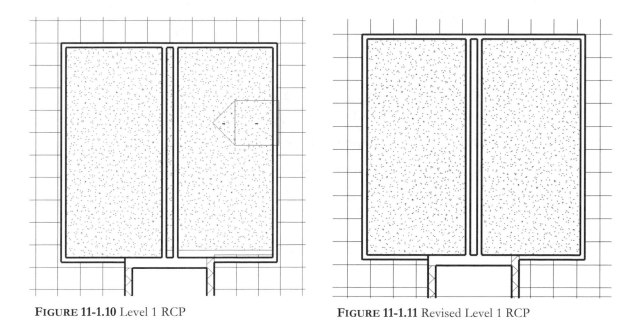

FIGURE 11-1.10 Level 1 RCP   **FIGURE 11-1.11** Revised Level 1 RCP

Exercise 11-2:
Cabinets

In this exercise you will look at adding cabinets and casework to your project. As usual, Revit provides several families to be placed into the project.

Placing Cabinets

You will add base and wall cabinets in a break room on Level 1.

1. Open ex11-1.rvt and **Save As ex11-2.rvt**.

2. Switch to **Level 1** view and zoom into the area shown in **Figure 11-2.1**.

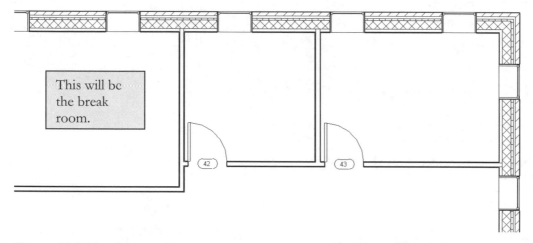

FIGURE 11-2.1 Level 1 – north-east corner

3. *Load* the following components into the project
   (*all local files – folder listed after filename*):

   a. **Counter Top w Sink Hole** (Casework\Counter Tops)

   b. **Base Cabinet-Double Door Sink Unit** (Casework\Base Cabinets)

   c. **Base Cabinet-Single Door** (Casework\Base Cabinets)

   d. **Base Cabinet-4 Drawers** (Casework\Base Cabinets)

   e. **Upper Cabinet-Double Door-Wall** (Casework\Wall Cabinets)

   f. **Sink Kitchen-Single** (*Plumbing\Architectural\Fixtures folder*)

   g. **Refrigerator** (*Specialty Equipment\Domestic folder*)

You are now ready to place the cabinets into your floor plan.

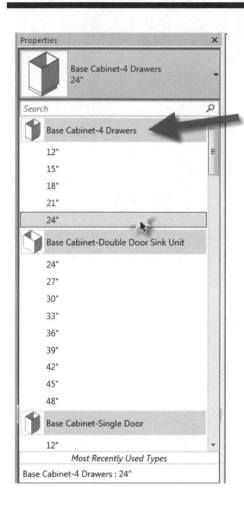

FYI: As with other components (i.e., doors and windows), Revit loads several types to represent the most valuable/useful sizes available. Cabinets typically come in 3″ increments with different types (i.e., single or double door unit) having maximum and minimum sizes.

4. With the *Component* tool selected, pick <u>*Base Cabinet-4 Drawers:*</u> **24″** from the *Element Type Selector*.

5. Place the cabinet as shown in **Figure 11-2.2**.

TIP: The control arrows are on the front side of the cabinet; the cursor is on the back. Press the spacebar to rotate while placing (i.e., before picking).

6. Place the other two base cabinets as shown in **Figure 11-2.3**, with a 24″ single door base cabinet in the middle and a 48″ sink base to the north end.

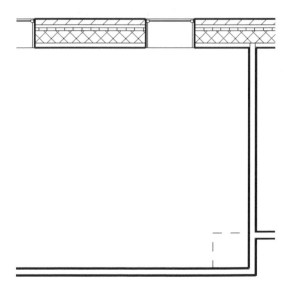

FIGURE 11-2.2 First cabinet placed

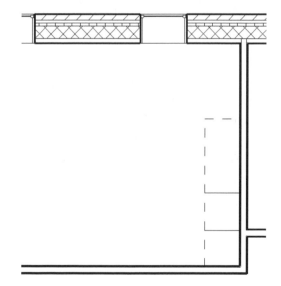

FIGURE 11-2.3 Three base cabinets placed

Add the remaining items as shown in Figure 11-2.4; be sure to use snaps.
(E.g., place sink to one side, use move and snap to a mid-point of the sink bowl and then to the mid-point of the same line representing the hole in the countertop).

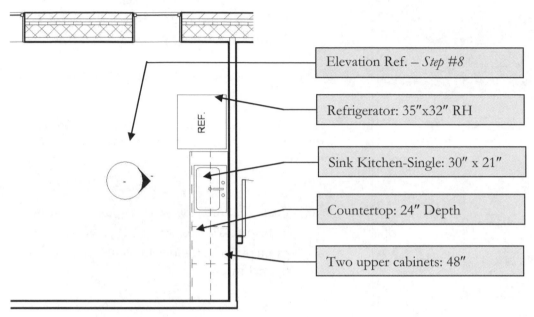

Elevation Ref. – *Step #8*

Refrigerator: 35″x32″ RH

Sink Kitchen-Single: 30″ x 21″

Countertop: 24″ Depth

Two upper cabinets: 48″

FIGURE 11-2.4 Completed plan view

7. Add an interior elevation tag to set up the interior elevation view (Figure 11-2.4).

8. Rename the new elevation view to **Break Room (east)**.

9. Switch to the new view, **Break Room (east)**.

10. Adjust the **Crop Region** so the slab on grade is not visible.

11. Your view should look like **Figure 11-2.5**.

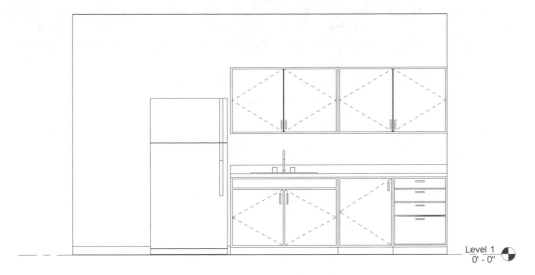

FIGURE 11-2.5 Interior elevation

You will add notes and dimensions to the elevation. You can also add 2D line work to the elevation.

12. Set the *View Scale* to ½″ = 1′-0″.

13. Add the notes and dimensions per **Figure 11-2.6**.

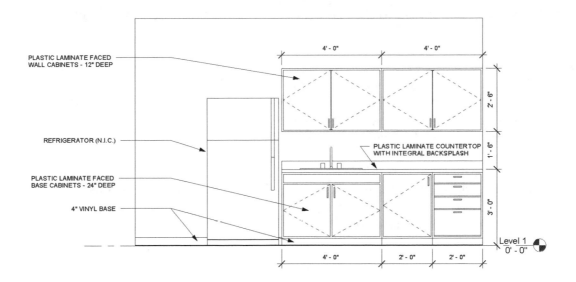

FIGURE 11-2.6 Interior elevation with annotations

14. Use the **Model Line** tool, on the *Architecture* tab, to draw the line on the wall behind the refrigerator indicating the vinyl base.

When you select the **Model Line** tool, Revit will ask you what plane you want to draw on (Figure 11-2.7). This will allow Revit to restrict all your line work to a particular plane. Otherwise you would not know exactly at what depth the lines would be drawn on.

15. Select **Pick a Plane** (Figure 11-2.7) and then pick the wall in the elevation view (use the TAB key and make sure the tool tip lists the wall before picking the plane).

> **TIP:** *If you are prompted to switch to a different view, you selected the wrong plane. Click cancel and try again.*

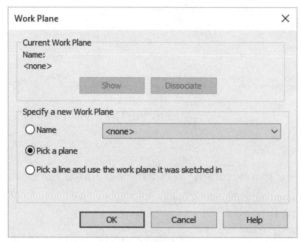

FIGURE 11-2.7 Work Plane prompt

16. Draw the line, and snap to the endpoint of the base cabinet toe kick.

> **FYI:** *The main difference between the Model Line tool on the Architecture tab and the Detail Line tool on the Annotate tab is this: any line work drawn with the Model Line tool will show up on other views which see that surface. On the other hand, any line work drawn with the Detail Line tool will only show up in the view in which they were created. So in this example, you may want to draw the lines using the Detail Line tool so the base does not show up in building and wall sections. Do not change this at this time, however.*

17. Select the line and in the *Type Selector*, change the *Line Style* to **Thin Lines**.

18. **Save** your project as **ex11-2.rvt**.

Exercise 11-3:
Furniture

This lesson will cover the steps required to lay out office furniture. The processes are identical to those previously covered for toilets and cabinets. Various manufacturers are beginning to provide Revit content; for example, take a look at www.haworth.com.

Loading the Necessary Families

1. Open ex11-2.rvt and **Save As ex11-3.rvt**.

2. Select the **Component** tool and load the following items into the current project:

   <u>**Local Files**</u> *(i.e., on your hard drive)*
   a. **Work Station Cubicle.rfa** (Furniture System)
   b. **Work Station Desktops** (Furniture System)
   c. **Sofa-Pensi** (Furniture\Seating)
   d. **Chair-Breuer** (Furniture\Seating)
   e. **Chair-Executive** (Furniture\Seating)
   f. **Chair-Task Arms** (Furniture\Seating)
   g. **Table-Round** (Furniture\Table)

   <u>**Online Files**</u> *(i.e., SDC Publications)*
   h. **Copier-Floor**

These files represent various families that will be used to design the offices.

TIP:
You can set the View mode for the Open dialog box (which is displayed when you click Load Family). One option is Thumbnail mode; this displays a small thumbnail image for each file in the current folder. This makes it easier to see the many symbols and drawings that are available for insertion. You can make the preview images larger by holding down the Ctrl key and then spinning the wheel on your mouse.

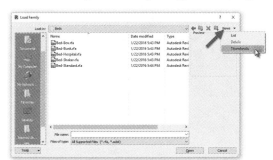
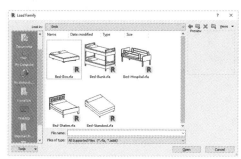

View set to List mode   View set to Thumbnail mode

Designing the Office Furniture Layout

3. Switch to the **Level 3** view.

4. Place the furniture as shown in Figure 11-3.1.

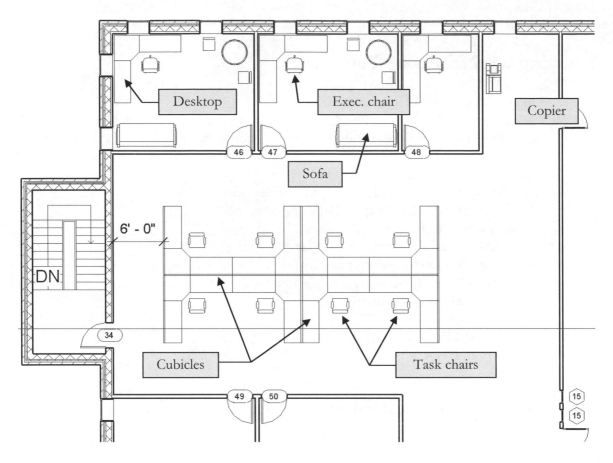

FIGURE 11-3.1 Level 3 – furniture layout

3D View of Office Layout

Next you will look at a 3D view of your office area. This involves adjusting the visibility of the roof and skylights.

5.   Switch to the *Default* **3D** view.

6.   Make sure nothing is selected and the *Properties Palette* is open.

7.   Click **Edit** next to the *Visibility/Graphic Override* parameter.

8.   Uncheck the *Roof* category and click **OK**.

The roof should not be visible now. However, you should still see the skylights floating in space. You will make those disappear next.

9.   Select one of the skylights floating above the office area.

10. Click the **Temporary Hide/Isolate** from the *View Control Bar*.

You should see the menu shown in **Figure 11-3.2** show up next to the *Temporary Hide/Isolate* icon. This allows you to isolate an object (so it's the only thing on the screen) or hide it (so the object is temporarily not visible).

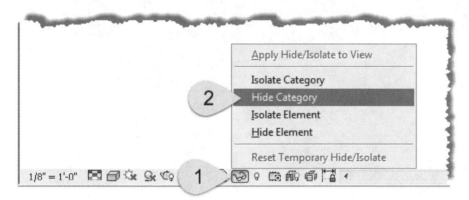

FIGURE 11-3.2 Hide/Isolate popup menu

11. Click **Hide Category** in the menu (Figure 11-3.2).

> **FYI:** *This makes all the skylights hide; you could leave it at "Hide Element" and select each skylight.*

12. Adjust your 3D view to look similar to **Figure 11-3.3** by clicking and dragging your mouse on the *ViewCube*.

You will now restore the original visibility settings for the 3D view.

13. Click the *Hide/Isolate* icon and then select **Reset Temporary Hide/Isolate** from the pop-up up menu.

    NOTE: The Temporary Hide/Isolate feature is just meant to be a temporary control of element visibility while you are working on the model. If you want permanent results, you can click the "Apply Hide/Isolate to View" option. Also, to the right of the Hide/Isolate icon is the Reveal Hidden Elements icon (the light bulb icon), which will clearly show any elements that have been previously hidden.

14. Reset the 3D view's visibility settings so the roof is visible.

15. **Save** your project as **ex11-3.rvt**.

    Notice the furniture and toilet rooms are represented in 3D.

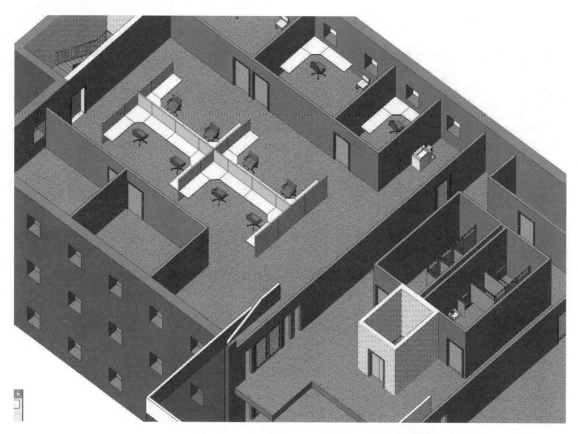

FIGURE 11-3.3 3D view with roof not visible

Online Content

A few locations on the internet provide additional content for use in Revit. Some are free and some are not. Hopefully more product manufacturers will start providing content based on the products they make, making it easier for people to include that manufacturer's product in their project (both the virtual and real projects).

The following sites also contain content that can be downloaded:
- www.revitcity.com
- http://revit.autodesk.com/library/html/ (old Revit content library)
- https://www.turbosquid.com/revit

You should occasionally search the internet to see if additional content becomes available. You can do an internet search for "revit content"; make sure to include the quotation marks. The rendering content, such as that offered by www.archvision.com, will be covered in Chapter 13.

Exercise 11-4:
Adding Guardrails

This lesson will cover the steps required to lay out guardrails. The steps are similar to drawing walls; you select your style and draw its path.

Adding a Guardrail to the Atrium

1. Open ex11-3.rvt and **Save As ex11-4.rvt**.

2. Switch to **Level 2** view.

3. Select **Architecture → Circulation → Railing → Sketch Path**.

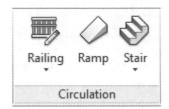

4. **Zoom** into the Atrium area (South of the elevator).

At this point you will draw a line representing the path of the guardrail. The railing is offset to one side of the line, similar to walls. However, you do not have the *Location Line* option as you do with the *Wall* tool, so you have to draw the railing in a certain direction to get the railing to be on the floor and not hovering in space just beyond the floor edge.

5. Draw a line along the edge of the floor as shown in **Figure 11-4.1**.

> ***TIP:*** *Select Chain from the Options Bar to draw the railing with fewer picks.*

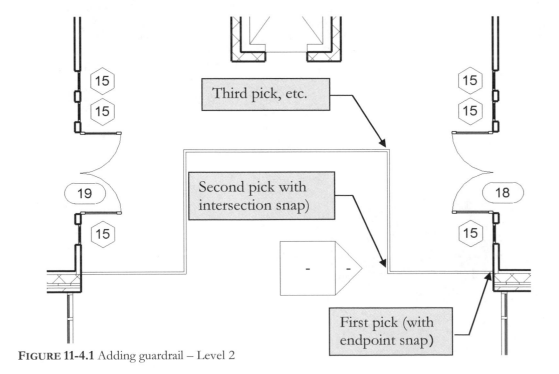

FIGURE 11-4.1 Adding guardrail – Level 2

6. Click the **green check mark** on the *Ribbon* to finish the Railing tool.

The railing has now been drawn. In the next step you will switch to a 3D view and see how to quickly change the railing style. This will also involve changing the height of the railing. Most building codes require the railing height be 42″ when the drop to the adjacent surface is more than 30″; this is called a guardrail.

7. Switch to the *Default* **3D** view.

8. Zoom into the railing shown on Level 2, looking at it through the curtain wall. (Notice the railing style – Figure 11-4.2.)

FIGURE 11-4.2 Added railing – 3D view

9. Select the railing. You may have to use the *Tab* key to cycle through the various selection options.

10. With the railing selected, select the various railing types available in the *Type Selector* on the *Ribbon*. When finished make sure **Railing: Guardrail – Pipe** is selected (Figure 11-4.3).

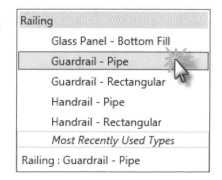

FIGURE 11-4.3 Options for selected railing

Your railing should now look like Figure 11-4.4. Notice that a handrail was added to the railing. You should also notice that the handrail is on the wrong side of the guardrail. You will adjust that next.

FIGURE 11-4.4 Railing with new style

11. Switch to the **Level 2** plan view and select the railing.

12. Click on the **Control Arrows** to flip the railing orientation. (See the *FYI* below for additional information.)

13. You can switch back to the **3D view** to see the change.

14. Finally, from the Level 2 plan view, **Copy** the railing to the clipboard and **Paste** it into the Level 3 view.

15. **Save** your project as **ex11-4.rvt**.

FYI: The last modification, using the control arrows, flipped the railing about the sketch lines, so the railing is now hanging out in space. You can fix this by selecting the railing, clicking "edit path" on the Ribbon, and then moving the sketch lines in to compensate.

You can Copy/Paste a railing style from this drawing into one of your project files. Then you select your railing and pick the newly imported one from the *Type Selector*. This process was done to achieve the image below (Figure 11-4.5). Notice the glass railing with brackets.

FIGURE 11-4.5 Optional railing configuration

Self-Exam:

The following questions can be used as a way to check your knowledge of this lesson. The answers can be found at the bottom of this page.

1. The toilet room fixtures are preloaded in the template file. (T/F)

2. You do not need to be connected to the internet when loading content from the online web library. (T/F)

3. Revit content is not always in compliance with codes. (T/F)

4. You can draw 2D lines on the wall in an interior elevation view. (T/F)

5. Use the _____ tool to copy fixtures to other floors.

Review Questions:

The following questions may be assigned by your instructor as a way to assess your knowledge of this section. Your instructor has the answers to the review questions.

1. Revit provides several different styles of toilet stalls for placement. (T/F)

2. Most of the time Revit automatically updates the ceiling when walls are moved, but occasionally you have to manually make revisions. (T/F)

3. It is not possible to draw dimensions on an interior elevation view. (T/F)

4. Cabinets typically come in 6" increments. (T/F)

5. Base cabinets automatically have a countertop on them. (T/F)

6. What can you adjust so the concrete slab does not show in section?

   _____

7. How does Revit determine where to place 2D lines in an elevation view (based on the example in this lesson)?

   _____

8. What is the current size of your Revit Project? _____

9. What should you use to assure accuracy when placing furniture?

   _____

10. You use the _____ tool to make various components temporarily invisible.

Lesson 12
SCHEDULES:

You will continue to learn the powerful features available in Revit. This includes the ability to create parametric schedules; you can delete a door number on a schedule and Revit will delete the corresponding door from the plan.

Exercise 12-1:
Room and Door Tags

This exercise will look at adding room tags and door tags to your plans. As you insert doors, Revit adds tags to them automatically. However, if you copy or mirror a door you can lose the tag and have to add it.

Adding Room Tags

You will add a Room Tag to each room on your Level 1 floor plan.

Room

1.   Select **Architecture → Room & Area → Room**.

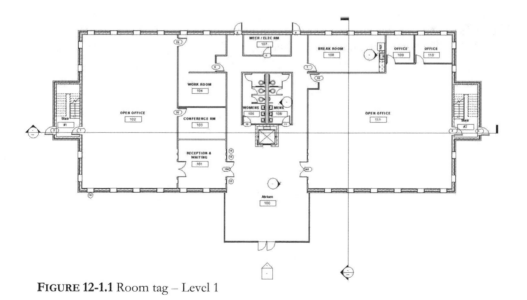

FIGURE 12-1.1 Room tag – Level 1

TIP: Change the height of the Room element in the stairs to go up to Level 3 + 10'-0" via the Properties Palette – and then just "tag" the room at Levels 2 and 3 rather than adding new room objects.

Placing a *Room/Room Tag* is similar to placing a ceiling in the reflected ceiling plan; as you move your cursor over a room, the room (perimeter) highlights. When the room you want to place a *Room* element in is highlighted, you click to place it.

2. Click your cursor within the atrium area to place a **Room**, which will also automatically place a room tag (Figure 12-1.1).

By default, Revit will simply label the space "Room" and number it "1." You will change these to something different.

3. Press **Esc** or select **Modify** to cancel the *Room* command.

4. Click on the *Room Tag* you just placed to select it.

5. Now click on the room name text to change it; enter **Atrium**.

6. Now click on the room number to change it; enter **100**.

7. Add *Room Tags* (using the **Room** tool) for each room on Level 1, incrementing each room number by 1 (Figure 12-1.1).

The stair shafts typically are numbered Stair #1, Stair #2, etc. The same number is then placed on each level. This is because stair shafts are really one tall room and the finishes would apply to the entire shaft, not each floor. When you try to place a tag with the same name and number, Revit will warn you; see the tip on the previous page to avoid this problem.

8. Add Room/Tags to Levels 2 and 3. The numbering for Level 2 should start with 200 and Level 3 should start with 300. (For Level 2, see Figure 12-1.2; for Level 3, see Figure 12-1.3.)

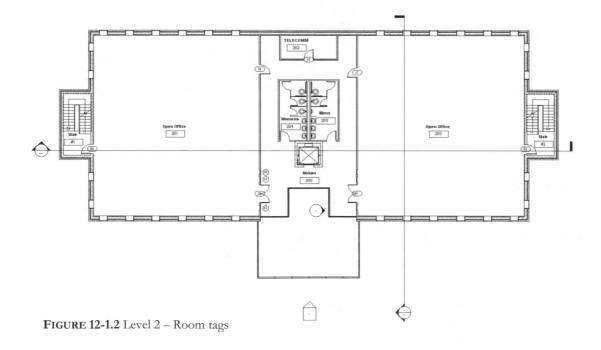

FIGURE 12-1.2 Level 2 – Room tags

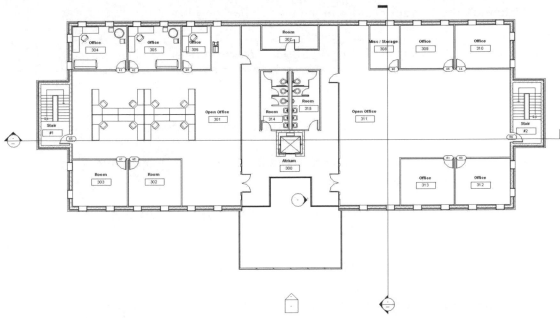

FIGURE 12-1.3 Level 3 – Room tags

Adding Door Tags

Next you will add Door Tags to any doors that are missing them. Additionally, you will adjust the door numbers to correspond to the room numbers.

Revit numbers the doors in the order they are placed into the drawing. This would make it difficult to locate a door by its door number if door number 1 was on Level 1 and door number 2 was on Level 3, etc. Typically, a door number is the same as the room that the door swings into. For example, if a door swung into an office numbered 304, the door number would also be 304. If the office had two doors into it, the doors would be numbered 304A and 304B.

9. Switch to **Level 1** view.

10. Select **Annotate → Tag → Tag by Category** button on the *Ribbon* (Figure 12-1.4).

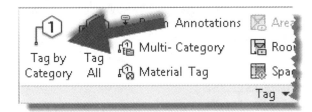

FIGURE 12-1.4 Annotate tab, Tag panel

Notice as you move your cursor around the screen Revit displays a tag, for any elements that can have tags, when the cursor is over it. Revit actually places a tag when you click the mouse.

11. **Uncheck** the **Leader** option on the *Options Bar*.

12. Place a door tag for each door that does not have a tag; do this for each level.

13. Renumber all the door tags to correspond to the room they open into; do this for each level (Figure 12-1.5).

> **REMEMBER:** *Click Modify, select the Tag and then click on the number to edit it.*

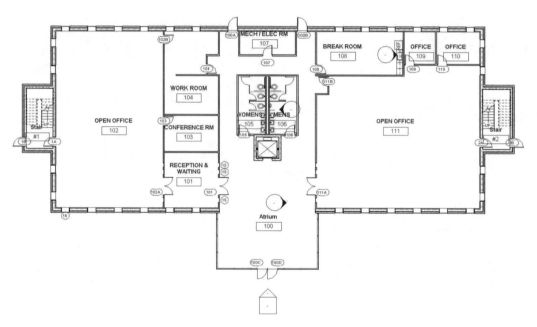

FIGURE 12-1.5 Level 1 – door tags (scale changed to make tags larger on this page)

TIP: TAG ALL...
This tool allows you to quickly tag all the elements of a selected type (e.g., doors) at one time. After selecting the tool, you select the type of element from a list and specify whether or not you want a leader. When you click OK, Revit tags all the untagged doors in that view.

14. **Save** your project as **ex12-1.rvt**.

Exercise 12-2:
Generate a Door Schedule

This exercise will look at creating a door schedule based on the information currently available in the building model (i.e., the tags).

Create a Door Schedule View

A door schedule is simply another view of the building model. However, this view displays numerical data rather than graphical data. Just like a graphical view, if you change the view it changes all the other related views. For example, if you delete a door number from the schedule, the door is deleted from the plans and elevations.

1. Open ex12-1.rvt and **Save As ex12-2.rvt**.

2. Select **View → Create → Schedule/Quantities** button from the *Ribbon*.

3. Select **Doors** under *Category*, make sure *Phase* is set to **New Construction** and then click **OK** (Figure 12-2.1).

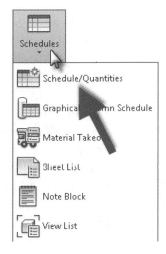

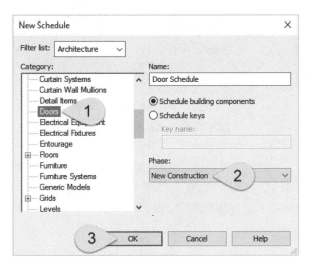

FIGURE 12-2.1 New Schedule dialog

You should now be in the *Schedule Properties* dialog where you specify what information is displayed in the schedule, how it is sorted and the text format.

4. On the **Fields** tab, add the information you want displayed in the schedule. Select the following (Figure 12-2.2):

   a. Mark

      *TIP: Click the **Add** →*
      button each time.

   b. Width
   c. Height
   d. Frame Material
   e. Frame Type
   f. Fire Rating

As noted in the dialog, the fields added to the list on the right are in the order they will be in the schedule view. Use the *Move Up* and *Move Down* buttons to adjust the order.

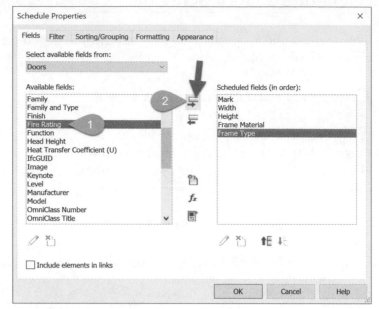

FIGURE 12-2.2 Schedule Properties – Fields

5. On the **Sorting/Grouping** tab, set the schedule to be sorted by the **Mark** (i.e., door number) in ascending order (Figure 12-2.3).

   *TIP: **The Formatting and Appearance tabs** allow you to adjust how the schedule looks. The formatting is not displayed until the schedule is placed on a plot sheet. Also, you cannot print a schedule unless it is on a sheet. Only views and sheets can be printed – not schedules or legends.*

6. Click the **OK** button to generate the schedule view.

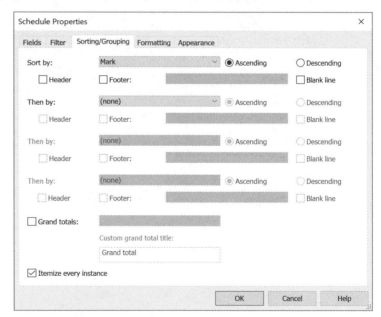

FIGURE 12-2.3 Schedule Properties – Sorting

You should now have a schedule similar to Figure 12-2.4.

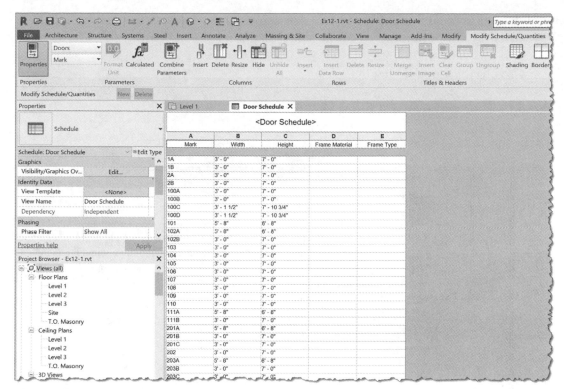

FIGURE 12-2.4 Door schedule view

TIP #1: Use **CTRL + Scroll Wheel** to zoom in and out in schedules. Press **CTRL + 0** to rest the view to the default.

TIP #2: You can select **File Tab → Export → Reports → Schedule** to create a text file (*.txt) that can be used in other programs such as MS Excel.

The example below is from a real-world Revit project; notice the detailed header information.

DOOR AND FRAME SCHEDULE												
DOOR NUMBER	DOOR				FRAME		DETAIL				FIRE RATING	HDWR GROUP
	WIDTH	HEIGHT	MATL	TYPE	MATL	TYPE	HEAD	JAMB	SILL	GLAZING		
1000A	3' - 8"	7' - 2"	WD		HM		11/A8.01	11/A8.01					
1046	3' - 0"	7' - 2"	WD	D10	HM	F10	11/A8.01	11/A8.01 SIM				34	
1047A	6' - 0"	7' - 10"	ALUM	D15	ALUM	SF4	6/A8.01	6/A8.01	1/A8.01 SIM	1" INSUL		2	CARD READER N. LEAF
1047B	8' - 0"	7' - 2"	WD	D10	HM	F13	12/A8.01	11/A0.01 SIM		60 MIN		85	MAG HOLD OPENS
1050	3' - 0"	7' - 2"	WD	D10	HM	F21	8/A8.01	11/A8.01		1/4" TEMP		33	
1051	3' - 0"	7' - 2"	WD	D10	HM	F21	8/A8.01	11/A8.01		1/4" TEMP		33	
1052	3' - 0"	7' - 2"	WD	D10	HM	F21	8/A8.01	11/A8.01		1/4" TEMP		33	
1053	3' - 0"	7' - 2"	WD	D10	HM	F21	8/A8.01	11/A8.01		1/4" TEMP		33	
1054A	3' - 0"	7' - 2"	WD	D10	HM	F10	8/A8.01	11/A8.01		1/4" TEMP	-	34	
1054B	3' - 0"	7' - 2"	WD	D10	HM	F21	8/A8.01	11/A8.01		1/4" TEMP	-	33	
1055	3' - 0"	7' - 2"	WD	D10	HM	F21	8/A8.01	11/A8.01		1/4" TEMP	-	33	
1056A	3' - 0"	7' - 2"	WD	D10	HM	F10	9/A8.01	9/A8.01			20 MIN	33	
1056B	3' - 0"	7' - 2"	WD	D10	HM	F10	11/A8.01	11/A8.01			20 MIN	34	
1056C	3' - 0"	7' - 2"	WD	D10	HM	F10	20/A8.01	20/A8.01			20 MIN	33	
1057A	3' - 0"	7' - 2"	WD	D10	HM	F10	8/A8.01	11/A8.01			20 MIN	34	
1057B	3' - 0"	7' - 2"	WD	D10	HM	F30	9/A8.01	9/A8.01		1/4" TEMP	20 MIN	33	
1058A	3' - 0"	7' - 2"	WD	D10	HM	F10	9/A8.01	9/A8.01			-	33	

Image courtesy LHB (www.LHBcorp.com)

Next you will see how adding a door to the plan automatically updates the door schedule. Likewise, deleting a door number from the schedule deletes the door from the plan.

7. Switch to the **Level 1** view.

8. Add a door as shown in **Figure 12-2.5**; number the door **111C**.

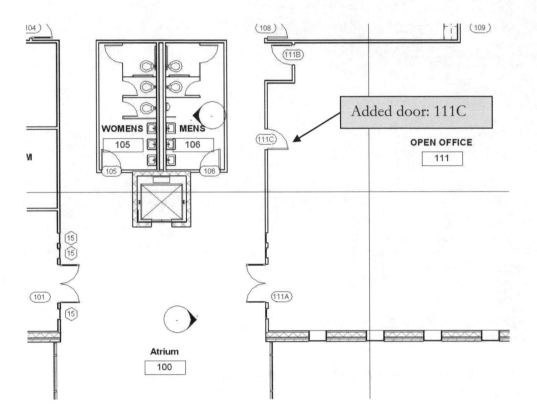

FIGURE 12-2.5 Level 1 – door added

9. Switch to the **Door Schedule** view, under Schedules/Quantities in the *Project Browser*. Notice door 111C was added (Figure 12-2.6).

109	3' - 0"	7' - 0"
110	3' - 0"	7' - 0"
111A	6' - 0"	7' - 0"
111B	3' - 0"	7' - 0"
111C	3' - 0"	7' - 0"
201A	6' - 0"	7' - 0"
201B	3' - 0"	7' - 0"

FIGURE 12-2.6 Updated door schedule

Next you will delete door 111C from the door schedule view.

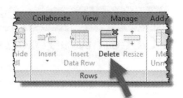

10. Click in the cell with the number **111C**.

11. Now click the **Delete** button from the *Ribbon* (Figure 12-2.7).

FIGURE 12-2.7
Ribbon for the door schedule view

You will get an alert. Revit is telling you that the actual door will be deleted from the project model (Figure 12-2.8).

12. Click **OK** to delete the door (Figure 12-2.8).

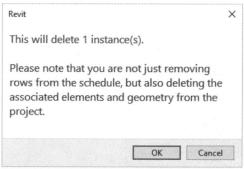

13. Switch back to the **Level 1** view and notice that door 111C and its tag have been deleted from the project model.

14. **Save** your project as **ex12-3.rvt**.

FIGURE 12-2.8 Revit alert message

TIP: You can also change the door number in the schedule and even the size; however, changing the size actually changes the door family which affects all the doors of that type.

The image below shows a real-world Revit project with several doors; notice the door numbers match the room numbers. Also, the shaded walls are existing (Revit can manage phases very well).

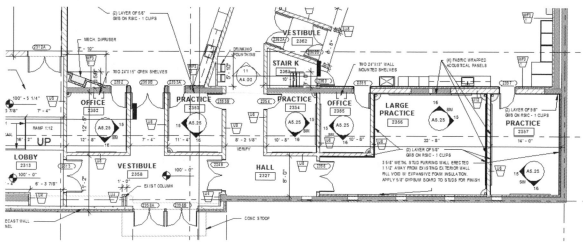

Image courtesy LHB (www.LHBcorp.com)

Exercise 12-3:
Generate a Room Finish Schedule

In this exercise you will create a room finish schedule. The process is similar to the previous exercise. You will also create a color-coded plan based on information associated with the *Room* element.

Create a Room Finish Schedule

1. Open ex12-2.rvt and **Save As ex12-3.rvt**.

2. Select **View → Create → Schedule/Quantities** button from the *Ribbon*.

3. Select **Room** under *Category* and then click **OK** (Figure 12-3.1).

4. In the **Fields** tab of the *Schedule Properties* dialog, add the following fields to be scheduled (Figure 12-3.2):

   a. Number
   b. Name
   c. Base Finish
   d. Floor Finish
   e. Wall Finish
   f. Ceiling Finish
   g. Area

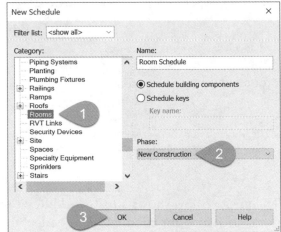

FIGURE 12-3.1 New Schedule dialog

Area is not typically listed on a room finish schedule. However, you will add it to your schedule to see the various options Revit allows.

5. On the **Sorting/Grouping** tab set the schedule to be sorted by the **Number** field.

6. On the **Appearance** tab, select ¼″ **Arial** for the *Title Text* (Figure 12-3.3).

7. Select **OK** to generate the **Room Schedule**.

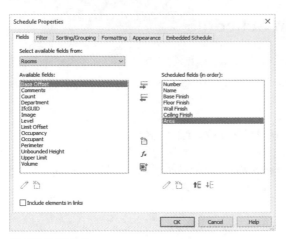

FIGURE 12-3.2
Schedule Properties – Fields

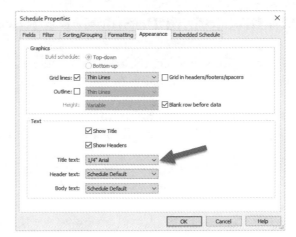

FIGURE 12-3.3
Schedule Properties – Appearance

Your schedule should look similar to the one to the left. (Figure 12-3.4)

8. Resize the *Name* column so all the room names are visible. Place the cursor between the *Name* and *Base Finish* and drag to the right until all the names are visible (Figure 12-3.4).

The formatting (i.e., Bold header text) will not show up until the schedule is placed on a plot sheet.

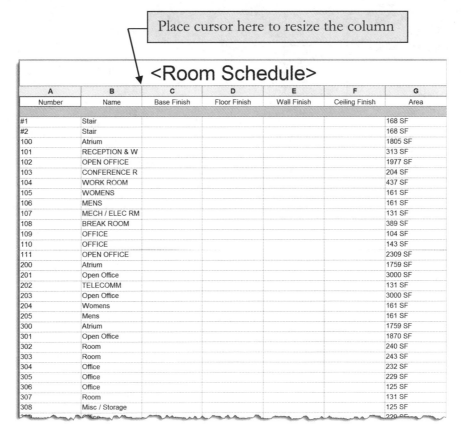

Place cursor here to resize the column

<Room Schedule>

	A	B	C	D	E	F	G
	Number	Name	Base Finish	Floor Finish	Wall Finish	Ceiling Finish	Area
	#1	Stair					168 SF
	#2	Stair					168 SF
	100	Atrium					1805 SF
	101	RECEPTION & W					313 SF
	102	OPEN OFFICE					1977 SF
	103	CONFERENCE R					204 SF
	104	WORK ROOM					437 SF
	105	WOMENS					161 SF
	106	MENS					161 SF
	107	MECH / ELEC RM					131 SF
	108	BREAK ROOM					389 SF
	109	OFFICE					104 SF
	110	OFFICE					143 SF
	111	OPEN OFFICE					2309 SF
	200	Atrium					1759 SF
	201	Open Office					3000 SF
	202	TELECOMM					131 SF
	203	Open Office					3000 SF
	204	Womens					161 SF
	205	Mens					161 SF
	300	Atrium					1759 SF
	301	Open Office					1870 SF
	302	Room					240 SF
	303	Room					243 SF
	304	Office					232 SF
	305	Office					229 SF
	306	Office					125 SF
	307	Room					131 SF
	308	Misc / Storage					125 SF

FIGURE 12-3.4 Room Schedule view

Modifying and Populating a Room Schedule

Like the door schedule, the room schedule is a tabular view of the building model. So you can change the room name or number on the schedule or in the plans.

9.   In the **Room Schedule** view, change the name for room **307** (this should be the room directly north of the toilet rooms) to **MECH/ELEC RM**.

> ***TIP:*** *Click on the current room name and then click on the down-arrow that appears. This gives you a list of all the existing names in the current schedule, otherwise you can type a new name.*

10.   Switch to the **Level 3** view to see the updated room tag.

You can quickly enter finish information to several rooms at one time. You will do this next.

11.   In the Level 3 plan view, select the *Rooms* (not the room tags) for all private offices – 9 total (Figure 12-3.5).

> ***REMEMBER:*** *Hold the Ctrl key down to select multiple elements.*

> ***TIP:*** *Move the cursor near the room tag but not over it to select the room – the large "X" will appear when the Room is selectable (see image below).*

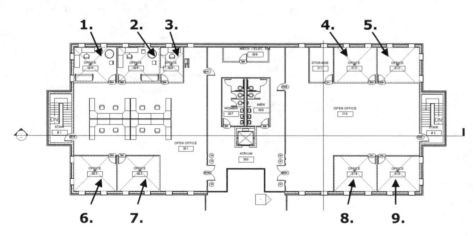

FIGURE 12-3.5 Level 3 – selected room

12.   Next you will make changes in the *Properties Palette*; type *PP* to open it if needed.

The *Parameters* listed here are the same as the *Fields* available for display in the room schedule. When more than one tag is displayed and a parameter is not the same (e.g., different names), that value field is left blank. Otherwise, the values are displayed for the selected *Room* element. Next you will enter values for the finishes.

13. If the *Name* field is blank, enter **OFFICE** so the nine rooms are labeled office.

14. Enter the following for the finishes (Figure 12-3.6):

   a. *Base Finish:* **Wood**
   b. *Ceiling Finish:* **ACT 1** *(ACT = acoustic ceiling tile)*
   c. *Wall Finish:* **VWC 1** *(VWC = vinyl wall covering)*
   d. *Floor Finish:* **Carpet 1**

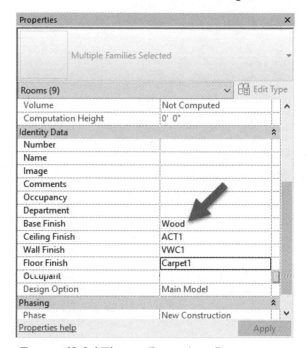

FIGURE 12-3.6 Element Properties – Room

15. Click **Apply**.

16. Switch back to the **Room Schedule** view to see the automatic updates (Figure 12-3.7).

You can also enter data directly into the Room Schedule view.

17. Enter the following data for the Men's and Women's toilet rooms:

   e. *Base:* **COVED CT**
   f. *Ceiling:* **Gyp. Bd.**
   g. *Wall:* **CT**
   h. *Floor:* **CT**

Hopefully, in the near future, Revit will be able to enter the finishes based on the wall, floor and ceiling types previously created!

TIP:
You can add fields and adjust formatting anytime by selecting one of the edit buttons in the Properties Palette. This gives you the same options that were available when you created the schedule.

204	WOMENS					161 SF
205	MENS					161 SF
300	Room					1786 SF
302	OFFICE	Wood	Carpet 1	VWC 1	ACT 1	229 SF
303	OFFICE	Wood	Carpet 1	VWC 1	ACT 1	232 SF
304	OPEN OFFICE					1896 SF
305	OFFICE	Wood	Carpet 1	VWC 1	ACT 1	232 SF
306	OFFICE	Wood	Carpet 1	VWC 1	ACT 1	229 SF
307	OFFICE	Wood	Carpet 1	VWC 1	ACT 1	125 SF
308	MECH / ELEC RM					98 SF
309	WOMENS					161 SF
310	MEN					161 SF
311	MISC. / STORAGE					125 SF
312	OFFICE	Wood	Carpet 1	VWC 1	ACT 1	229 SF
313	OFFICE	Wood	Carpet 1	VWC 1	ACT 1	232 SF
314	OPEN OFFICE					1896 SF
315	OFFICE	Wood	Carpet 1	VWC 1	ACT 1	232 SF
316	OFFICE	Wood	Carpet 1	VWC 1	ACT 1	229 SF
# 1	STAIR					168 SF

FIGURE 12-3.7 Partial Room Schedule with new data

Setting Up a Color-Coded Floor Plan

With the *Rooms* in place you can quickly set up color-coded floor plans. These are plans that indicate (with color) which rooms are Offices, Circulation, Public, etc., based on the room name in our example.

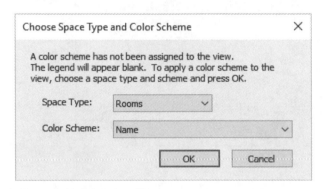

18.   Switch to **Level 3** view.

19.   Select **Annotate → Color Fill → Color Fill Legend**.

20.   Click just below the floor plan on the right side.

21.   Select **Name** and then **OK** to the following prompt (Figure 12-3.8).

You now have a color-coded plan where the colors are assigned by room <u>name</u>; e.g., all the rooms named "Office" have the same color (Figure 12-3.9).

FIGURE 12-3.8 Color fill prompt

*TIP: You can use a **Room Separation Line**, drawn along the railing, to stop the Atrium "Room" from extending out over the multi-story atrium area (i.e. no floor area). This will also make the "area" correct.*

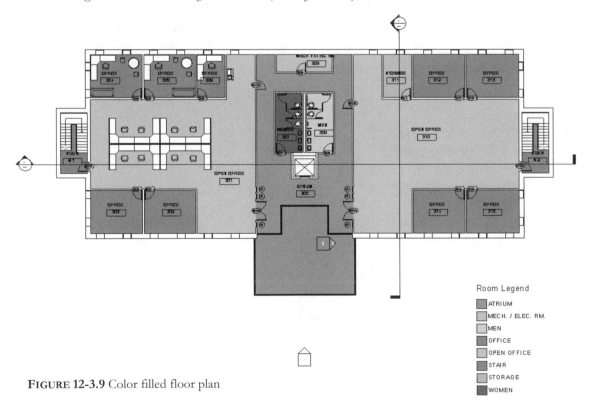

FIGURE 12-3.9 Color filled floor plan

22.   Select the *Room Legend* shown in **Figure 12-3.9**.

23.   Click **Edit Scheme** on the *Ribbon* (Figure 12-3.10).

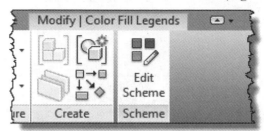

FIGURE 12-3.10 Ribbon;
Color Legend key selected

Each unique room name will get a different color. Before you finish you will change one *Color* and one *Fill Pattern*.

24.   Click on the *Color* for the **Atrium**.

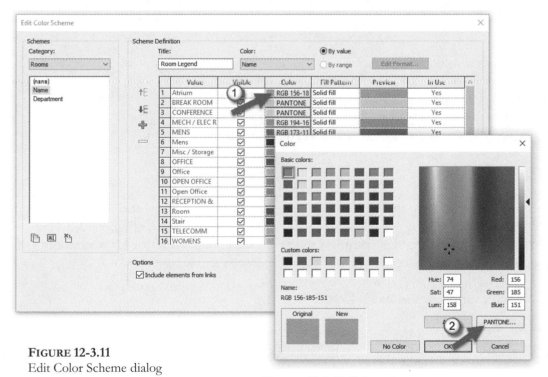

FIGURE 12-3.11
Edit Color Scheme dialog

FIGURE 12-3.12 Color selector

25.   Click the **PANTONE...** button to select a standard *Pantone* color (Figure 12-3.12).

26. Pick any color you like (Figure 12-3.13).

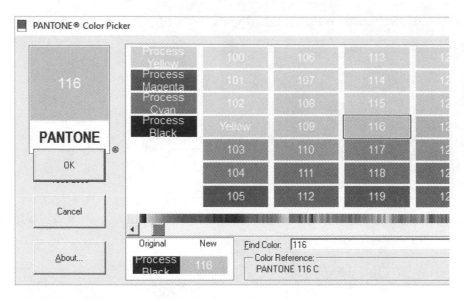

FIGURE 12-3.13 PANTONE Color Picker

27. Click **OK** to accept.

28. Now click on the *Fill Pattern* for the **MECH /ELEC RM**.

29. Click the down-arrow and select **Vertical-small** from the list.

30. Click **OK**.

Your plan should now have the new color you selected for the Atrium and a hatch pattern in the mechanical room. The color legend can also sort by *Department* (see Figure 12-3.11) in addition to many other variables common to the *Room* element. (Interior Designers can create a color filled plan based on the floor finishes, for example.)

31. **Save** your project as **ex12-4.rvt**.

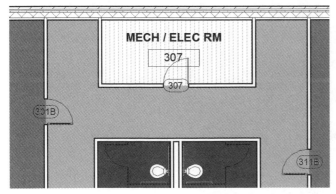

TIP: *In the Properties Palette you can set the Color Scheme Location to "foreground" if you do not want 3D objects like chairs and toilets to obscure the color beneath them.*

Self-Exam:

The following questions can be used as a way to check your knowledge of this lesson. The answers can be found at the bottom of this page.

1. Revit is referred to as a Building Information Modeler (BIM). (T/F)

2. The area for a room is calculated when a *Room* element is placed. (T/F)

3. Revit can tag all the doors not currently tagged on a given level with the *"Tag All"* tool. (T/F)

4. You can add or remove various fields in a door or room schedule. (T/F)

5. Use the _____ tool to add color to the rooms in a plan view.

Review Questions:

The following questions may be assigned by your instructor as a way to assess your knowledge of this section. Your instructor has the answers to the review questions.

1. You can add a door tag with a leader. (T/F)

2. You can export your schedule to a file that can be used in MS Excel. (T/F)

3. A door can be deleted from the door schedule. (T/F)

4. The schedule formatting only shows up when you place the schedule on a sheet. (T/F)

5. It is not possible to add the finish information (i.e., base finish, wall finish) to multiple rooms at one time. (T/F)

6. When setting up a color scheme, you can adjust the color and the _____ pattern in the *Edit Scheme* dialog.

7. Use the _____ palette to adjust the various fields associated with each room in a plan view.

8. Most door schedules are sorted by the _____ field.

9. Revit provides access to the industry standard _____ color library.

Notes:

Lesson 13
SITE and RENDERING:

In this chapter you will take a look at Revit's photo-realistic rendering abilities as well as the basic site development tools. Rather than reinventing the wheel, Autodesk makes several high-end rendering programs like *Autodesk 3DS Max*, *Autodesk Maya* and *Autodesk Renderer*, which work with Revit models in various ways; the integration options improve with each new release.

Exercise 13-1:
Site tools

This lesson will give the reader a quick overview of the site tools available in Revit. The site tools are not intended to be an advanced site development package. Autodesk has other programs much more capable of developing complex sites such as AutoCAD Civil 3D 2021. These programs are used by professional Civil Engineers and Surveyors. The contours generated from these advanced civil CAD programs can be used to generate a topography object in Revit.

In this lesson you will create a topography object from scratch – you will also add a sidewalk.

Once the topography object (the topography object, or element, is a 3D mass that represents part *or* all of the site) is created the grade line will automatically show up in building and wall sections, exterior elevations and site plans. The sections even have the earth pattern filled in below the grade line.

As with other Revit elements you can select the object after it is created and set various properties for it, such as surface material, Phase, Etc. One can also return to Sketch mode to refine or correct the surface – this is done in the same way most other Sketched objects are edited – by selecting the item and clicking *Edit* on the *Options Bar*.

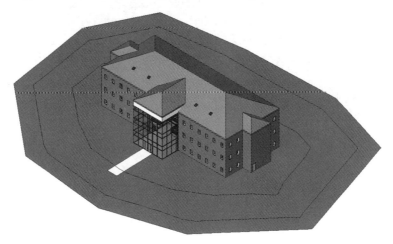

Overview of Site tools located on the Ribbon:

Below is a brief description of what the site tools are used for. After this short review you will try a few of these tools on your office project.

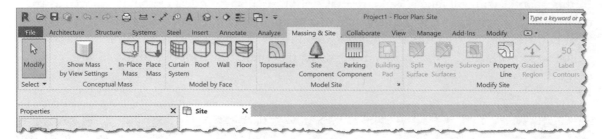

Toposurface: Creates a 3D surface by picking points (specifying the elevation of each point picked) or by using linework, within a linked AutoCAD drawing, that were created at the proper levels.

Subregion: Allows an area to be defined within a previously drawn *Toposurface* – the result is an area within the *Toposurface* that can have a different material than the *Toposurface* itself. The *Subregion* is still part of the *Toposurface* and will move with it when relocated. If a *Subregion* is selected and deleted the original surface/properties for that area are revealed.

Split Surface: This tool is similar to the *Subregion* tool in every way except that the result is separate surfaces that can intentionally, or accidentally, be moved apart from each other. If a split surface is selected and deleted it results in a subtraction or a void relative to the original *Toposurface*.

Merge Surfaces: After a surface has been split into one or more separate surfaces you can merge them back together. Only two surfaces can be merged together at a time. The two surfaces to be merged must share a common edge or overlap.

Graded Region: This tool is used to edit the grade of a *Toposurface* that represents the existing site conditions and the designer wants to use Revit to design the new site conditions. This tool is generally only meant to be used once; when used it will copy the existing site conditions to a new phase and set the existing site to be demolished in the new construction phase. The newly copied site object can then be modified for the new site conditions.

Property Line: Creates property lines (in plan views only).

Building Pad: Used to define a portion of site to be subtracted when created below grade (and added when created above grade). An example of how this might be used is to create a *Pad* that coincides with a basement floor slab, which would remove the ground in section above the basement floor slab (otherwise the basement would be filled with the earth fill pattern). Several pads can be imposed on the same *Toposurface* element.

<u>Parking Component:</u> These are parking stall layouts that can be copied around to quickly lay out parking lots. Several types can be loaded which specify both size and angle.

<u>Site Component:</u> Items like benches, dumpsters, etc. that are placed directly on the *Toposurface* at the correct elevation at the point picked.

<u>Label Contours:</u> Adds an elevation label to the selected contours.
FYI: Contours are automatically created based on the Toposurface.

Site Settings:

The *Site Settings* dialog controls a few key project wide settings related to the site – below is a brief description of these settings.

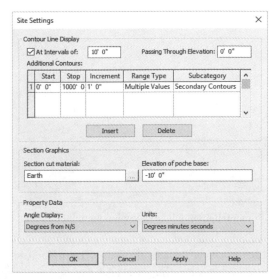

The **Site Settings** dialog is accessed from the *arrow* link in the lower-right corner of the *Model Site* panel. You should note that various tools under the Settings menu affect the entire project, not just the current view.

<u>Contour Line Display:</u> This controls if the contours are displayed when the *Toposurface* is visible (via the check box) and at what interval. If the *Interval* is set to 1'-0" you will see contour lines that follow the ground's surface and each line represents a vertical change of 1'-0" from the adjacent contour line (the contour lines alone do not tell you what direction the surface slopes in a plan view – this is where contour labels are important). The *Passing Through Elevation* setting allows control over where the contour intervals start from. This is useful because architects usually base the first floor of the building on elevation 0'-0" (or 100'-0") and the surveyors and Civil Engineers will use the distance above sea level (e.g. 1009.2'). So this feature allows the contours to be reconciled between the two systems. The Additional Contours section allows for more "contour" detail to be added within a particular area (vertically) – on a very large site you might want 1'-0" contours only at the building and 10'-0" contours everywhere else. However, this only works if the building is in a distinctive set of vertical elevations. If the site is relatively level or an adjacent area shares the same set of contours you will have undesirable results.

<u>Section Graphics:</u> This area controls how the earth appears when it is shown in section (e.g. exterior elevations, building sections, wall sections, etc.). Here is where the pattern is selected that appears in section (the pattern is selected from the project "materials" similar to the process for selecting the pattern to be displayed within a wall when viewed in section).

<u>Elevation of poche base:</u> Controls the depth of the pattern in section views relative to the grade line.

<u>Property Data:</u> This section controls how angles and lengths are displayed for information describing property lines.

Creating topography in Revit

1. Open file 12-3.rvt and *Save-As* **13-1.rvt**.

2. Switch to the **Site** plan view.

   a. Adjust the category visibility so your view matches Figure 13-1.1. Also, set the *Detail Level* to *Fine*. Hide elevation and section tags.

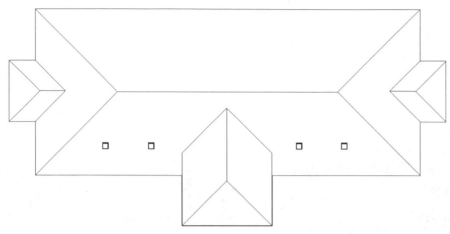

FIGURE 13-1.1 Site plan view

Here you will basically see what appears to be a roof plan view of your project (Figure 13-1.1). This view has the visibility set such that you see the project from above and the various "site" categories turned on, so they are automatically visible once they are created. Next you will take a quick look at the *View Range* for this view so you understand how things are set up.

3. Make sure nothing is selected and a command is not active by clicking the **Modify** tool. This will ensure you have access to the *View Properties* via the *Properties Palette*.

4. In the *Properties Palette*, scroll down and select **Edit** next to *View Range*.

Here you can see the site is being viewed from 200′ above the first floor level, so your building/roof would have to be taller than that before it would be "cut" like a floor plan. The *View Depth* could be a problem here: on a steep site, the entire site will be seen if part of it passes through the specified *View Range*. However, items completely below the *View Range* will not be visible.

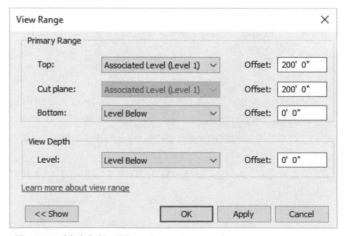

FIGURE 13-1.2 Site Plan view range settings

5. In the *View Depth* section, set the *Level* to **Unlimited**.

This change will ensure everything shows up on a steeper site.

6. Click **OK**.

You are now ready to create the site object. The element in Revit which represents the site is called *Toposurface*.

TIP: Structural engineers can use the toposurface tool to model ledge rock, which would aid in determining how deep foundations need to be.

7. Select **Massing & Site** → **Model Site** → **Toposurface** from the *Ribbon*.

Toposurface

By default the *Place Point* tool is selected on the *Ribbon*; you are now in sketch mode. This tool allows you to specify points within the view at various elevations – Revit will generate a 3D surface based on those points, so the more points you provide the more accurate the surface. Notice on the *Options Bar* (Figure 13-1.3) that you can enter an elevation for each point as you click to place them on the screen.

| Modify | Edit Surface | Elevation | 0' 0" | Absolute Elevation ▼ |

FIGURE 13-1.3 Options bar for Toposurface tool

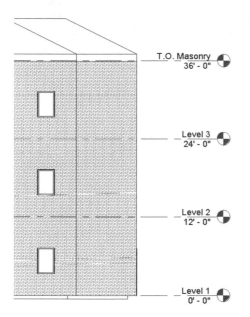

T.O. Masonry
36' - 0"

Level 3
24' - 0"

Level 2
12' - 0"

Level 1
0' - 0"

FIGURE 13-1.4 Partial South Elevation

What elevation should I enter on the Options Bar?

The elevation you enter for each point should relate to the level datums set up in your project – for example, look at your South exterior elevation (Figure 13-1.4). Recall that the first floor is set to 0'-0".

As you can see in the South elevation, the only place you would want to add points at an elevation of 0'-0" would be at the doors. Everywhere else should be at about -4" so the grade does not rise higher than the top of the foundation wall and come into contact with the brick, and possibly block the weeps and brick vents.

8. With the *Elevation* set to **0'-0"** pick the six points shown in Figure 13-1.5; these points are at the exterior door locations.

TIP: You can switch to the Level 1 Floor Plan view while in the Toposurface tool; just click the Place Point tool again on the Ribbon.

9. Change the **Elevation** to **-0'-4"** (don't forget the minus sign) on the *Option Bar* and then pick all the inside and outside corners along the perimeter of the building which have not yet been selected.

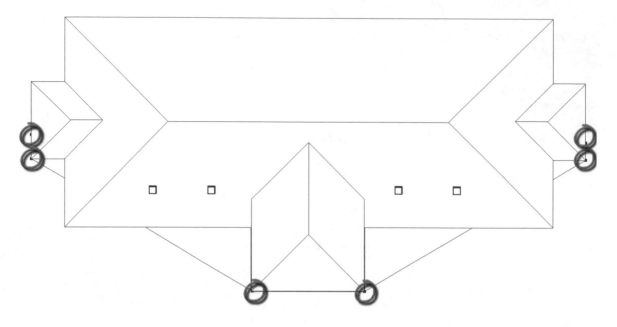

FIGURE 13-1.5 Site plan view – 6 points to be selected (elevation -0'-0")

10. Set the **Elevation** to **-2'-6"** and point to the ten points shown in Figure 13-1.6 which define the extents of the *Toposurface*.

FYI: The elevations selected will generally provide a positive slope away from the building – this will be visible in elevations and sections.

Sketchy Lines

Within a view's Graphic Display Options you can turn on Sketchy Lines and adjust the sliders for two related settings: *Jitter* and *Extensions*. The first determines how wavy the lines are and the latter controls how far the lines cross at corners.

This option is set per view. You will also want to turn on "smooth lines" for this view.

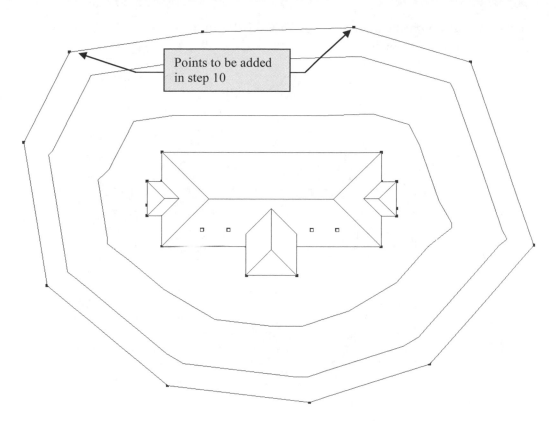

Points to be added
in step 10

FIGURE 13-1.6 Site plan view – 10 points to be selected (elev – 2'-6")

11. Select the **green check mark** on the *Ribbon* to finish the *Toposurface*.

Revit has now created the *Toposurface* based on the points you specified. Again, the more points you add the more refined and accurate the surface will be.

Ideally you would use the *Toposurface* tool to create a surface from a surveyor's points file or contour lines drawn in an AutoCAD file – Revit can automatically generate surfaces from these sources, rather than you picking points. This process is beyond the scope of this tutorial.

12. Switch to the **3D view** to see your new ground surface (Figure 13-1.7).

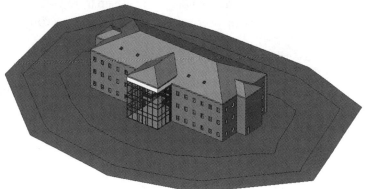

FIGURE 13-1.7 Toposurface seen in 3D view

The toposurface is automatically added in sections and elevations (Figure 13-1.8). It may need to be turned off in other views.

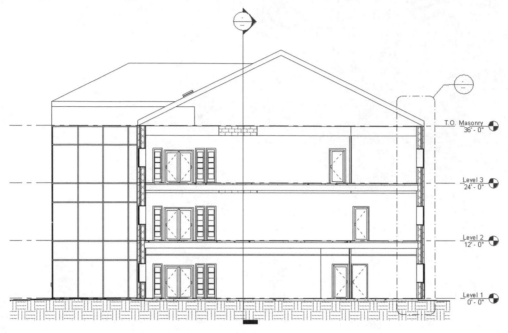

FIGURE 13-1.8 Section with ground pattern added

If your building has a basement or space partially below grade you will need to add a **Building Pad** to "stamp" out an area of the *Toposurface*. To add one you select the tool, draw an outline in a plan view and then specify the depth, or height, of the pad. The *Toposurface* then removes, or adds, ground below the *Building Pad*. If you have a stepped basement you will need multiple Building Pads. Their edges can align but they cannot overlap.

Next you will quickly create a sidewalk to wrap up this lesson.

You can use the *Subregion* tool or the *Split Surface* tool to create the driveway and sidewalks. The *Subregion* tool defines an area that is still part of the main site object – the *Split Surface* tool literally breaks the *Toposurface* into separate elements. Splitting a surface can create problems when trying to move or edit the site so you will use the *Subregion* tool.

13. Switch back to the **Site** plan view.

14. Select **Massing & Site** → **Modify Site** → **Subregion** tool and sketch the lines for the sidewalk (Figure 13-1.9).

15. Click the **green check mark** on the *Ribbon* to finish the *Subregion*.

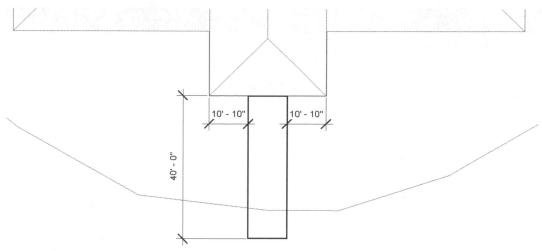

FIGURE 13-1.9 Partial site plan – closed sketch lines for subregion

16. Switch to 3D view to see your sidewalk (Figure 13-1.10). Notice the 2D lines sketched in the site plan view have been projected down onto the surface of the 3D site object!

Notice that the shade and render material is still the same as the main ground surface. You will learn how to change this in a moment.

When you select the *Toposurface* and click *Edit Surface* on the *Ribbon* you can select existing points and edit their elevation to refine the surface.

17. In the 3D view, select the sidewalk *Subregion*.

18. In the *Properties Palette*, set the *Material* to **Concrete – Cast-in-place concrete**.

19. Set the main site element's *Material* to **Site – Grass**.

20. Make sure the *Visual Style* for your 3D view is set to **Shaded** so you can see the "shaded" version of the materials.

   TIP: *Set via View Control bar.*

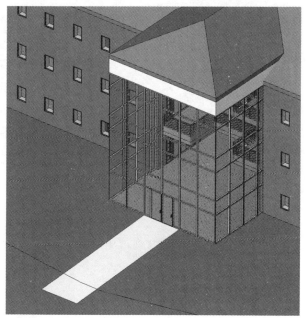

FIGURE 13-1.10 3D view with sidewalk added

That concludes this overview of the site tools provided within Revit. You now have everything modeled in your project so you can start to develop rendered images of your project for presentations.

Exercise 13-2:
Creating an Exterior Rendering

The first thing you will do to prepare a rendering is set up a view. You will use the *Camera* tool to do this. This becomes a saved view that can be opened at any time from the *Project Browser*. A *Camera* view differs from the default 3D view in that it is a perspective view.

Creating a Camera View

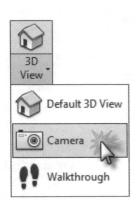

1. Open file 13-1.rvt and *Save-As* to **13-2.rvt**.

2. Open the **Level 1** view and **Zoom to Fit**, so you can see the entire plan.

3. Select **View → Create → 3D Views → Camera**.

4. Click the mouse in the lower right corner of the screen to indicate the camera eye location.

    NOTE: Before you click, Revit tells you it wants the eye location first on the Status Bar.

5. Next click near the atrium curtain wall; see Figure 13-2.1.

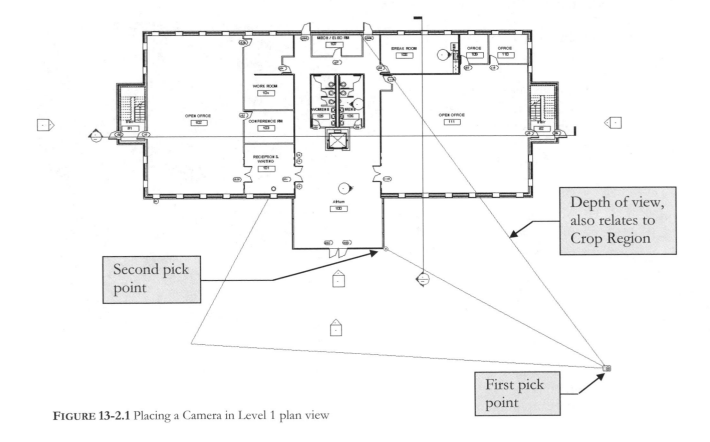

FIGURE 13-2.1 Placing a Camera in Level 1 plan view

Revit will automatically open a view window for the new camera. Take a minute to look at the view and make a mental note of what you see and don't see in the view (Figure 13-2.2).

6. Switch back to the **Level 1** plan view.

7. Adjust the camera, using its grips, to look similar to Figure 13-2.3.

> ***TIP:*** *If the camera is not visible in plan view, right click on the 3D view name in the Project Browser (3D View 1) and select Show Camera (do this while still in the plan view).*

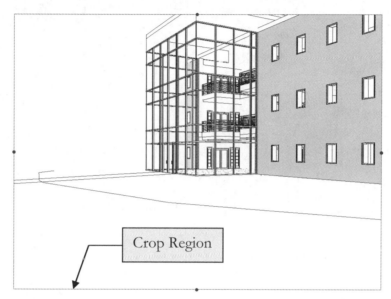

FIGURE 13-2.2 Initial Camera view

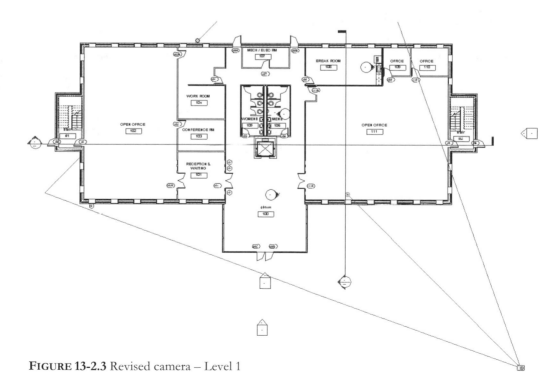

FIGURE 13-2.3 Revised camera – Level 1

8. Now switch to *3D View 1* and adjust the **Crop Region** to look similar to **Figure 13-2.4**. Also, uncheck **Far Clip Active** in the *Properties Palette*.

This will be the view we render later in this exercise.

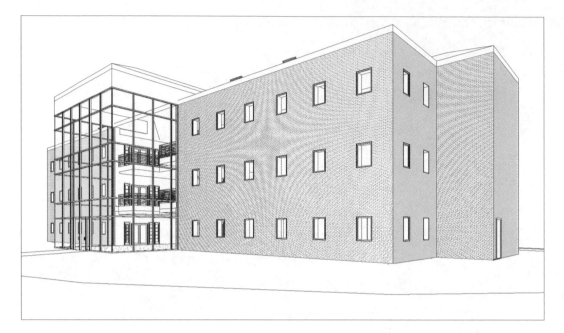

FIGURE 13-2.4 Revised camera – 3D View 1

Assigning Materials to Objects

Materials are scanned images or computer-generated representations of the materials your building will be made of.

Typically, materials are added while the project is being modeled. For example, when you create a material (using the *Materials* command on the *Manage* tab), you can assign a material at that time. Of course, you can go back and add or change it later. Next you will change the material assigned for the exterior brick wall.

9. Switch to **Level 1** plan view.

10. Select an exterior wall somewhere in plan view.

11. Click **Edit Type** in the *Properties Palette* and then click **Edit** *structure*.

12. Notice the material selected for the exterior finish is **Brick, Common**; click in that cell (Figure 13-2.5). **FYI:** In the provided starter project, the material is named Masonry – Brick, which is based on an old Autodesk naming convention.

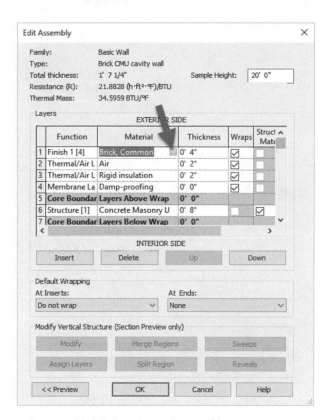

FIGURE 13-2.5 Exterior wall assembly

13. Click the "**…**" icon to the right of the label **Brick, Common**.

Now you will take a look at the definition of the material *Brick, Common*.

You are now in the *Materials* dialog. You should notice that a material is already selected on the left. Next you will select a different brick material.

14. On the Appearance tab (Fig. 13-2.6) click on the **Replace Asset** icon (Step 1 in Figure 13-2.7).

You can browse through the list and select any material in the list to be assigned to the *Brick, Common* material in Revit. The material does not have to be brick but would be confusing if something else were assigned to the *Brick, Common* material.

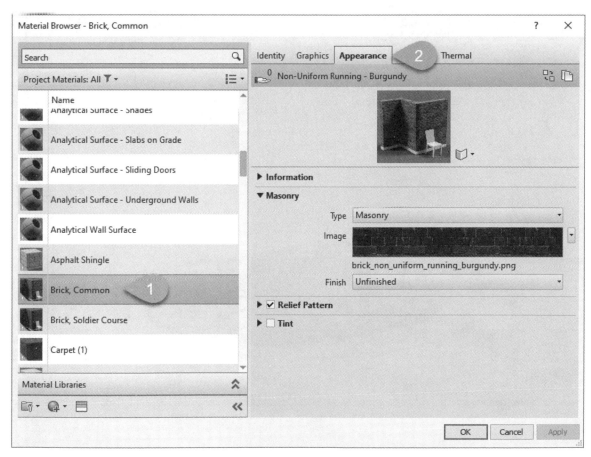

FIGURE 13-2.6 Materials dialog

15. Scroll down in the *Masonry* section and double-click **Non-uniform Running - Red**, and then click the red **X** to close the *Asset Browser* (Figure 13-2.7).

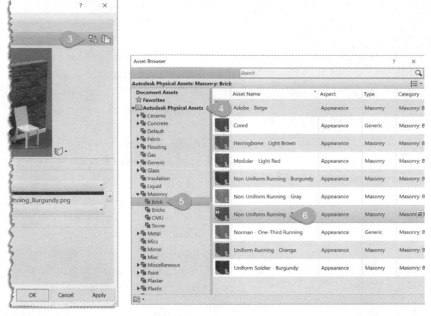

Notice the material listed is now updated.

16. Click **OK** to close *all open* dialog boxes.

Now, when you render any element (wall, ceiling, etc.) that has the material *Non-uniform Running - Red* associated with it, will have a red brick appearance.

If you need more than one brick color, you assign that material to another wall type.

FIGURE 13-2.7 Revit's Asset Library

Sun Path

It is possible to turn on a feature called *Sun Path*. This shows the path of the sun over a day and a year. This is accessed from the *View Control Bar* while in a 3D view.

Once on, you can click and drag the sun along its daily or yearly path!

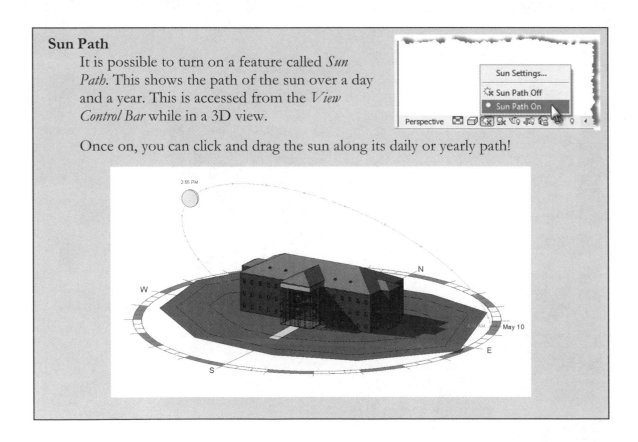

Project Location

Next you will specify the project location on the earth. This will make the daylight accurate in your renderings.

Location

17. Select **Manage → Project Location → Location** from the *Ribbon*.

18. In the *Project Address* field enter **Minneapolis, MN**.

    a.   You may enter your location if you wish.

    b.   It is also possible to enter the actual street address.

19. Press **Enter**.

You should now see an internet based map as shown below, including latitude and longitude.

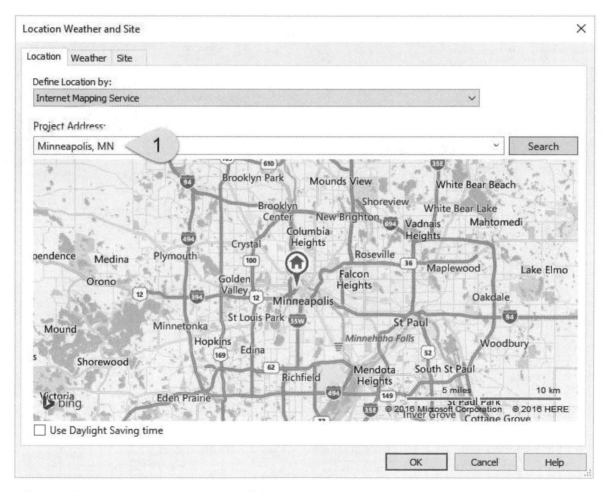

FIGURE 13-2.8 Location Weather and Site dialog

20. Click **OK** to save the location settings.

Sun Settings

Next you will define the sun and shadow settings. You will explore the various options available.

21. Select **Manage → Settings → Additional Settings → Sun Settings** from the *Ribbon*.

22. Make the following changes (Figure 13-2.9).

    a. *Solar Study:* **Still**.

    b. Uncheck *Ground Plane* at Level.

    c. Select **Summer Solstice** on the left.

    d. Change the year to the current year.

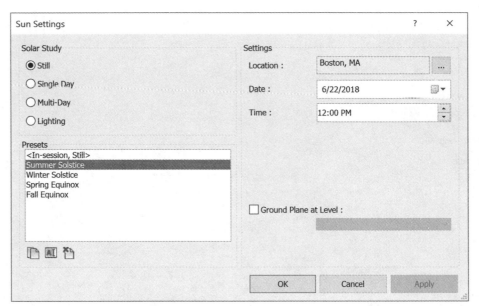

FIGURE 13-2.9 Sun Settings dialog

23. Click **OK** to close the dialog.

Setting Up the Environment

You have limited options for setting up the building's environment. If you need more control than what is provided directly in Revit you will need to use another program like Autodesk 3DS Max 2021 which is designed to work with Revit and can create extremely high quality renderings and animations; it even has day lighting functionality that helps to validate LEED® (Leadership in Environmental and Energy Design) requirements. You can adjust the lighting and the background.

24. Switch to your camera view **3D View 1** and then select the **Show Rendering Dialog** icon on the *View Control Bar*; it looks like a teapot (Figure 13-2.10).

> *FYI: This icon is only visible when you are in a 3D view, the same as the Navigation Wheel and ViewCube.*

The *Rendering* dialog box is now open (Figure 13-2.11). This dialog box allows you to control the environmental settings you are about to explore and actually create the rendering (which you will do soon!).

FIGURE 13-2.10 Render dialog icon

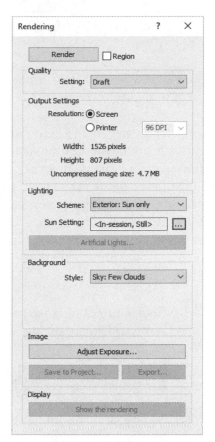

FIGURE 13-2.11 Rendering dialog

25. In the *Lighting* section, click the down-arrow next to *Scheme* to see the options. Select **Exterior: Sun and Artificial** when finished (Figure 13-2.12).

The lighting options are very simple choices: is your rendering an interior or exterior rendering and is the light source Sun or Artificial – or both? You may have artificial lights (i.e., light fixtures like the ones you placed in the office) but still only desire a rendering solely based on the light provided by the sun.

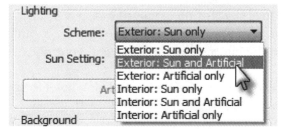

FIGURE 13-2.12 Lighting options

26. In the *Lighting* section, set the *Sun* set to **Summer Solstice**. (This relates to the settings in Figure 13-2.9.)

> *FYI: In the Sun Settings dialog box, Revit lets you set up various "scenes" which control time of day. Two examples would be:*
>   • *Daytime, summer*
>   • *Nighttime, window*

Looking back at Figure 13-2.9, you would click *Duplicate* and provide a name. This name would then be available from the *Sun* drop-down list in the *Render* dialog box.

27. Click on the **Artificial Lighting** button (Figure 13-2.13).

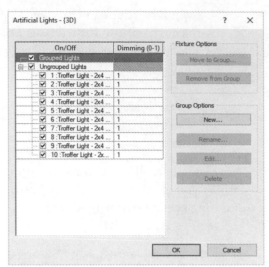

You will now see a dialog similar to the one shown to the right (Figure 13-2.13).

You will see several 2x4 light fixtures. The light fixtures relate to the fixtures you inserted in the reflected ceiling plans. It is very convenient that you can place lights in the ceiling plan and have them ready to render whenever you need to (i.e., render and cast light into the scene!). Here you can group lights together so you can control which ones are on (e.g., exterior and interior lights).

FIGURE 13-2.13 Scene Lighting dialog

28. Click **Cancel** to close the *Artificial Lighting* dialog.

29. Click the down-arrow next to *Style* in the *Background* area (Figure 13-2.14).

30. Select **Sky** (should be the default).

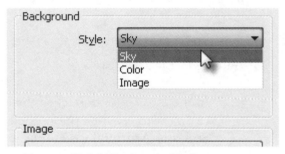

FIGURE 13-2.14 Background style

Notice that for the background, one option is Image. This allows you to specify a photograph of the site or one similar. It can prove difficult getting the perspective just right, but the end results can look as if you took a picture of the completed building.

31. Close the *Renderings* dialog by clicking the "X" in the upper right; all your render settings will be saved for this view.

Next you will place a few trees into your rendering. You will adjust their exact location so they are near the edge of the framed rendering so as not to cover too much of the building.

32. Switch to **Site** plan view and select **Architecture → Build → Component → Place a Component** from the *Ribbon. (Close the Render dialog if it is still open.)*

33. Pick *RPC Tree – Deciduous:* **Largetooth Aspen 25′** from the *Type Selector* on the *Ribbon.*

    FYI: *If the tree is not listed in the* Type Selector, *click* Load Family *and load the deciduous tree family from the* Plantings *folder.*

34. Place three trees as shown in **Figure 13-2.15**. (You will make one smaller in a moment.)

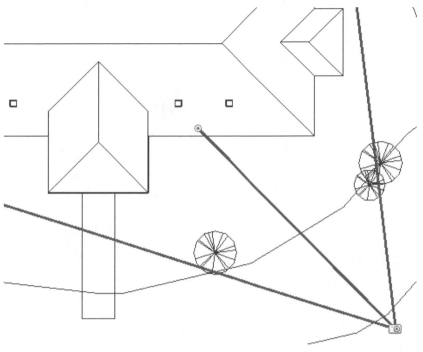

FIGURE **13-2.15** Level 1 with trees added

35. Adjust the trees in plan view, reviewing the effects in the 3D View 1 view, so your 3D view is similar to Figure 13-2.16.

36. In the Level 1 plan view, select the tree that is shown smaller in **Figure 13-2.14**.

37. Select **Edit Type** on the *Properties Palette*. Click **Duplicate** and enter the name *RPC Tree – Deciduous:* **Largetooth Aspen 18′**.

Images courtesy of LHB (www.LHBcorp.com)

38. Change the *Height* to **18′** (from 25′) and then click **OK** to close the open dialog box.

The previous three steps allow you to have a little more variety in the trees being placed. Otherwise, they would all be the same height, which is not very natural.

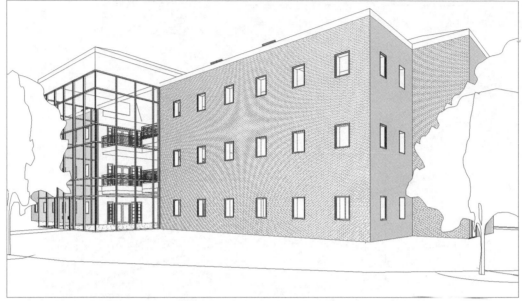

FIGURE 13-2.16 3D View 1 – with trees

39. Open the **3D View 1** camera view.

40. Open the **Rendering** dialog again.

41. Make sure the *Quality* is set to **Draft**.

>   ***FYI:*** *The time to process the rendering increases significantly as the quality level is raised.*

42. Click **Render** from the *Rendering* dialog box.

You will see a progress bar while Revit is processing the rendering (Figure 13-2.17).

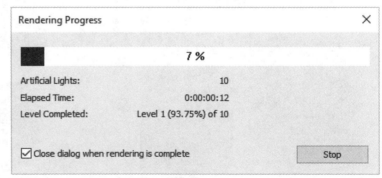

FIGURE 13-2.17 Rendering progress

After a few minutes, depending on the speed of your computer, you should have a rendered image similar to Figure 13-2.18 below. You can increase the quality of the image by adjusting the quality setting in the *Render* dialog. However, these higher settings require substantially more time to generate the rendering. The last step before saving the Revit project file is to save the rendered image to a file.

FYI: Each time you make changes to the model, you will have to re-render the view to get an updated image. Depending on exactly how your view was set up, you may be able to see light from one of the light figures in the office off the Atrium. Also, notice the railing through the curtain wall; Revit has the glass in the windows set to be transparent!

FIGURE 13-2.18 Rendered view

43. From the *Rendering* dialog select **Export**.

> *FYI: The "Save to Project" button saves the image within the Revit Project for placement on Sheets. This is convenient, but it makes the project size larger, so you should delete old ones!*

44. Select a *location* and provide a *file name*.

45. Set the *Save As* type to **JPEG**.

46. Click **Save**.

Render   Render   Render
in Cloud   Gallery

47. If you are a student or work for a company who has Autodesk Subscription, try using the **Render in Cloud** tool on the *View* tab. Adjust the settings to the maximum resolution to see how fast it will render. Click the **Render Gallery** button to see the results! **Note:** You must log into *Autodesk 360* first.

The image file you just saved can now be inserted into MS Word or Adobe Photoshop for editing. Try rendering again with the quality and printer DPI increased, e.g., Medium and 300dpi.

Exercise 13-3:
Rendering an Isometric in Section

This exercise will introduce you to a view tool called *Section Box*. This tool is not necessarily related to renderings, but the two tools together can produce some interesting results. The Section Box works in any 3D, even camera views. This can be used throughout the design process to better visualize your model.

Setting up the 3D View

1. Open file ex13-2.rvt and **Save As ex13-3.rvt**.

2. Switch to the *Default* **3D** view via the 3D icon on the *QAT (not the 3D View 1 from Exercise 13-2)*.

3. Make sure nothing is selected so the *Properties Palette* is showing the **View Properties**.

4. Activate the **Section Box** parameter and then click **OK**.

You should see a box appear around your building, similar to Figure 13-3.1. When selected, you can adjust the size of the box with its grips. Anything outside the box is not visible. This is a great way to study a particular area of your building while in an isometric view. You will experiment with this feature next.

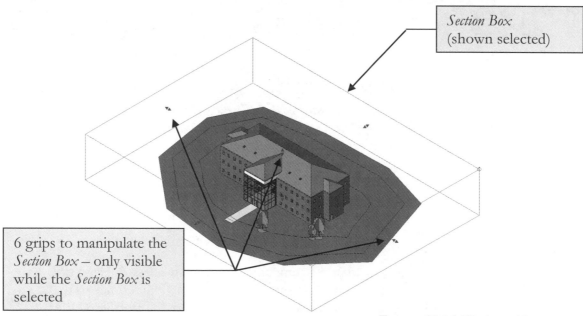

Section Box (shown selected)

6 grips to manipulate the *Section Box* – only visible while the *Section Box* is selected

FIGURE 13-3.1 3D view with Section Box activated

5. To practice using the **Section Box**, drag the grips around until your view looks similar to **Figure 13-3.2**.

    TIP: This will require the ViewCube tool as well.

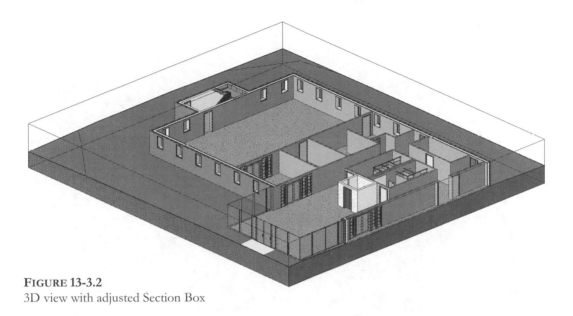

FIGURE 13-3.2
3D view with adjusted Section Box

This creates a very interesting view of the Level 1 – West Wing. What client would have trouble understanding this drawing? Notice the ground in section as well.

6. Now re-adjust the **Section Box** to look similar to **Figure 13-3.3**. Notice toilet room is visible.

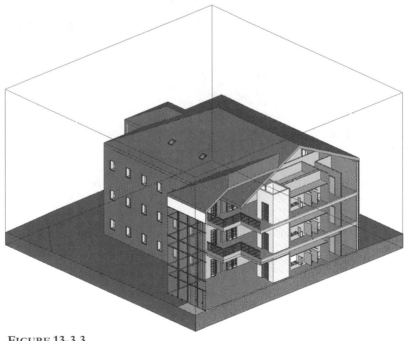

FIGURE 13-3.3
3D view

7. From the *Manage* tab, select *Additional Settings***Sun Settings** and change the following *Sun* settings (Figure 13-3.4):
   a. Click **Duplicate** (name: 28 February 8am)
      (Must select something other than <In-session> first)
   b. *Date:* **2/28/2016**
   c. *Time:* **8:00am**

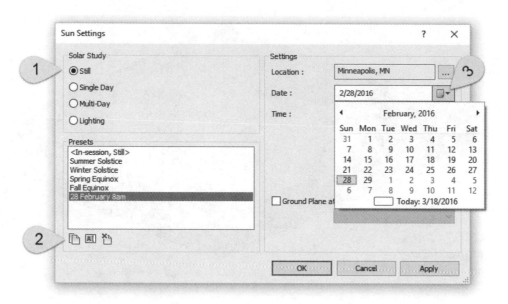

FIGURE 13-3.4 Modified Sun settings

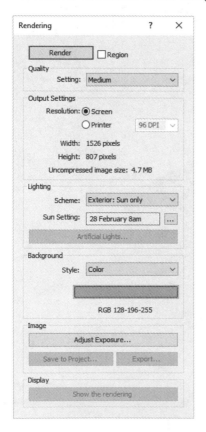

8. Select the **Render** icon, set the *Quality* to **Medium**, select the *Scheme* **Exterior: Sun only**, and then set *Sun* to **23 February 8am**.

9. Set the *Background Style* to **Color**. Leave the default color as is.

10. Select the **Region** option and then adjust the "render region crop" that appears to indicate the area to be rendered.

    TIP: *This tool is nice for checking a material before rendering the entire building, which takes longer.*

11. Click **Render** to generate a rendered image.

The image will take a few minutes to render (again, depending on the speed of your computer). When finished it should look similar to **Figure 13-3.5**. The image looks much better on the screen or printed in color.

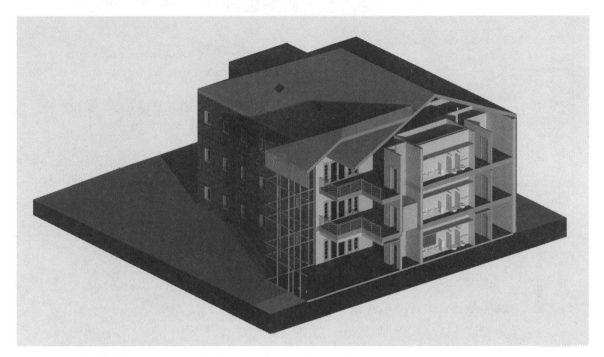

FIGURE 13-3.5 Rendered isometric view

TIP: 3D views can be toggled between Isometric and Perspective via the view's properties.

Adjusting an Element's Material

As previously mentioned, most elements already have a material assigned to them. This is great because it allows you to quickly render your project to get some preliminary images. However, they usually need to be adjusted. You will do this next.

12. Switch to **Level 1** plan view and zoom in on the toilet rooms.

13. Select one of the toilet partitions.

14. Click the **Edit Type** button in the *Properties Palette*.

Notice the toilet partition material is set to *Toilet Partition*. This value is a material; you will change this next (Figure 13-3.6).

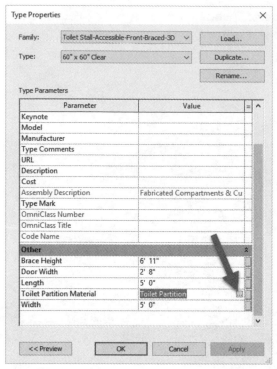

FIGURE 13-3.6 Toilet Partition properties

15. **Click** on the Toilet Partition value (which will cause a "**...**" icon to display to the right); click the icon.

You are now in the *Materials* dialog box where you can create and edit *Materials*.

16. Select **Toilet Partition** (if not already selected on the left).

17. Change the *Render Appearance* to *Stone - Granite*, **Polished - Black** (Figure 13-3.7).

    TIP: Try using the search box near the top, and type "Granite."

18. Close the open dialog boxes.

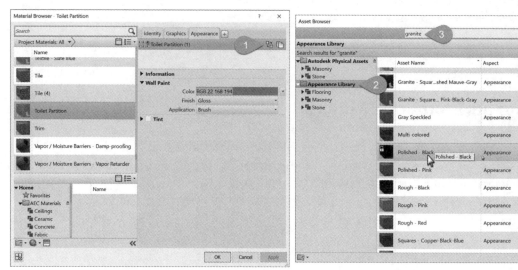

FIGURE 13-3.7 Material selection

19. You can now re-render the 3D view and see the results.

20. **Save** your project as ex **13-3.rvt**.

Duplicating a Material

It is important to know how to properly duplicate a *Material* in your model so you do not unintentionally affect another *Material*. The information on this page is mainly for reference and does not need to be done in your model.

If you **Duplicate** a *Material* in your model, the **Appearance Asset** will be associated to the new *Material* AND the *Material* you copied it from! For example, in the first image (the right), we will right-click on Carpet (1) and duplicate it. Before we duplicate it, notice the *Appearance Asset* named "RED" is not shared (arrow #3).

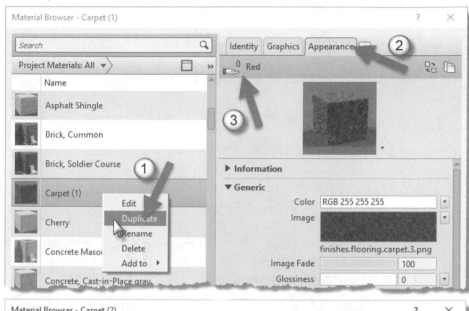

Once you have duplicated a *Material*, notice the two carpet materials, in this example (second image), now indicate they both share the same *Appearance Asset*. Changing one will affect the other. Click the **Duplicate this asset** icon in the upper right (arrow #3).

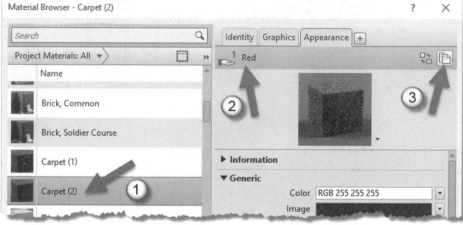

Once the *Appearance Asset* has been duplicated (third image), you can expand the information section and rename the asset. You can now make changes to this material without affecting other materials.

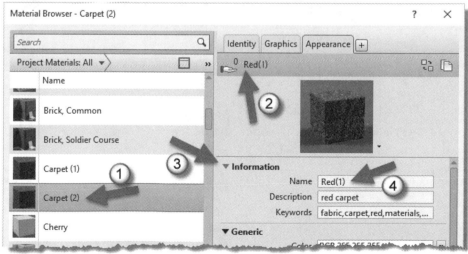

Exercise 13-4:
Creating an Interior Rendering

Creating an interior rendering is very similar to an exterior rendering. This exercise will walk through the steps involved in creating a high quality interior rendering.

Setting up the Camera View:

1. Open ex13-3.rvt and **Save As ex13-4.rvt**.

2. Open **Level 2** view.

3. From the *QAT*, select **3D View → Camera**.

4. Place the *Camera* as shown in **Figure 13-4.1**.

Revit uses default heights for the camera and the target. These heights are based on the current level's floor elevation. These reference points can be edited via the camera properties.

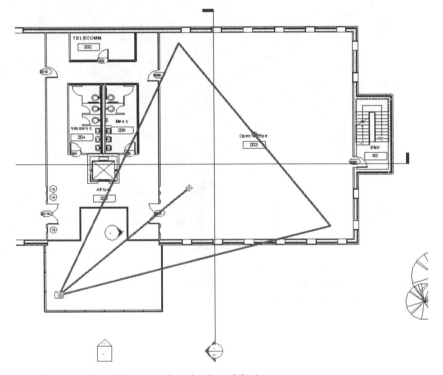

FIGURE 13-4.1 Camera placed – Level 2 view

Revit will automatically open the newly generated camera view. Your view should look similar to **Figure 13-4.2**.

FYI: *Make sure you created the camera on Level 2 and picked the points in the correct order.*

FIGURE 13-4.2 Material selector

5. Using the **Crop Region** rectangle, modify the view to look like **Figure 13-4.3**.

> *TIP: You will have to switch to plan view to adjust the camera's depth of view to see the trees.*

> *REMEMBER: If the camera does not show in plan view, right-click on the camera view label in the project browser and select Show Camera. If you did not add a ceiling to the third floor lobby previously, you should do that now.*

> *FYI: A ceiling was added at the second and third levels in this image to "clean" things up for the rendering.*

FIGURE 13-4.3 Modified interior camera

6. Switch back to **Level 2** to see the revised *Camera* view settings; if you cannot see the camera, right-click on the camera view name in the *Project Browser* and select **Show Camera** from the pop-up menu that appears.

Notice the field of view triangle is wider based on the changes to the Crop Region (Figure 13-4.4).

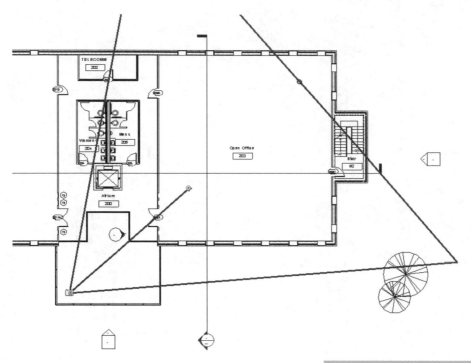

FIGURE 13-4.4 Modified camera – Level 2

7. Select the *Camera* in the level 2 plan view.

8. Change the **Eye Elevation** to **5'-6"** (Figure 13-4.5).

9. Click **Apply**.

Your interior camera view should now look similar to **Figure 13-4.6**. This would be a person standing on Level 1 looking up. The vertical lines are distorted due to the wide field of view (crop region). This is similar to what a camera with a 10-15mm lens would get in the finished building.

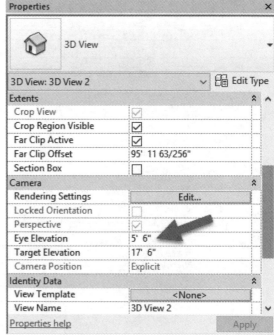

FIGURE 13-4.5 Camera properties

Creating the rendering

Next you will render the view.

10. Select **Show Render Dialog** from the *View Control Bar.*

11. Set the *Scheme* to **Interior: Sun only**.

12. Set the *Sun Setting* to **Lighting: Sunlight from Top Right**.

13. Click **Render** to begin the rendering process.

This will take several minutes depending on the speed of your computer. When finished, the view should look similar to Figure 13-4.7.

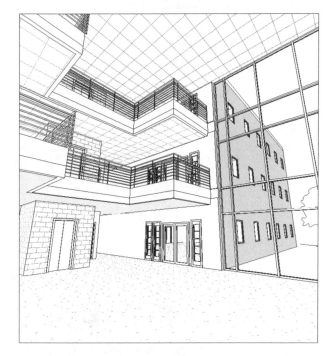

FIGURE 13-4.6 Interior camera view

14. Click **Export** from the *Rendering* dialog box to save the image to a file on your hard drive. Name the file **Atrium.jpg** (jpeg file format).

You can now open the *Atrium.jpg* file in Adobe Photoshop or insert into MS Word to manipulate or print.

To toggle back to the normal hidden view, click **Show the Model** from the *Rendering* dialog box.

There are many things you can do to make the rendering look even better. You can add interior light fixtures and props (e.g., pictures on the wall, items on the countertop, and lawn furniture). Once you add interior lights, you can adjust the Sun setting to nighttime and then render a night scene.

TIP: Setting the output to Printer rather than Screen allows you to generate a higher resolution image. Thus, between the Quality setting and the output setting you can create an extremely high quality rendering, but it might take hours, if not days, to process!

Revit also gives you the ability to set a material to be self-illuminating. This will allow you to make a button on the dishwasher look like it is lit up or, if applied to the glass on the range door, like the light in the oven is on! You can also set a lamp shade to glow when a light source has been defined under it so it looks more realistic.

FIGURE 13-4.7 Rendered view (draft quality)

Image courtesy of LHB (www.LHBcorp.com)

Rendering a Night Scene

One more variation we will look at is rendering the interior atrium view at nighttime. This involves adjusting the sun settings so the Sun is not considered and making sure you have the correct number of light fixtures to light the space being rendered.

15. Add Ceilings and light fixtures to Levels 2 and 3 per steps covered in previous chapters.

16. Open the *Rendering* dialog box.

17. Make the adjustments shown in the image to the left. Make sure Scheme is set to **Interior: Artificial Only**.

18. While in the camera view for the Atrium, click on the **Render** button.

When the rendering is completed you will have a night view of your interior atrium. This clearly shows the effect the 2x4 light fixtures have on the rendering, as they are the primary light source for this rendering. Your image should look similar to **Figure 13-4.8**.

You can also try this (especially if you have placed light fixtures for the entire building) on your exterior camera view. Nighttime renderings can be very dramatic.

The image to the right is a pool created and rendered in Revit. The rendering is only using artificial lights and the "water" material is distorting the pool light and the striping on the bottom of the pool.

Image courtesy of LHB (www.LHBcorp.com)

FIGURE 13-4.8 Rendered nighttime view

TIP: *Setting the Visual Style to realistic can produce nice results without doing a rendering. Clicking the Graphic Display Options, see image to the left, allows you to turn on a feature called Show Ambient Shadows and/or Cast Shadows. Both of these settings help to produce a more realistic look without doing a rendering. This can make the view slow but not other views.*

You will learn how to add people in the next lesson!

Notice you can see reflections in the curtain wall glass. Revit accurately renders reflective surfaces like glass and shiny or polished metal (like the elevator doors). This creates a more realistic rendering.

19. **Save** your project as **13-4.rvt**.

Exercise 13-5:
Adding People to the Rendering

Revit provides a few RPC (Rich Photorealistic Content) people to add to your renderings. These are files from a popular company that provides 3D photo content for use in renderings (http://www.archvision.com). You can buy this content in groupings (like college students) or per item. In addition to people, they offer items like cars, plants, trees, office equipment, etc.

Loading Content into Current Project

1. Open ex13-4.rvt and **Save As 13-5.rvt**.

2. Switch to **Level 2** view.

3. Select **Component** → **Place a Component**.

4. Click the **Load Family** button on the *Ribbon*.

5. Browse to the **Entourage** folder and select both the **RPC Male** and **RPC Female** files (using the *Ctrl* key to select both at once) and click **Open**.

6. Place one **Male** and one **Female** as shown in **Figure 13-5.1**.

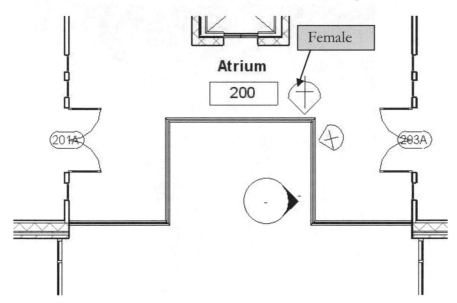

FIGURE 13-5.1 Level 2 – RPC people added

The line in the circle (Figure 13-5.1) represents the direction a person is looking. You simply rotate the object to make adjustments.

7. Switch to **Level 1** view.

8. Place a few of the other people available (similar to Fig. 13-5.3).

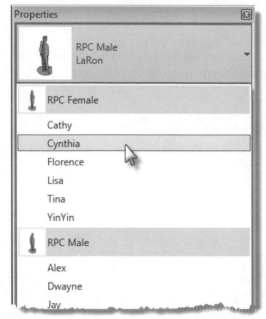

FIGURE 13-5.2 Element Type Selector

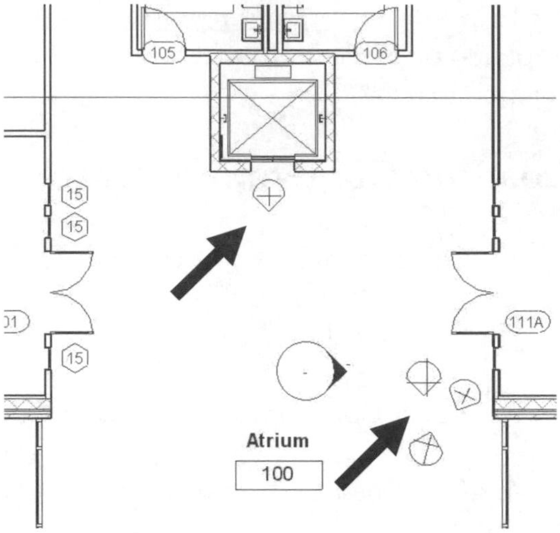

FIGURE 13-5.3 Level 1 – people added

9. Switch to your interior atrium camera view.

10. Render the Atrium view with the daytime settings previously used.

Your rendering should now have people in it and look similar to **Figure 13-5.4**.

Adding people and other "props" gives your model a sense of scale and makes it look a little more realistic. After all, architecture is for people. These objects can be viewed from any angle. Try a new camera view from a different angle to see how the people adjust to match the view and perspective, maybe from the third floor looking down to Level 1.

FIGURE 13-5.4 Interior Atrium view with people added; rendered view

11. **Save** your project as **ex13-5.rvt**.

Revit also has several settings which can make your camera view look more realistic without doing a rendering. The image to the right is an example. It has the *Visual Style* set to **Realistic** (notice even the RPC people are visible), which creates a nice bright image. The view also has **Ambient Shadows** and **Photographic Exposure** enabled in the view's *Graphic Display Options* dialog. Another option is to set the *Visual Style* to **Ray Trace**.

Revit Extras:

There is an add-in for Revit called **Enscape**, a real-time rendering environment which can also be used for videos, VR and Google Cardboard. This add-in is free for students. Check it out at www.enscape3d.com.

FIGURE 13-5.5
Interior Atrium view with people added; Revit camera view set to Realistic

The image below shows all the transportation content available through **Archvision**. When using their subscription option, you have access to all of their content (including people, trees and more). Using the Archvision Dashboard program, you simply drag and drop content into Revit. Students and teachers should contact Archvision about special student options. These elements work in Realistic views, renderings and A360 Cloud renderings.

Self-Exam:

The following questions can be used as a way to check your knowledge of this lesson. The answers can be found at the bottom of this page.

1. Creating a camera adds a view to the *Project Browser* list. (T/F)

2. Materials are defined in Revit's *Materials* dialog box. (T/F)

3. After inserting a light fixture, you need to adjust several settings before rendering and getting light from the fixture. (T/F)

4. You cannot create a nighttime rendering as the sun is always on. (T/F)

5. Use the _____ tool to remove a large portion of the model in 3D.

Review Questions:

The following questions may be assigned by your instructor as a way to assess your knowledge of this section. Your instructor has the answers to the review questions.

1. You cannot "point" the people in a specific direction. (T/F)

2. You cannot get accurate lighting based on day/month/location. (T/F)

3. Adding *Families* to your project does not make the project file bigger. (T/F)

4. Creating photo-realistic renderings can take a significant amount of time for your computer to process. (T/F)

5. The RPC people can only be viewed from one angle. (T/F)

6. The RPC components do not cast shadows. (T/F)

7. Adjust the _____ to make more of a perspective view visible.

8. You use the _____ tool to load and insert RPC people.

9. You can adjust the *Eye Elevation* parameter of the camera via the camera's

   _____.

10. What is the file size of (completed) Exercise 13-5? _____ MB

Lesson 14
CONSTRUCTION DOCUMENTS SET:

This lesson will look at bringing everything you have drawn thus far together onto sheets. The sheets, once set up, are ready for plotting. Basically, you place the various views you have created on sheets. The scale for each view is based on the scale you set while drawing that view, which is important to have set correctly because it affects the text and symbol sizes. When finished setting up the sheets, you will have a set of drawings ready to print, individually or all at once. See the view on sheets for additional information.

Exercise 14-1:
Setting Up a Sheet

Creating a Sheet View

1. Open ex13-5.rvt and **Save As** 14-1.rvt.

2. Select **View →** **Sheet Composition →** **New Sheet**.

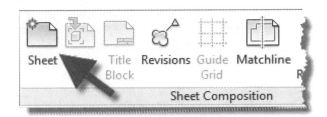

Next Revit will prompt you for a Titleblock to use. The template file you started with only has one; that's the one you will use (Figure 14-1.1).

3. Click **OK** to select the **E1 30x42 Horizontal** titleblock.

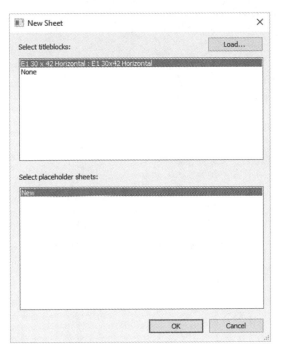

NOTE: A new item shows up in the Project Browser under the heading Sheets. Once you get an entire drawing set ready, this list can be very long.

FIGURE 14-1.1 Select a Titleblock

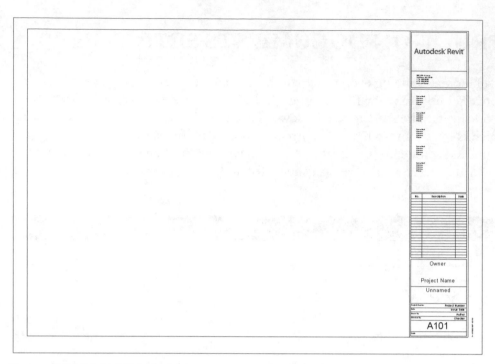

FIGURE 14-1.2 Initial Titleblock view

4. **Zoom** into the sheet number area (lower right corner).

5. Adjust the text to look similar to **Figure 14-1.3**.

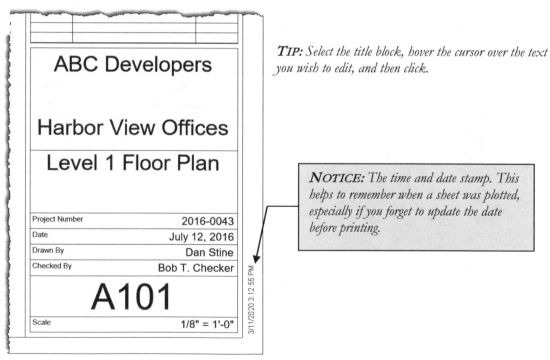

TIP: Select the title block, hover the cursor over the text you wish to edit, and then click.

NOTICE: The time and date stamp. This helps to remember when a sheet was plotted, especially if you forget to update the date before printing.

FIGURE 14-1.3 Revised Titleblock data

6. **Zoom out** so you can see the entire sheet.

7. With the sheet fully visible, click and drag the **Level 1** label (under floor plans) from the *Project Browser* onto the sheet view.

You will see a box that represents the extents of the view you are placing on the current sheet.

8. Move the cursor around until the box is somewhat centered on the sheet (this can be adjusted later at any time).

Your view should look similar to **Figure 14-1.4**.

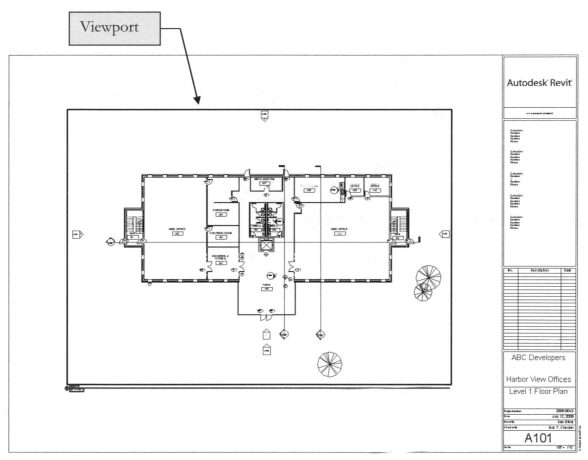

FIGURE 14-1.4 Sheet view with Level 1 added

9. Click the mouse in a "white" area (not on any lines) to deselect the Level 1 view. Notice the box goes away (unless the Crop Region is set to display in the view being placed).

10. **Zoom In** on the lower left corner to view the drawing identification symbol that Revit automatically added (Figure 14-1.5).

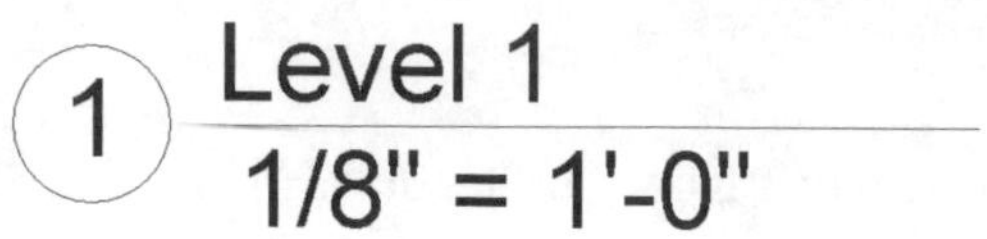

$$\begin{array}{c} \text{Level 1} \\ \hline \text{1/8" = 1'-0"} \end{array}$$

(1)

NOTE: *The drawing number for this sheet is added automatically. The next view you add will be number 2. The view name is listed. This is another reason to rename the elevation and section views as you create them. Also notice that the drawing scale is listed. Again, this comes from the scale setting for the Level 1 view.*

FIGURE 14-1.5 Drawing title

11. **Zoom Out** to see the entire sheet again.

12. Add two more sheets and set up Levels 2 and 3 on them:

   a.   Sheet A102 → Level 2 Floor Plan

   b.   Sheet A103 → Level 3 Floor Plan

NOTE: *When you create a new sheet, most of the titleblock is filled in and the number has increased by 1. This information can be changed if needed.*

TIP: *Back in the view, if you select the Crop Region, you can select Edit Crop to adjust the perimeter to be L-shaped if needed. This also works with the Callout tool on the View tab.*

Setting Up Exterior Elevations:

Next you will set up the exterior elevations on the A200 series sheets.

13. Create a new Sheet and adjust the title block data:

   a.   Sheet Title: Exterior Elevations

   b.   Sheet Number: A200

14. Drag the **South** elevation view onto the sheet. Place the drawing near the lower right.

15. Drag the **North** elevation view onto the same sheet. Place the drawing so that the drawing title tag is aligned. (Revit will snap to this position vertically.)

   TIP: *When placing a view on a sheet, Revit will display a dashed reference line when the model portion of the view aligns with an adjacent model view. This allows you to ensure views align with each other on a sheet.*

Your drawing should look similar to Figure 14-1.6. These same steps can be applied to the floor plan views to hide the trees if desired.

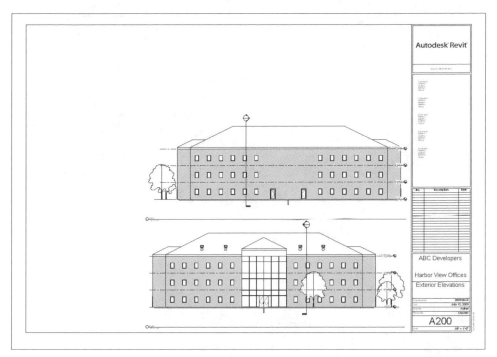

FIGURE 14-1.6 North and South exterior elevations

Next you will turn off the trees in the south view. Normally you would turn them off in all views. However, you will only turn them off in the south view to show that you can control visibility per view on a sheet.

16. Click anywhere on the South elevation view to select it.

17. Now **Right-Click** and select **Activate View** from the pop-up menu (see image to right). *TIP: you can also double-click within the view.*

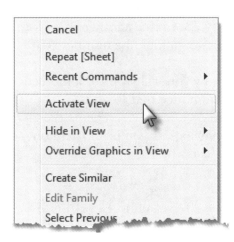

At this point you are in the viewport and can make changes to the project model to control visibility, which is what you will do next.

18. Type **VV** on the keyboard (do not press **Enter**).

19. In the visibility dialog **Uncheck Planting**.

20. Close the open dialog box.

21. Right-click anywhere in the drawing area and select **Deactivate View** from the pop-up menu. *TIP: you can also double-click outside of the viewport to deactivate.*

Now the trees are turned off for the South Elevation but not the North.

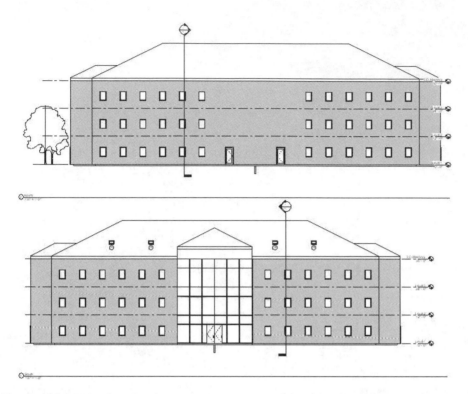

FIGURE 14-1.7 North & South exterior (trees removed from south view)

22. Create another sheet for the other two exterior elevations (East and West); the sheet should be number **A201**.

Now you will stop for a moment and notice that Revit is automatically referencing the drawings as you place them on sheets.

23. Switch to **Level 1** (see Figure 14-1.8).

Notice in Figure 14-1.8 that the number A200 represents the sheet number that the drawing can be found on. The number one (1) is the drawing number to look for on sheet A200.

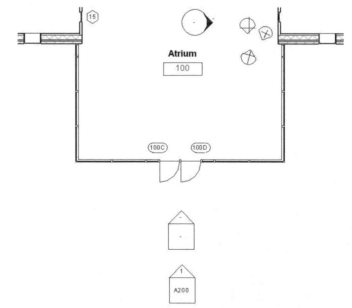

FIGURE 14-1.8 Level 1 – elev. Reference tag filled-in

Setting Up Sections

24. Create a sheet numbered **A300** and titled **Building Sections**.

25. Add a cross section, in plan view, through the atrium area.

26. Add the three building sections as shown in **Figure 14-1.9**.

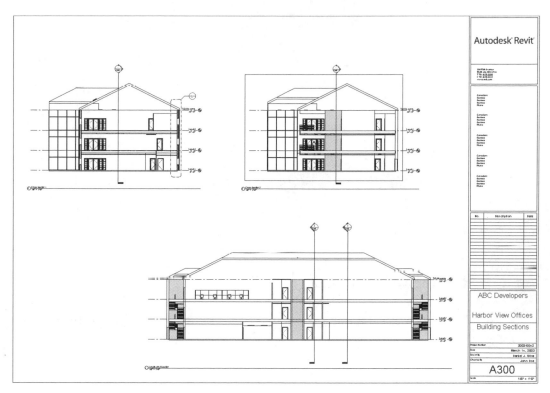

FIGURE 14-1.9 Sheet A300 building sections

27. Switch to _Level 1_ plan view and zoom into the area shown in Figure 14-1.10.

Notice, again, that the reference bubbles are automatically filled in when the referenced view is placed on a sheet. If the drawing is moved to another sheet, the reference bubbles are automatically updated.

You can also see in Figure 14-1.9 (above) that the reference bubbles on the building sections are filled in.

Keep in mind that the _View Scale_ for each view controls the size of the text, dimensions and symbols. This ensures all the annotation is the same size on the printed sheet no matter what the scale of the drawing/view is. So if you added a 3″=1′-0″ detail in the open space, in the lower left, on sheet A300 above; the annotation would be the same size as the adjacent building section views.

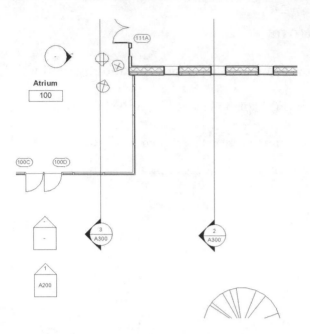

FIGURE 14-1.10 Level 1 – Section refs filled in

Setting Up Remaining Sheets

Next you set up sheets for the remaining views that have yet to be placed on a sheet (except for the 3D views).

Create the following sheets and place the appropriate views on them:

- A111   Level 1 Reflected Ceiling Plan
- A112   Level 2 Reflected Ceiling Plan
- A113   Level 3 Reflected Ceiling Plan
- A400   Wall Sections
- A500   Interior Elevations
- A600   Details (see Figure 14-1.13)
- A800   Schedules

Notice that the new sheets, just created, can be found under *Sheets (all)* on the *Project Browser* (Figure 14-1.11).

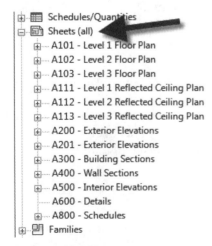

FIGURE 14-1.11
Project Browser; Sheets (all)

Question: On a large project with hundreds of views, how do I know for sure if I have placed every view on a sheet?

Answer: Revit has a feature called *Browser Organization* that can hide all the views that have been placed on a sheet. You will try this next.

28. Take a general look at the *Project Browser* to see how many views are listed. (See Figure 14-1.13 on page 14-10.)

29. Select **Views (all)** at the top of the *Project Browser* (Figure 14-1.12).

30. In the *Type Selector*, pick **not on sheets** (Figure 14-1.12).

31. Notice the list in the *Project Browser* now has fewer views listed. (See Figure 14-1.15 on page 14-10.)

The *Project Browser* now only shows views which have not been placed onto a sheet. Of course, you could have a few views that do not need to be placed on a sheet, but this feature will help eliminate errors.

Next you will reset the *Project Browser*.

32. Using the process just described, set the *Project Browser* back to **All** (Figure 14-1.12).

The image to the right, Figure 14-1.13, shows the details created in Chapter 8. If you recall, these are 2D Drafting Views not directly tied to the 3D model.

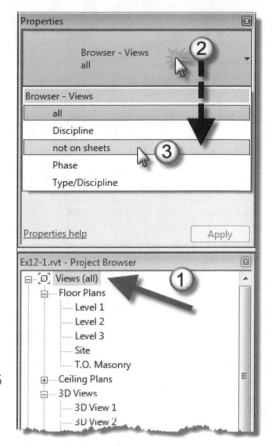

FIGURE 14-1.12 Project Browser and Type Selector

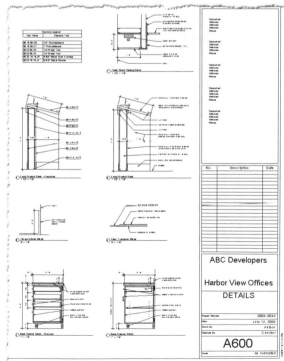

FIGURE 14-1.13 Drafting Views placed on sheet

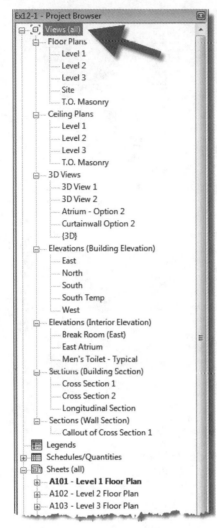

FIGURE 14-1.14
Project Browser; Views (all)

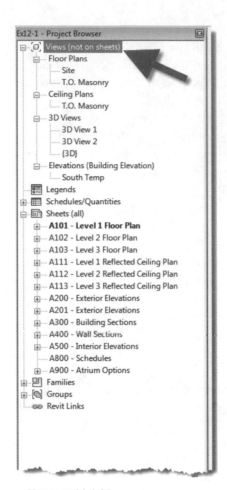

FIGURE 14-1.15
Project Browser; Views (not on sheets)

Sheets with Design Options

Finally, you will set up a sheet to show two of the atrium *Design Options*.

33. Create a *Sheet* named **Atrium Options** and number it **A900**.

34. Open both *Atrium – Option 2* and *Curtainwall Option 2 views* and change the scale to
    ⅛″ = 1′-0″.

35. Place the two views, mentioned above, on sheet A900. (See Figure 14-1.16.)

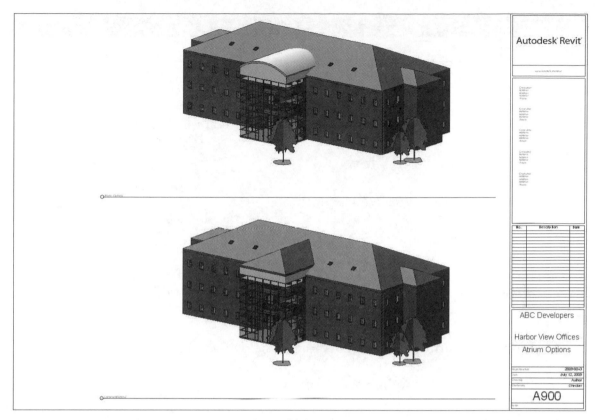

FIGURE 14-1.16 Atrium Options Sheet; two views with Design Options added

Design Options can be used to present alternates to the client or the contractor.

36. **Save** your project as **ex14-1.rvt**.

The sheet border can be modified to match the design firms sheet design, including adding a scanned logo; however, creating the logo in a native Revit format would be better. As you saw, the default sheet size in the Revit template is 30″x42″. However, most design firms have standardized on 22″x34″ as half size sets can easily be printed on 11″x17″ paper. Half size sets are a nice size for a reference set at one's desk. Plus it is one-fourth the paper of a full size set.

FIGURE 14-1.17 Standard sheet size of 22″x34″ can be printed to 11″x17″ at half scale

Exercise 14-2:
Sheet Index

Revit has the ability to create a sheet index automatically. The sheet index is placed on the first sheet in the set of drawings to help the contractor find specific information when it is needed. Some drawing sets can have hundreds of sheets. You will study this now.

Creating a Sheet List View

1. Open ex14-1.rvt and **Save As** ex14-2.

2. Select **View → Create → Schedule → Sheet List**.  Sheet List

You are now in the *Sheet List* dialog box. Here you specify which fields you want in the sheet index and how to sort the list (Figure 14-2.1). This process is identical to the steps required to set up a schedule.

3. Add **Sheet Number** and **Sheet Name** to the right *(click Add →)*.

4. On the *Sorting* tab, sort by **Sheet Number**.

5. Click **OK**.

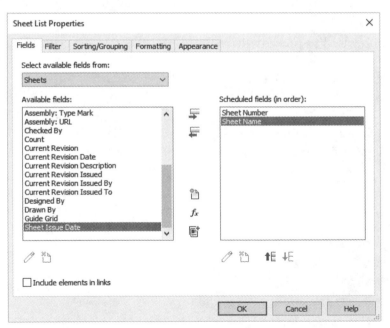

FIGURE 14-2.1
Sheet List Properties Dialog; sheet number and name "added"

Now you should notice that the *Sheet Names* are cut off because the column is not wide enough. You will adjust this next.

6. Move your cursor over the right edge of the *Drawing List* table and click-and-drag to the right until you can see the entire name (Figure 14-2.2).

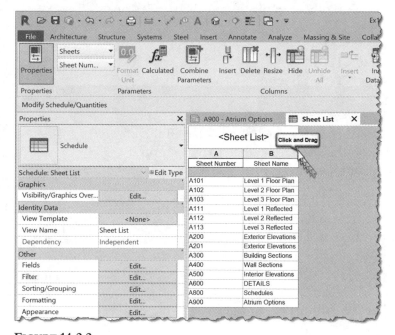

FIGURE 14-2.2
Sheet List view; notice sheet names are cut off in the right column

TIP: *Via the Properties palette you can adjust the settings on the Sorting/Grouping tab to sort the sheets by sheet number.*

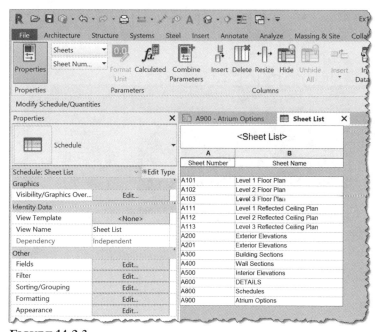

FIGURE 14-2.3
Sheet List view; sheet names are now visible

Setting Up a Title Sheet

Now you will create a title sheet to place your sheet index on.

7. Create a new Sheet:
   a. *Number:* **T001**
   b. *Name:* **Title Sheet**

8. From the *Schedules/Quantities* category of the *Project Browser*, place the view named **Sheet List** on the *Title Sheet*. Once on the sheet, drag the "triangle" grip to adjust the column width (Figure 14-2.4).

Next you will place one of your rendered images that you saved to file (raster image). If you have not created a raster image, you should refer back to Lesson 11 and create one now (otherwise you can use any BMP or JPG file on your hard drive if necessary).

9. Select **Insert → Import → Image**.

10. Browse to your JPG or BMP raster image file, select it and click **Open** to place the Image.

11. Click on your title sheet to locate the image.

Your sheet should look similar to Figure 14-2.5.

Sheet List	
Sheet Number	Sheet Name
A101	Level 1 Floor Plan
A102	Level 2 Floor Plan
A103	Level 3 Floor Plan
A111	Level 1 Reflected Ceiling Plan
A112	Level 2 Reflected Ceiling Plan
A113	Level 3 Reflected Ceiling Plan
A200	Exterior Elevations
A201	Exterior Elevations
A300	Building Sections
A400	Wall Sections
A500	Interior Elevations
A600	DETAILS
A800	Schedules
A900	Atrium Options
T001	Title Sheet

FIGURE 14-2.4
Sheet list on a sheet

Image

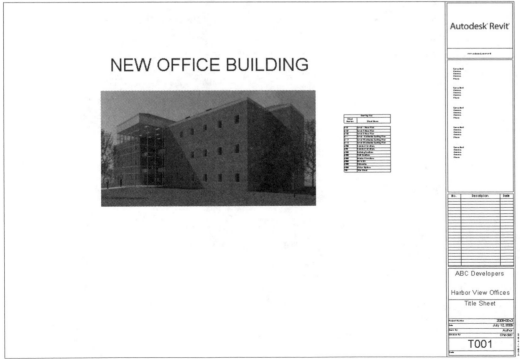

FIGURE 14-2.5 Sheet View: Title Sheet with sheet list, text and image added

12. Use the **Text** command to add the title shown in Figure 14-2.5.

    a.   Create a new text style named **1″ Arial**.

When you have raster images in your project, you can manage them via the *Raster Images* dialog.

13. Select **Insert → Link → Manage Links**, click the **Images** tab.

You now see the *Raster Image* list, which gives you some information about the image and allows you to delete it from the project (Figure 14-2.6). You can also "swap" the image out with another image by using the *Reload From…* button, or place another instance with the *Add…* button.

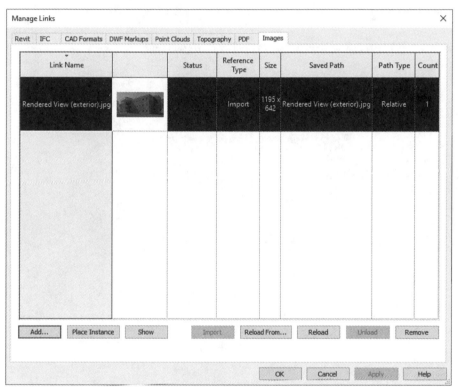

FIGURE 14-2.6 Raster Image dialog

14. Click **OK** to close the *Raster Images* dialog.

15. **Save** your project as **ex14-2.rvt**.

TIP: Appears in Sheet List

If you have a presentation drawing setup on a sheet and don't want it to show up in the sheet index (i.e. sheet list), simply uncheck the **Appears in Sheet List** in the *Properties Palette*.

Exercise 14-3:
Printing a Set of Drawings

Revit has the ability to print an entire set of drawings in addition to printing individual sheets. You will study this now.

Printing a Set of Drawings

1. **Open ex14-2.rvt.**

2. Select **File Tab → Print**.

3. In the *Print range* area, click the option **Selected views/sheets** (Figure 14-3.1).

4. Click the **Select…** button within the *Print range* area.

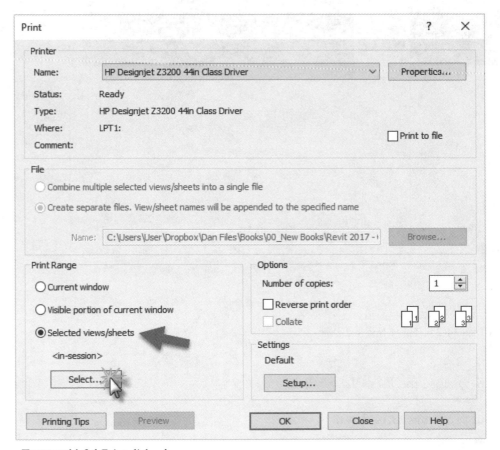

FIGURE 14-3.1 Print dialog box

You should now see a listing of all views and sheets (Figure 14-3.2).

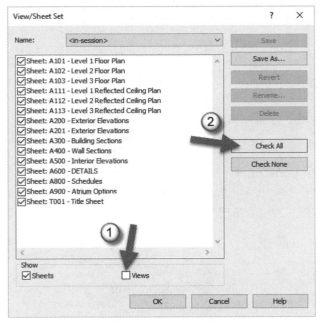

FIGURE 14-3.2 Selecting tool for printing

Notice at the bottom you can **Show** both **Sheets** and **Views**, or each separately. Because you are printing a set of drawings you will want to see only the sheets.

5. **Uncheck** the **Views** option.

The list is now limited to just sheets set up in your project.

6. Select all the drawing Sheets.

7. Click **OK** to close the **View/Sheet Set** dialog.

FYI: Once you have selected the sheets to be plotted you can click Save. This will save the list of selected drawings to a name you choose. Then, the next time you need to print those sheets, you can select the name from the drop-down list at the top (Figure 14-3.2). On very large projects (e.g., with 20 floor plan sheets) you could have a "Plans" list saved, a "Laboratory Interior Elevations" list saved, etc.

8. IF YOU ACTUALLY WANT TO PRINT A FULL SET OF DRAWINGS, you can do so now by clicking OK. Otherwise click **Cancel**.

9. You do not need to save the file at this time.

Autodesk User Certification Exam
Be sure to check out Appendix A for an overview of the official Autodesk User Certification Exam and the available study book and software from SDC Publications and this author.

Successfully passing this exam is a great addition to your resume and could help set you apart and help you land a great job. People who pass the official exam receive a certificate signed by the C.E.O. of Autodesk, the makers of Revit.

Self-Exam:

The following questions can be used as a way to check your knowledge of this lesson. The answers can be found at the bottom of this page.

1. You have to manually fill in the reference bubbles after setting up the sheets. (T/F)

2. You cannot control the visibility of objects per viewport. (T/F)

3. It is possible to see a listing of only the views that have not been placed on a sheet via the *Project Browser*. (T/F)

4. You only have to enter your name on one titleblock, not all. (T/F)

5. Use the _____ tool to create another drawing sheet.

Review Questions:

The following questions may be assigned by your instructor as a way to assess your knowledge of this section. Your instructor has the answers to the review questions.

1. You need to use a special command to edit text in the titleblock. (T/F)

2. The template you started with has several titleblocks to choose from. (T/F)

3. You only have to enter the project name on one sheet, not all. (T/F)

4. The scale of a drawing placed on a sheet is determined by the scale set in that view's properties. (T/F)

5. You can save a list of drawing sheets to be plotted. (T/F)

6. Use the _____ tool to edit the model from a sheet view.

7. The reference bubbles will not automatically update if a drawing is moved to another sheet. (T/F)

8. On new sheets, the sheet number on the titleblock will increase by one from the previous sheet number. (T/F)

Lesson 15
Introduction to Revit Content Creation:

This chapter will introduce you to many of the basic concepts which relate to using and creating families. The reader will benefit from learning to create custom content as the need often arises. An entire book could be written just on creating *Families*. This chapter is intended to provide a very basic introduction to help get you started.

Exercise 15-1:
Basic Family Concepts

Kinds of Families

Autodesk Revit has three primary types of *Families*:
- *System Families*
- *Loadable Families*
- *In-Place Families*

This book will mainly focus on the second, that being loadable families. Below is a brief explanation of each, followed by a graphic to help tell the story.

<u>System Families:</u>
Autodesk describes *System Families* generally as the portion of the building that is **constructed on site**. Things like walls, floors, ceilings, roofs, stairs, wiring, ductwork and piping are system families. These families can only be defined and exist within the *Project Environment*.

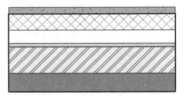

Unlike loadable families (doors, windows, furniture), System families cannot exist separately outside the project environment. This image shows a wall in plan with its various layers of material.

System Families have the ability to **host** *Loadable Families*. The concept of *Hosted Families* will be covered in more detail later in this section, but here is a simplified explanation: a wall hosted *Family*, such as a window, can only exist within a wall; it automatically moves with the wall and is deleted when the wall is deleted.

It is too bad that *System Families* cannot exist outside the project in individual files as it would be helpful to have several construction types predefined and load them when needed, as opposed to having them predefined within the Revit project template. Therefore, your options are to have the *System Family* predefined in a template file or Copy/Paste them between projects. This author prefers to load relevant content and not have extraneous items cluttering the selection lists. There are always tricks and workarounds to facilitating this concept but it is not as simple as *Loadable Families* which are described next.

<u>Loadable Families:</u>
Building components which are **constructed in a factory** and shipped to the project site are what mainly make up *Loadable Families*. These are typically just referred to as *Families* rather than "Loadable" Families; this book will mainly use the term *Family* and mean "*Loadable*" *Family*. Building elements such as doors, windows, furniture, casework, columns, beams, appliances, electrical devices (i.e., outlets and switches), electrical panels, mechanical equipment (i.e., VAV boxes, air handling units and water heaters), duct fittings and plumbing fixtures (i.e., toilets and sinks) constitute *Loadable Families*.

Plumbing fixtures, such as this toilet, are "loadable," that is, they can exist in a file outside of the project environment.

Image credit: Stabs, Wingate

As the name implies, these *Families* can be stored outside of the *Project Environment* within individual files (*.rfa is the family extension, versus *rvt which is the project file extension). Using a "Load Family" command (found on the Insert tab), one is able to bring one or more Families into the Project Environment for placement.

<u>In-Place Families:</u>
Using a special command (*Architecture → Build → Component → Model In-Place*) within the *Project Environment* it is possible to create what is called an *In-Place Family*. This feature allows you to create a *Family* within the context of your project.

The *In-Place Family* has a couple of rather significant warnings that go along with it. <u>First</u>, it is meant for one-off items like a reception desk or a unique built in cabinet. If you copy an *In-Place Family* it is really just making another independent instance of the family – which makes the Revit project file grow in size. If you know something will occur more than once it should be created as a *Loadable Family*, not

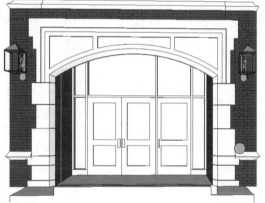

Stone trimmed opening created as an in-place family to interface with project conditions.
Image courtesy of LHB, Inc.

an *In-Place Family*. <u>Second</u>, *In-Place* families cannot ever exist outside the *Project Environment*, similar to System Families, so the item may not be added to your firm's library for use on another project. One last point is from Autodesk's "Revit Platform Performance Document": *In-Place Families* tend to reduce system performance within the Project Environment, especially on larger projects.

So with these points in mind it is best to use *Loadable Families* whenever possible. However, they are still acceptable in some situations.

Creating *In-Place Families* will not specifically be covered in this book. However, the process is almost identical in several ways to creating *Loadable Families* so you should be able to create them without too much trouble when needed.

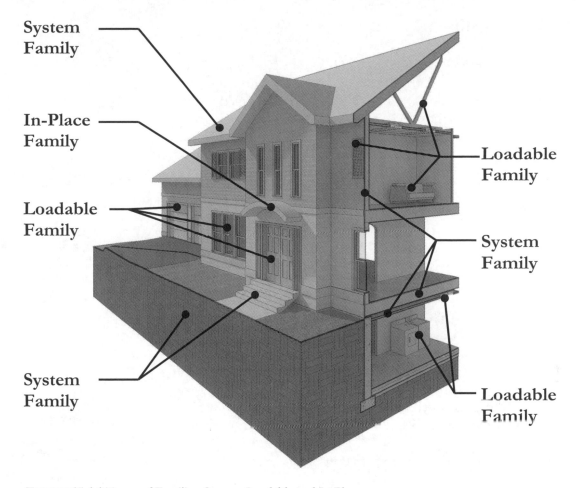

System
Family

In-Place
Family

Loadable
Family

Loadable
Family

System
Family

System
Family

Loadable
Family

FIGURE 15-1.1 Types of Families; System, Loadable and In-Place

Nested Families

It is possible to insert one or more *Families* into another *Family*; this is called nesting a family. One way in which this is useful might be two double-hung windows, as shown in the image above. Although it is possible to simply place the two windows close to each other in the project, this does not represent the reality of how the window will need to be documented in the project and how it will be delivered to the job site. The two windows would typically be mulled together by the window manufacturer and shipped and installed as one unit. By nesting one family into another and then copying it, you can quickly create a two window unit, which can also be tagged as one window. Lastly, nested families can typically have their position controlled more easily by parameters because they act as one element rather than several, or even hundreds.

Hosted Families

Loadable Families can be created to be Hosted or Non-hosted. As briefly mentioned above, the host is a *System Family* (e.g., a wall, floor, ceiling or roof) and when the host is moved, copied or deleted the Hosted Family is moved, copied or deleted.

This feature helps solve some coordination problems but also creates a few new problems.

Problems solved:
When a wall is moved, things like cabinets, toilet fixtures, windows, specialty equipment (e.g., paper towel dispensers, grab bars, mirrors, etc.), doors and electrical devices (e.g., light switches and outlets) all move with the wall. This significantly helps with last minute coordination.

Problems created:
A few problems have surfaced when using hosted content. For example, when designing a kitchen the wall cabinets are hosted. It is difficult to quickly mirror one's kitchen design. Also, when a supply air diffuser is hosted by a ceiling there is occasionally a problem with the ductwork when the ceiling is moved vertically.

Oftentimes the benefits outweigh the problems created.

> ***TIP:*** *One technique is creating your content initially as non-hosted and then nest that family into a hosted family, giving you access to both types, which is convenient as it is not possible to convert from one type to the other. Quick examples of why you would want to do this: wall cabinets in a kitchen are sometimes hung from rods above a peninsula countertop. Also, light fixtures which are attached to a ceiling are sometimes suspended within mechanical rooms or basements which do not have ceilings.*

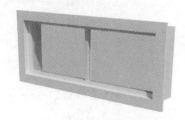

Wall hosted family - Window.
Image credit: Stabs, Wingate

Roof hosted family – Solar
Conforms to slope of roof.
Image credit: Stabs, Wingate

Ceiling hosted family – Light Fixture
Image credit: Stabs, Wingate

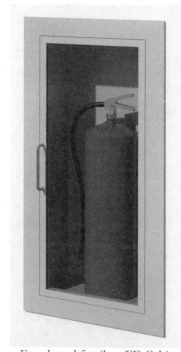

Face based family – FE Cabinet
Image courtesy of LHB, Inc.

Family Types

Revit *Families* have the ability to hold information in placeholders called parameters. These parameters are defined, when created, to hold a specific kind of information; for example, text, integer, number, length, currency, yes/no, etc. This helps to validate data when it is entered. You cannot type text into a parameter that is set to currency. Finally, some of the parameters, such as lengths, can be associated with actual dimensions in the *Family*. Adjusting a length parameter value actually changes the size of the *Family*.

Wall Cabinet:
With the information from the previous paragraph in mind, it is time to discuss how slight variations are dealt with for content that is geometrically the same; that is, for example, a wall cabinet is a rectangular box that has four different heights, two different depths and several widths.

It is possible to create 100 individual *Families*, each of which represents one wall cabinet. However, this would be very cumbersome to manage; if one change was required you would have to open 100 files and make the same change 100 times. This is where *Family Types* are employed.

A single *Family* can have several *Types* defined. A *Type* is simply a saved state of all the parameter settings. For example, the wall cabinet may have two *Types* defined as follows:

Type A		**Type B**	
Parameter	*Value*	*Parameter*	*Value*
Width	**24″**	Width	**27″**
Depth	14″	Depth	14″
Height	24″	Height	24″
Model	**W241424**	Model	**W271424**
Manufacturer	Merillat	Manufacturer	Merillat
LEED – 500mi	Yes/No	LEED – 500mi	Yes/No

Normally the *Type* name would be more meaningful than Type A and Type B. As you can see, the various parameters can be edited within each type. Therefore, one wall cabinet Family can represent 100 different sizes.

Some of the information in the table above has a direct impact on the 3D geometry; the Width, Depth and Height parameters can be associated with dimensions in the *Family*. When the parameter is changed the size of the *Family* actually changes. So one only has to change the *Family* type in a *Project Environment* to change the size of the placed *Family*. And, of course, some of the information has no effect on the 3D geometry. Rather it is used to track design decisions and for scheduling purposes. Shared parameters can show up in schedules.

There are always exceptions to the rule, but generally speaking, anytime the geometry changes you need to create a new *Family*. For instance, one Family can represent all the wall cabinets with a single door. But when the cabinet width dictates two doors a new *Family* is required. This change in geometry cannot easily be managed within a single *Family*.

When the geometry changes a new Family is typically required.
The example above has different handle, door and drawer locations.

Family Types within the Project Environment:

When utilizing content within the *Project Environment*, the *Family* and *Types* are presented as shown in the image below (Figure 15-1.2). Notice _Chair-Desk_ only has one *Type* defined within that *Family* whereas _Cook Top-2 Unit_ has three. Within the project anyone on the design team can manually add or delete *Types* associated with any *Family*.

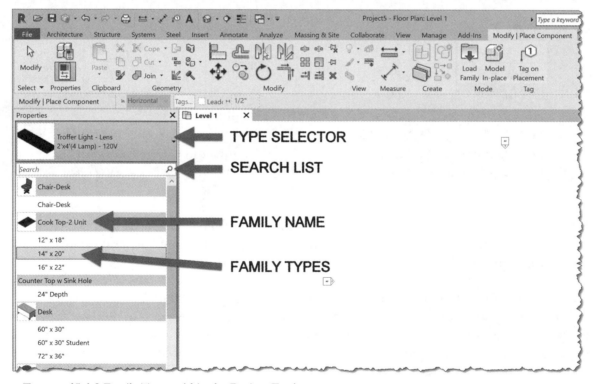

FIGURE 15-1.2 Family Types within the Project Environment

Type Catalog:

There is one last concept to point out about *Family Types* before moving on. When a *Family* is loaded into a project all of its *Types* are automatically loaded. Sometimes this is good and other times it is not. The image on the previous page (Figure 15-1.2) shows manageable lists of *Families*. However, you can imagine how this list would get rather long if you loaded the wall cabinet *Family*, described on the previous pages, that had 100 types. Not only would the list be long, but you are likely only using one depth and two heights.

To better manage this situation Revit has a feature for *Families* called a *Type Catalog*. Rather than loading all of the *Types* into the project with the *Family*, the user is prompted to select the *Types* desired. This process could be repeated until all of the *Types* are ultimately loaded for a specific *Family*.

The image below (Figure 15-1.3) shows the dialog box the user is presented with when loading the steel wide flange beam *Family*.

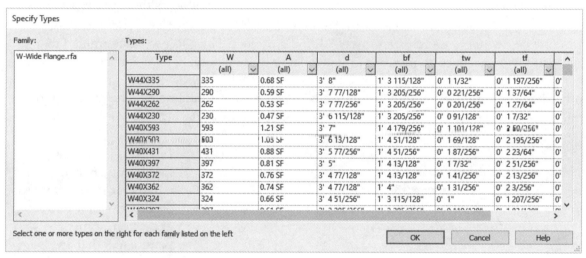

FIGURE 15-1.3 Selecting specific *Types* when loading a *Family* via the *Type Catalog*

The implementation and use of *Type Catalogs* will be covered later in the book.

The first *Family* you create, in the next lesson, will not even have a *Type* as they are optional.

Instance and Type Parameters

When creating *Parameters* for use in a *Family* you need to decide if it is an *Instance* or a *Type* parameter. Remember, parameters are placeholders for information in which some of them can be linked to geometry size and visibility.

An **Instance Parameter** only impacts the one or more *Families* that are selected, or about to be created within the *Project Environment*, whereas the **Type** parameter affects all instances of the *Family* within the project.

A good example to describe how Instance/Type parameters are used in a project is to consider a window. Looking at the image below you can see this window *Family* has been set up so that the width and height parameters are *Type* and the <u>Sill Height</u> is an *Instance* parameter. ***FYI:*** *You only know this here by the labels and arrows provided in the image (Figure 15-1.4).*

This works with the standard industry convention where all windows that are the same size (i.e., width and height) are labeled with a common mark or tag. For example, the 2'x4' window below might be referred to as "W1" in a project. Thus, if it was determined that all W1 window types need to increase in size to let in more natural light, changing one window's *Type Properties* would change the size of all "W1" windows in the project. On the other hand, the <u>Sill Height</u> dimension needs to be an *Instance* parameter. The "W1" windows might be 2'-4" above the first floor and 2'-8" above the second.

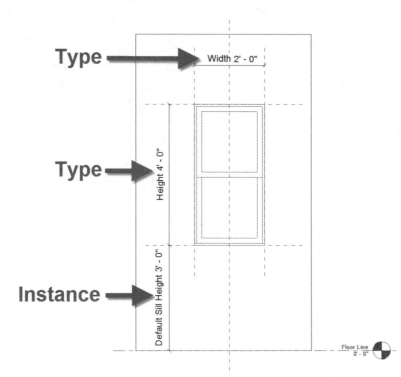

It is possible to edit a family and change a parameter from a *Type* to an *Instance*, with one exception: some parameters that are predefined in the template files are not editable. You cannot even change the name.

Final thought: a table *Family* might have its width and depth parameters set as *Type* or as *Instance* depending on how the design team wanted to use it in the project.

FIGURE 15-1.4 Exterior view of a window in the Family Editor

Family Templates

When starting a new *Family* you need to know what it is and how it will be used in a project. With this information you can then select a *Family Template* from which to begin the content creation process.

Selecting the correct template saves time setting up the most basic parameters and in some cases is critical to the *Family* working. For example, the door and the window templates have the following parameters setup:

Door Template			Window Template		
Parameter	*Value*		*Parameter*	*Value*	
Width	3'-0"		Width	3'-0"	
Height	7'-0"		Height	4'-0"	
Frame Width	3"		**Sill Height**	3'-0"	
Model	*blank*		Model	*blank*	
Manufacturer	*blank*		Manufacturer	*blank*	

Other templates, such as **Generic Model**, do not have a <u>Width</u> and <u>Height</u> parameter predefined. Additionally, the door and window templates are wall hosted whereas the *Generic Model* is not.

Revit comes with over 70 template files. This book will only provide a discussion on a handful of them, but many are self-explanatory and it should be possible to discern what most of them are for after completing this book.

Family Categories

Another item that needs to be covered at an overview level is the concept of a *Family Category*. Each *Revit Family* needs to be assigned to a specific category. This is done automatically by way of the template files for most *Families*. For example, starting a new *Family* from the window template file, the category has been set to window as shown in Figure 15-1.5.

The category setting does two basic things: it controls which command is used to place the *Family* and it relates directly to manipulating visibility within the project; turning off the window category for a specific view makes all the windows disappear in that view.

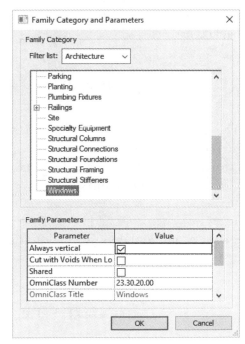

FIGURE 15-1.5 Family categories

Custom Family Libraries

Whether you are creating content for personal use or for a large firm, you should have a plan on how and where to store your custom content. There are probably an infinite number of solutions to this problem, but a relativly simple one will be suggested here.

It is best to store your custom Revit content in a separate folder and not in the folders that contain the Revit OOTB content (OOTB = out of the box, meaning the content that comes with the software). It is best to segregate the content that has been created or edited to comply with your firm's graphic or design standards. Additionally, each year, when a new version of Revit is released you would need to sift through all the OOTB content folders to find your custom content.

Revit has a specific set of folders created to house the OOTB content (see Figure 15-1.6). One suggestion is to create those same folders in a parallel folder and place your content there. Then, when content is needed, a user might first look through the OOTB location and then the firm location. At some point your custom content may be sufficient enough to reverse the order in which you look for content (i.e., Firm folder and then OOTB folder).

Of course, in a firm setting, the content should be located on a server so everyone has access to the same content, and so it gets backed up!

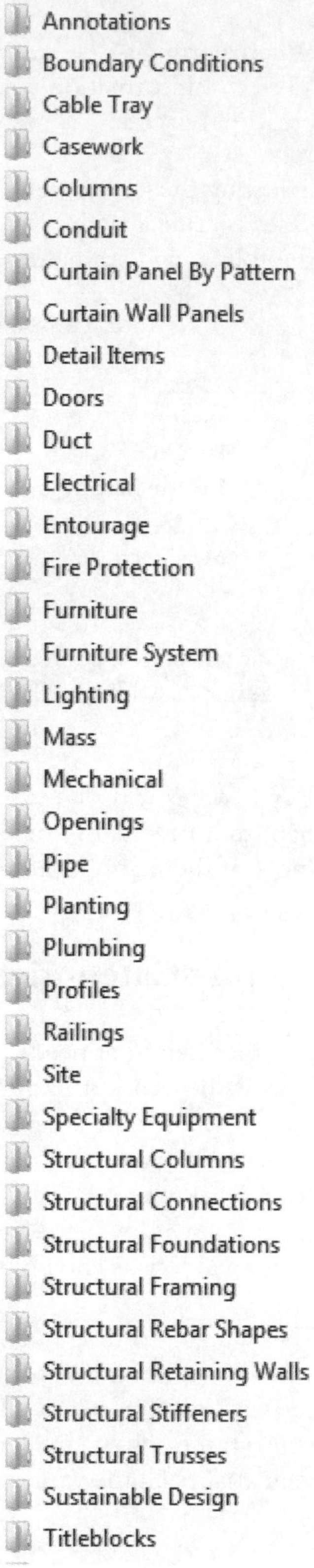

Naming Families

The name of the *Family* file (.rfa files) on your hard drive, or server, is also the name the users see in Revit. The following information is offered as a suggestion, as there is no hard and fast rule on *Family* naming.

The *Family* name should be concise and as short as possible, and it should not contain information that will appear in the *Type* name, or vice versa. For example:

<table>
<tr><td>*Family name*</td><td>*Type name*</td></tr>
<tr><td>Metal Locker – **Tall**</td><td>**Tall** 18″ x 18″</td></tr>
</table>

Annotations
Boundary Conditions
Cable Tray
Casework
Columns
Conduit
Curtain Panel By Pattern
Curtain Wall Panels
Detail Items
Doors
Duct
Electrical
Entourage
Fire Protection
Furniture
Furniture System
Lighting
Mass
Mechanical
Openings
Pipe
Planting
Plumbing
Profiles
Railings
Site
Specialty Equipment
Structural Columns
Structural Connections
Structural Foundations
Structural Framing
Structural Rebar Shapes
Structural Retaining Walls
Structural Stiffeners
Structural Trusses
Sustainable Design
Titleblocks

FIGURE 15-1.6
OOTB Family Folders

The *Family* name should be as generic as possible so others on the design team, or in the firm, can quickly ascertain what the various *Families* are. If only one type of locker is used then the *Family* name can be more generic. Similarly, the *Type* name within the *Family* should be easy to understand if possible; show the size of the locker rather than the model number. For example:

Generic naming convention:

<u>*Family name*</u>   <u>*Type name*</u>
Metal Locker – Tall   18″ x 18″

Detailed naming convention:

<u>*Family name*</u>   <u>*Type name*</u>
Arrow Locker – Tall   ALT181860

Locker family shown in perspective.

Remember, a *Family* often has several types defined within it. Thus, the example above would have several similarly named types which define several locker sizes.

The detailed naming convention above would be appropriate if a design firm used several locker manufacturers depending on the project type; maybe they specialize in sports facilities and design various types of locker rooms.

Another option, based on the example above, is to have a detailed *Family* name and a generic *Type* name:

<u>*Family name*</u>   <u>*Type name*</u>
Arrow Locker – Tall   18″ x 18″

A good naming convention will help promote efficiency in the project environment. Many of the *Families* are added to the Revit model via the catch-all tool: **Component: Place a Component** (on the *Architecture* tab). When the *Component* tool is active, the *Element Type Selector* lists all the *Families* that are loaded into the project—that is, all the *Families* that do not have their own insertion tool, as windows and doors do.

These *Families* are listed in alphabetical order in the *Element Type Selector*, within the project environment. Therefore, if the *Families* are not named properly, you might have a toilet at the bottom of the list, a Bathtub near the top and many other types of content sprinkled in between.

One solution to this problem would be to implement some sort of abbreviation or prefix naming system.

Families
- Annotation Symbols
- Casework
- Ceilings
- Columns
- Curtain Panels
- Curtain Systems
- Curtain Wall Mullions
- Detail Items
- Doors
- Floors
- Furniture
- Lighting Fixtures
- Parking
- Pattern
- Planting
- Plumbing Fixtures
- Profiles
- Railings
- Ramps
- Roofs
- Site
- Specialty Equipment
- Stairs
- Structural Beam Systems
- Structural Columns
- Structural Foundations
- Structural Framing
- Walls
- Windows

FIGURE 15-1.7 Project Browser within the Project Environment showing Family Categories

Each *Family* falls within a *Family Category* as previously discussed. These categories are listed within the *Project Browser*; note the list varies depending on the content currently loaded into the project. Creating a two letter prefix would sort the content by category within the *Element Type Selector*. For example:

CW	= Casework
FN	= Furniture
PF	= Plumbing Fixtures
SE	= Specialty Equipment

This naming convention would cause all the related content to be grouped together within the *Element Type Selector*.

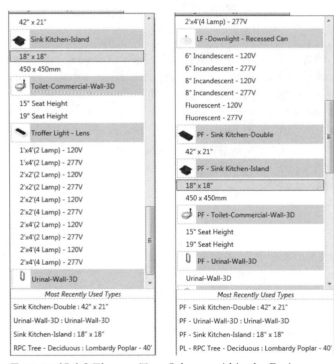

FIGURE 15-1.8 Element Type Selector within the Project Environment showing loaded Families available for placement via the Component tool. **Left**, default naming; **right**, prefix added to control sort order.

Utilizing a naming convention like this means a firm's name or initials cannot appear at the beginning of the name.

Family and Type naming is highly subjective, so one firm's solution may be completely different from another's. But, in any case, some sort of standard should be developed and agreed upon within each design firm.

Exercise 15-2:
The Box: Creating the Geometry

The emphasis will be placed on Revit features and techniques related to *Families* and not so much on geometry. A simple box will be created and then utilized to show many of the things that can be done with, and within, a *Family*. So, to keep things simple early on, a basic box will be used.

This exercise demonstrates how to start a *Family* and develop a 3D box that can be adjusted in size; i.e., a **parametric box**. Note that more detailed steps and graphics will be provided the first time a subject is covered. The subsequent steps for the same subject will be less detailed and may not contain graphics. It is therefore beneficial to work through the lessons in order.

Creating a New Family

1. Open Revit

    a. There is no difference to the user between 32bit and 64bit versions of the software. Therefore, it does not matter which one is used with this book.

2. **File Tab → New → Family**
   (Figure 15-2.1)

Assuming Revit was installed properly you should see a large list of files from which to choose. If not, you may download the *Family* template files from Autodesk's website.

The template names make it rather obvious as to what each template file is intended to be used for. When a template is selected, Revit makes a copy of the file and then opens it as an unnamed *Family* called **Family1** until saved.

FIGURE 15-2.1 Starting a new Family

3. Select **Generic Model.rft** and then click **Open** (Figure 15-2.2).

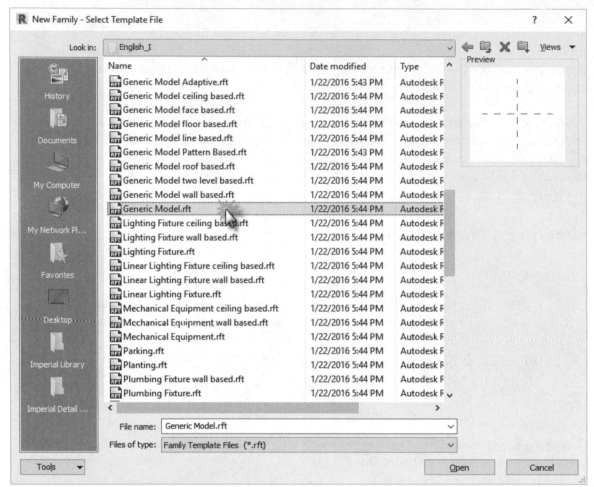

FIGURE 15-2.2 Selecting a Family template file

The initial view is a plan view from the top (Figure 15-2.3). As seen in the *Project Browser*, the following views have been established, via the template file:

- Floor Plan
- Ceiling Plan
- 3D View
- Elevations
  - o Back
  - o Front
  - o Left
  - o Right

FIGURE 15-2.3 Initial view shown reference planes

In the initial plan view, Figure 15-2.3, are two *Reference Planes*: one horizontal and one vertical. These two *Reference Planes* define the center of the content about to be created, in each direction. The intersection of the two also defines the origin. The origin is the insertion point, relative to your cursor, when placing the *Family* in a project. Ultimately the box to be created needs to be centered on the intersection of the reference planes.

The *Reference Planes* do not appear in the *Project Environment*. However, Revit is aware of them and can use them when dimensioning and aligning to other elements; this depends on how a few properties for the *Reference Planes* are set, but in any case the *Reference Planes* are never visible.

Creating the Framework for a New Family

A common method used when creating Revit content is to first add *Reference Planes*, make them parametric and then create the 3D geometry and lock its edges to the *Reference Planes*. This allows complex *Families* to be broken down into more manageable elements and makes controlling multiple 3D objects easier (i.e., moving one reference plane moves several 3D objects).

In the next few steps you will create *Reference Planes* that ultimately will control the size and location of a 3D box.

4. Select **Create → Datum → Reference Plane**.

To add a *Reference Plane* one simply picks two points in a view. A *Reference Plane* is a 3D plane that will appear in all views which cut through it, on end (or perpendicular to), or are within the Select/Elevation/Plan "view range." The edges of the *Reference Planes* are typically only visible unless "show" is toggled on under the *Create* tab on the *Ribbon*, and then only the current *Work Plane* is shown.

5. Add a horizontal *Reference Plane* as shown in Figure 15-2.4.

   a. The exact location and length does not matter at this time; it will be adjusted later.
   b. Make sure the *Reference Plane* is snapped to the horizontal plane before picking the second point. Watch for the cyan-colored dashed line and a *tooltip* which displays the word horizontal.

   FYI: The use of the word "horizontal" above is in reference to the computer screen.

The first *Reference Plane* has now been added to the new *Family*. Later in the book, a dimension will be added between the two parallel *Reference Planes* shown in Figure 15-2.4. This dimension can be locked so it does not move, or made to be parametric in which case the end user in the project environment can edit a value in the properties of a selected element, causing the *Reference Plane* to move accordingly.

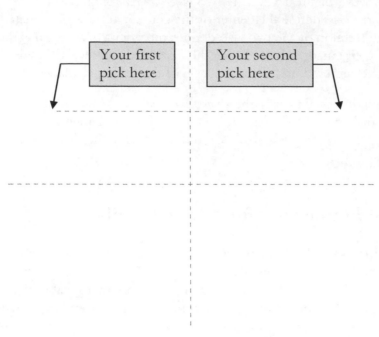

FIGURE 15-2.4 Adding a "horizontal" reference plane

6. Switch to the **Right** elevation view by double-clicking on it in the *Project Browser*.

Notice that the newly added *Reference Plane* is visible in this view, as well as the <u>Left</u> view. Again, this is because the *Reference Plane* is a 3D element.

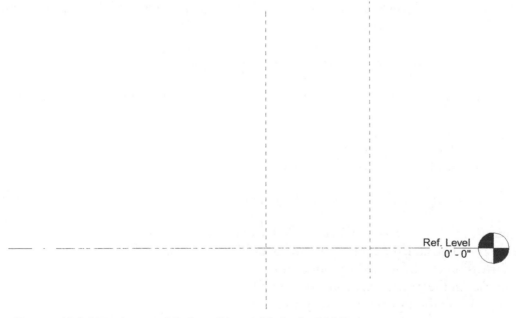

FIGURE 15-2.5 Newly created *Reference Plane* visible in the "side" views

7. Close the **Right** elevation view and switch back to the **Plan** view.

8. Draw three more *Reference Planes* approximately as shown in Figure 15-2.6.

9. Select **Modify**, on the *Ribbon*, to finish the *Reference Plane* command.

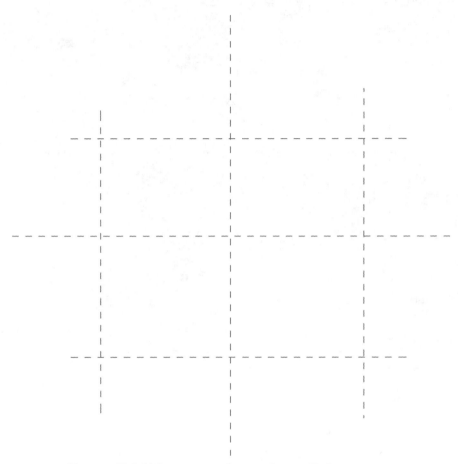

FIGURE 15-2.6 Three more reference planes added

The four *Reference Planes* just drawn will ultimately serve as the guides for the edge of the 3D box. The top and bottom edges of the box will be controlled by *Reference Planes* not visible in the current view.

Next, the newly added *Reference Planes* position will be adjusted relative to the center or origin *Reference Planes* that came with the template file. This can be done simply by selecting one *Reference Plane* at a time and editing the on-screen temporary dimension which appears.

10. Select the top horizontal *Reference Plane* (Figure 15-2.7).

11. Edit the *Temporary Dimension* to be **2′-0″** (Figure 15-2.7).

   a. If needed, adjust the ends of the *Reference Planes* so they cross each other and form a corner as shown below. Procedure: select a *Reference Plane* and then click and drag on the visible endpoint grip.

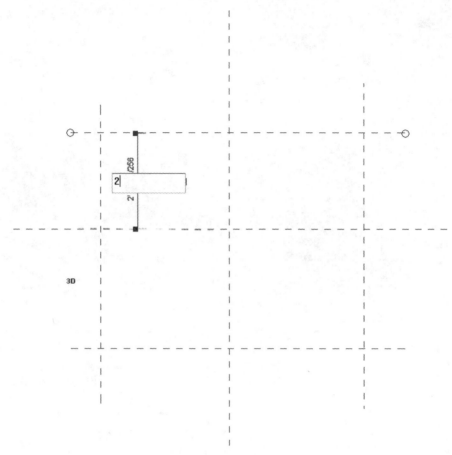

FIGURE 15-2.7 Top "horizontal" reference plane selected

12. Modify the remaining three *Reference Planes* to have the same **2′-0″** dimension off of the original centerline/origin *Reference Planes* that came with the template.

The four *Reference Planes* now define the extents of a 4′-0″ x 4′-0″ box. This only defines the sides of the box. Next the top and bottom of the box will be defined.

FIGURE 15-2.8 Four reference planes with position adjusted

13. In the *Project Browser*, double-click **Front** under *Elevations*.

The <u>Front</u> (or South elevation) shows two of the four *Reference Planes* created in this exercise. All *Reference Planes* that are perpendicular to the view are visible, as long as they are within view range. Additionally, the center (left/right) *Reference Plane* and one in line with the **Ref. Level** is visible, both of which came from the template.

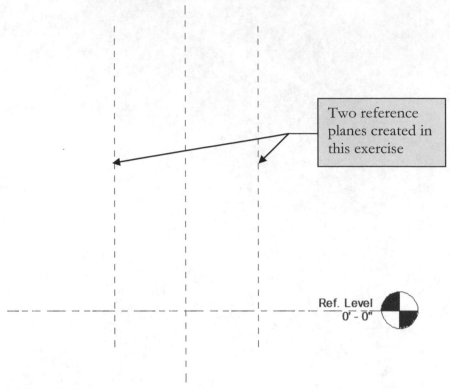

FIGURE 15-2.9 Front view

Next, a *Reference Plane* will be drawn as a guideline for the top of the 3D box. A *Reference Plane* is always drawn perpendicular to the "plane" that is being defined.

14. Draw a *Reference Plane* **2'-0"** above the Ref. Level (Figure 15-2.10).

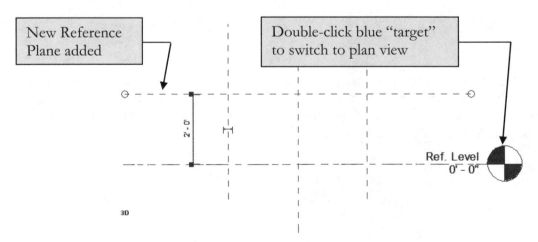

FIGURE 15-2.10 Front view; relocated ref. level

15. Switch back to the plan view, double-click **Ref. Level** under *Floor Plans* in the *Project Browser*, or Double-click the blue target; see Figure 15-2.10.

Creating the 3D Geometry

Now that the framework, or guidelines, have been established using Reference Planes, the 3D geometry can be created. A simple box will be created.

16. Select **Create** → **Forms** → **Extrusion** from the *Ribbon*.

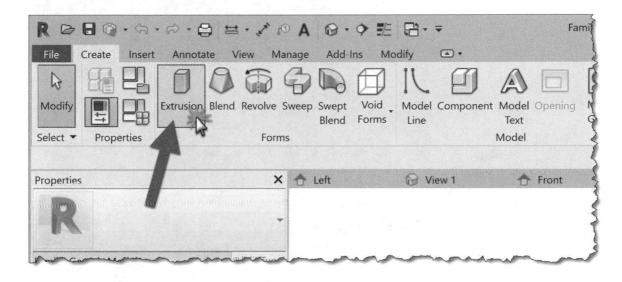

17. Draw a square within the outer *Reference Planes* (Figure 15-2.11).

    a. Select "rectangle" on the *Ribbon* in the *Draw* panel.

    b. Notice the depth is set to 1'-0" on the *Options Bar*; this is fine.

    c. The exact size does not matter here.

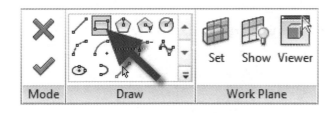

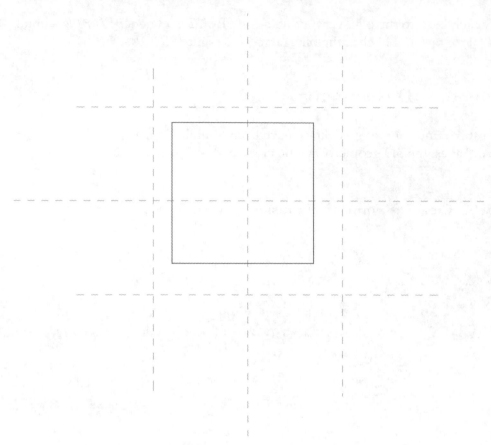

FIGURE 15-2.11 Sketch lines for 3D extrusion

When creating a *Solid Extrusion*, a simple 2D outline is sketched as in Figure 15-2.11. This defines the perimeter of the extrusion. When "finish extrusion" is selected, the 2D outline is extruded perpendicular to the sketch lines to a thickness (or depth) specified on the *Options Bar*.

The 2D sketch must be "clean," meaning no gaps or overlaps occur at the corners. Because the rectangle option was selected, the outline will automatically be "clean."

18. Select the **green check mark** on the *Ribbon*.

With the previous three simple steps a 3D box was created. This box will now serve as our "test subject" for an introduction on many of the basic options and settings that can be done with a *Family*.

The last step, in this exercise, is to align and lock the 3D geometry to the *Reference Planes*. Thus, whenever the *Reference Planes* move, so will the geometry. Revit provides a tool to easily do this; it is called **Align**. The tool brings two lines and/or surfaces into alignment. Once the *Align* tool has been employed, the opportunity to "lock" the relationship is available. This lock creates a parametric relationship within the *Family*. This is just one of the tools in which a *Family* can be made parametric.

19. Select **Modify → Align** on the *Ribbon*.

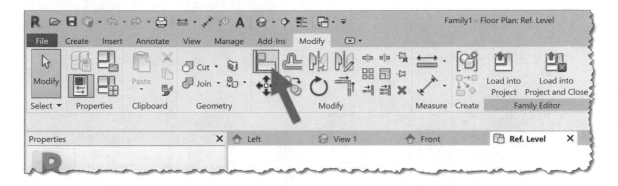

20. Select the vertical *Reference Plane* on the right (Figure 15-2.12).

21. Select the right-hand side of the box.

The right side of the box should now be aligned with the *Reference Plane* (Figure 15-2.12). If the *Reference Plane* moved rather than the edge of the box, click **Undo**. When using the *Align* tool, the element that does not move is selected first.

22. Click the **Padlock** icon that appears to lock the relationship between the edge of the box and the *Reference Plane*.

    a.   Figure 15-2.12 shows the icon in its initial position, unlocked.

    b.   Clicking the "unlocked" icon will change it to a "locked" icon.

    c.   When a *Reference Plane* is selected the "locked" icon will appear.

        i.   This lets the users know a "lock" exists.

        ii.   Clicking the "locked" icon will unlock the relationship.

    d.   Selecting the 3D box does not reveal the locked icon (except when in "edit sketch" mode).

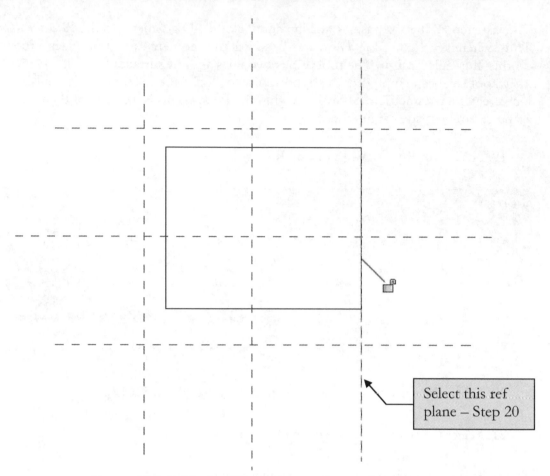

FIGURE 15-2.12 Aligning 3D geometry to reference planes

23. **Align** and **Lock** the other three sides of the box (Figure 15-2.13).

    a. Remember to select the *Reference Plane* first.

    b. Select *Undo* on the *Quick Access Toolbar* if you select things in the wrong order.

The four sides of the 3D box now have a parametric relationship to the four *Reference Planes* created in the plan view.

Next, the top and bottom of the 3D box will be *Aligned* and *Locked* to *Reference Planes*.

24. Switch to the **Front view** via the *Project Browser*.

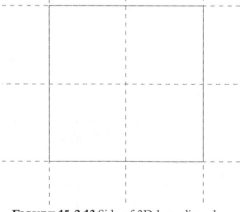

FIGURE 15-2.13 Side of 3D box aligned with *Reference Planes*

25. **Align** and **Lock** the top of the box, which is 1'-0" high, to the top *Reference Plane* (Figure 15-2.14).

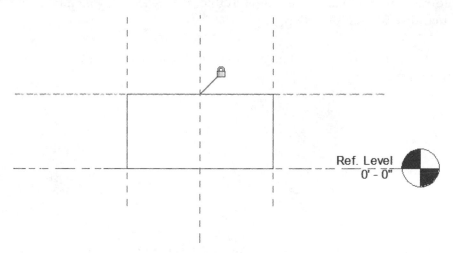

FIGURE 15-2.14 Aligning 3D geometry to reference planes

26. **Select the 3D box** to reveal the edit grips (Figure 15-2.15).

Notice the four triangle shaped grips on each side of the box. These grips can be dragged, which repositions the selected side. The one limitation with editing geometry with these grips is that there is no way to control the distance the edge is moved.

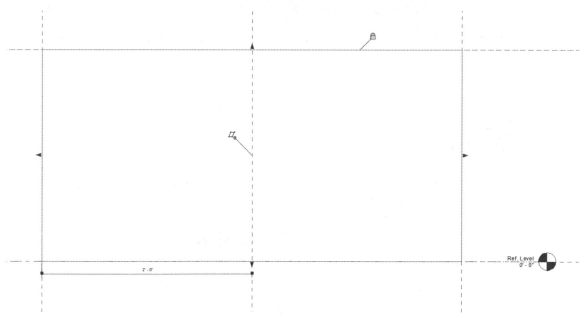

FIGURE 15-2.15 3D geometry selected, edit grips revealed

In the next step, the bottom of the box will be raised to make it easier to *Align* and *Lock* it to the bottom *Reference Plane*. It is not required that this be done. However, it is easier to select things in the correct order and assure the desired things are selected, which can be challenging when several things overlap.

27. Click and drag the bottom grip up as shown in Figure 15-2.16; the exact position does not matter.

FIGURE 15-2.16 Bottom of 3D box repositioned

28. *Align* and *Lock* the bottom edge of the 3D box with the bottom *Reference Plane*.

29. Switch back to the **Plan View**.

Flexing the Family

Now that the 3D box has been tied to the *Reference Planes*, any time the *Reference Planes* are moved the 3D box will move with it. This will be tested next.

30. Select the *Reference Plane* on the right and use the **Move** tool to move it **1'-0"** more to the right (Figure 15-2.17).
    a. The *Move* tool is only visible when something is selected.
    b. The exact distance here does not matter.

31. Notice the 3D box moves with the *Reference Plane* (Figure 15-2.17).

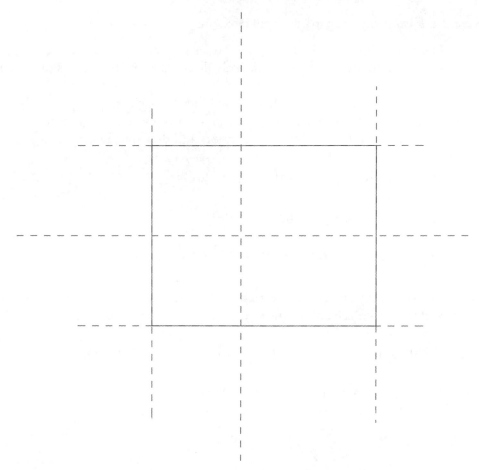

FIGURE 15-2.17 Right-hand reference plane repositioned

32. Click **Undo** on the *Quick Access Toolbar* to undo the previous step.

Now the *Family* will be loaded into a project to show how that process works. Once in the project, the content does not have any direct connection to the external *Family* created in this lesson.

33. Open a **new project**; see chapter one if needed.

    a.   Use the default template.

34. Switch back to the **Family Editor** by pressing **Ctrl + Tab**.

    a.   Ctrl+Tab cycles through open projects and views.

Before loading the *Family* into a project it should first be saved to a file. The main reason is to establish the *Family's* name. The *Family's* file name becomes the name of the content within a Revit project.

35. Save the *Family* to your hard drive as **Box.rfa**.

   a. The location where the file is saved does not matter; however, it would be a good idea to create a folder in which all the files created in this text are stored.

   b. It is highly recommended that all data files be backed up regularly. If a Revit file becomes corrupt a backup may be the only solution; another option would be to send the file to Autodesk support and they may be able to salvage it.

36. Click the **Load into Project** button on the *Ribbon*.

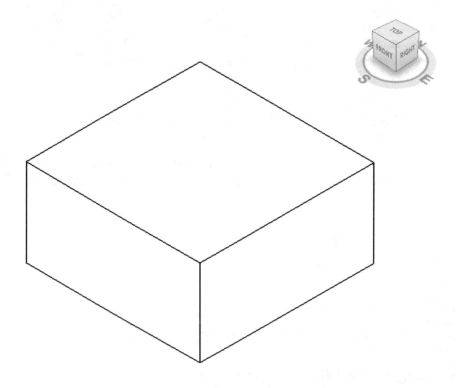

Revit automatically switched back to the Revit Project.

37. *In the project:* Select **Architecture → Build → Component**.

38. With the *Box* family current in the *Element Type Selector*, click somewhere within the floor plan view to place an instance of the box.

39. Click the **3D** icon on the *Quick Access Toolbar* to see the 3D box.

40. **Save** the *Family* as **Box** and the *Project* as **Box Project**.

Exercise 15-3:
The Box: Adding Parameters

In this exercise the steps required to make the box parametric will be covered. Parametric, in this case, means that certain parameters control the size of the box. Thus, in the project environment it is possible to select the box, go to its properties, edit a few parameters (e.g., width and height) and then the box will change size accordingly. This is a very powerful feature and can save time creating content, as a *Family* is not required for every size.

All parameters are either a ***Type*** parameter or an ***Instance*** parameter. An introduction to this concept was presented on page 15-8 of this chapter. This exercise will help to better explain the differences and how to implement them.

1. Open the ***Box*** *Family* created in the previous exercise.

    a. It is recommended that a copy of the "Box" *Family* be saved for each exercise in case problems arise and an older file is needed to revert back to. Maybe copy the "Box" *Family* file and rename it to Box 15-2 to save a copy of the *Family* from the previous exercise.

Adding Dimensions

The first step in making 3D geometry parametric is to add dimensions in a view. Two dimensions will be added in plan view, which will eventually be tied to parameters to control the width and depth of the 3D box.

2. In the *Ref. Level* plan view, select **Annotate → Dimension → Aligned**.

    Aligned

3. Click the vertical left and right *Reference Planes*, and then click a third point to position the dimension line (Figure 15-3.1).

    TIP: *The third click needs to be in a blank area, as clicking on something will add another segment to the dimension string.*

    TIP: *Be careful to select the reference plane and not the 3D box.*

4. Similar to the two previous steps, add another dimension for the other side of the box (Figure 15-3.2).

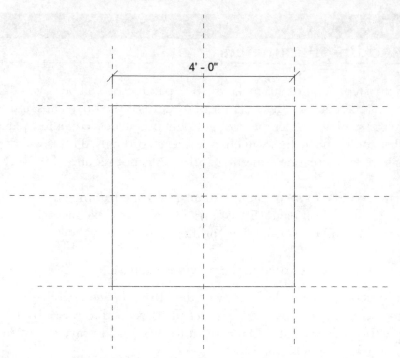

FIGURE 15-3.1 Dimension added to reference planes

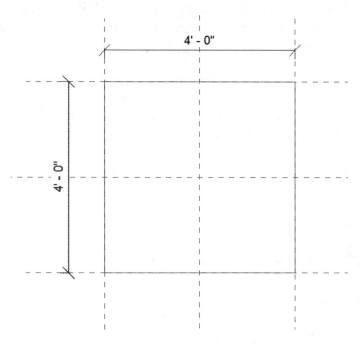

FIGURE 15-3.2 Second dimension added

Creating Parameters

Now that the dimensions are placed, they can be tied to parameters. The dimensions can be tied to previously created parameters or a new parameter can be created based on the selected dimension. The latter will be employed, that is, selecting a dimension and then creating a parameter for it.

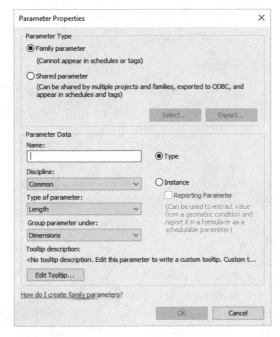

When defining a parameter, the user must specify the following:

- Name
- Parameter Type
  - ○ Family Parameter
  - ○ Shared Parameter
- Group parameter under
- Instance or Type
- Discipline
- Type of Parameter

You will gain a better understanding of how these various settings work once you start using them, but the following descriptions are offered as a primer for what is about to be covered in the next few steps.

Name: The *Parameter Name* should be descriptive. It can have spaces and symbols. However, dashes and other mathematical formula symbols should be avoided as they will confuse Revit when using them to do calculations. The name IS case sensitive. One common naming convention is to name all custom parameters with all uppercase letters to distinguish the Revit default parameters from the custom ones. The name can be changed at any time.

Parameter Type: The *Parameter Type* concept takes a little time to fully understand. A simple explanation will be provided here so as not to get too bogged down in the details early on.

Family parameters: cannot appear in schedules (e.g., a door schedule or a furniture schedule) or be used in tags (e.g., a door tag or a furniture tag). Thus, if one were to create a *Family Parameter* to keep track of the recycled content, that value (e.g., 80%) could not appear in a schedule, nor could it appear in a tag (e.g., a tag that listed the item number and directly below it the 80% value). This is the option that will be used initially in this book.

Shared parameters: can be shared by multiple projects and families, and appear in schedules and tags. This method requires the use of an external text file to manage the parameters.

Group parameter under: The *Group parameter under* setting simply specifies which section to place the parameter in when displayed in the properties dialogs in the project (see Figure 15-3.3). This setting can be changed at any time after the parameter has been created.

Instance vs. Type: The *Instance* vs. *Type* parameter has already been discussed; see page 15-8. This setting can be changed at any time after the parameter has been created.

Discipline: This setting is simply a way to manage the large number of value types as can be seen in the information below.

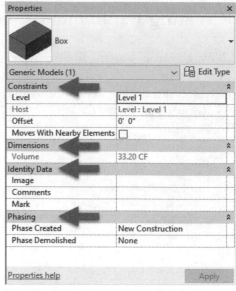

FIGURE 15-3.3 Instance Properties dialog for box family (in a project environment)

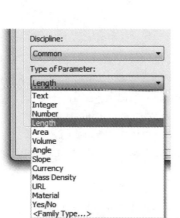

Type of Parameter: The *Type of Parameter* is how Revit knows what type of information will be stored in a specific placeholder, or parameter.

Looking at the images to the left, one can see the various disciplines and parameter types that can be used.

Setting this properly has the following benefits:

- *Input validation*: when a *Parameter* is set to the type *Integer*, a user working with the family in a project cannot enter a decimal number.

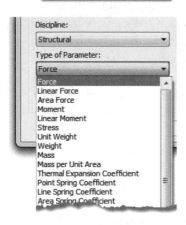

- *Proper formatting*:
  o Common, Currency = 1.50
  o Electrical, Wattage = 60 W
  o Structural, Force = 2.45 kip

FYI: Additional formatting can be done in the project via Project Units and also within the scheduling functionally.

This setting cannot be changed once the parameter has been created; it would have to be deleted and recreated.

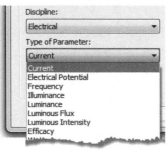

5. Select the horizontal dimension (see Figure 15-3.4).

Notice one of the options on the *Ribbon* is **Label Dimension**, and it has a drop-down list next to it. This might better be titled "Parameter" because the drop-down list presents *Parameters* that can be tied into the selected dimension and ultimately used to drive the dimensions from the *Properties* dialog from within a project. However, once a parameter has been tied into a dimension a *Label* does show up next to the dimension text as will be seen in a moment, so the *Label* title is not totally inappropriate.

6.   Click the **Create Parameter** icon on the *Ribbon* (see Figure 15-3.4).

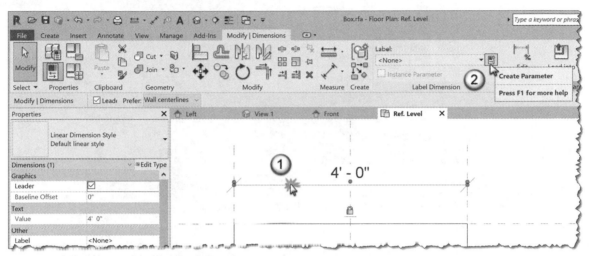

FIGURE 15-3.4 Adding a parameter to a previously drawn dimension

The information on creating parameters, covered on the previous two pages, can now be put to use. Notice, in Figure 15-3.5, that the *Discipline* and *Type* of *Parameter* have been automatically selected because there are no other options when tying a parameter into a dimension. Therefore, creating a parameter for a dimension using this method saves a little time.

7.   Enter the following in the *Parameter Properties* dialog (Figure 15-3.5):

    a.   *Parameter Type:*   **Family parameter**
    b.   *Name:*   **Width**
    c.   *Group under:*   **Dimensions**
    d.   *Instance vs. Type:*   **Type**

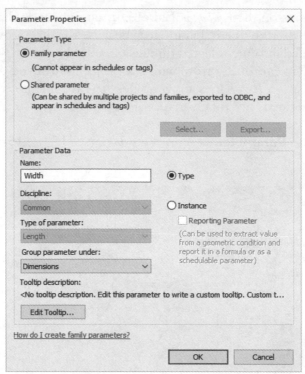

FIGURE 15-3.5 Creating the *Width* parameter

8. Repeat the previous steps to add a **Depth** parameter, or label, to the vertical dimension in the *Ref. Level* view.

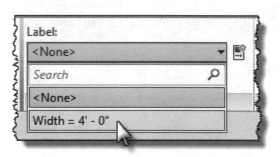

Notice, on the *Options Bar*, that previously created parameters are listed in the *Label* drop-down list. If the <u>Width</u> parameter were selected for the vertical dimension, one parameter would drive both the width and depth of the box.

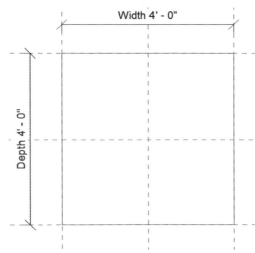

Looking at Figure 15-3.6, notice both dimensions have a *Label* associated with the dimension. This *Label* helps keep track of which dimensions are parametric. Another use for dimensions in a family is to "lock" it in order to maintain a relationship or spacing.

Even though the *Reference Planes* show up in the elevation views (front, left, etc.), the dimensions do not. So it is not visually obvious in the <u>Left</u> view that the *Depth* parameter has been created.

FIGURE 15-3.6 *Depth* parameter added

Next, an *Instance Parameter* will be created for the height of the box. Once the box family is loaded into a project, the height can vary from box to box (i.e., per instance). However, if the width (or depth) is changed, all boxes in the project will change. You will test this in a project momentarily.

9. Switch to the **Front** view.

Notice the *Width* dimension is not visible in this view.

10. Add a **dimension** from the *Reference Level* to the *Reference Plane* at the top of the box.

11. Add a **Height** parameter to the dimension (see Figure 15-3.7 and 15-3.8).

    a. Be sure to select the **Instance** option.

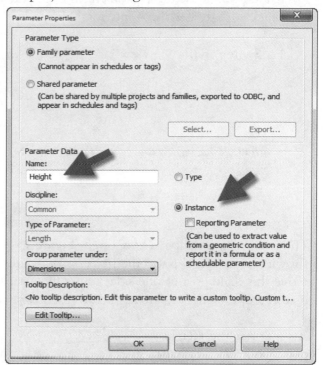

FIGURE 15-3.7 Creating the *Height* parameter

As can be seen in Figure 15-3.8, the Height parameter has been successfully added. This dimension can be seen in opposite views, the <u>Back</u> view in this case, but it cannot be seen in the side views, Left and Right in this case.

FIGURE 15-3.8 Front view; *Height* parameter added

15-35

Flexing the Family:

Now that parametric dimensions have been added to the *Family* it is necessary to flex them and make sure they work as intended before loading into a project.

12. Switch to the **3D view** via the *Quick Access Toolbar.*

13. Click the **Family Types** tool on the *Ribbon.*

> ***FYI:*** *Notice the last two tools on the* Ribbon *are repeated for each tab; this is for convenience within the Family Editor.*

The *Family Types* dialog is now open (Figure 15-3.9). Notice the three parameters created are showing up under the specified *Dimensions* heading. The easiest way to "flex" the *Family* is to move this dialog off to one side, adjust the dimension(s) and then press *Apply.* When *Apply* is selected, the changes are applied to the *Family* without the need to close the dialog box.

Also, notice the Height parameter has "(default)" next to its name. This is because it is an Instance parameter. The Height can vary, but when the Family is initially placed, the "default" value will be used.

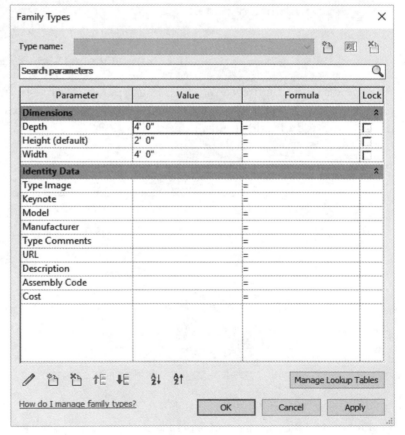

FIGURE 15-3.9 Family Types dialog

14. In the *Family Types* dialog, make the following adjustments:

    a.   Height: **6″**

    b.   Width: **1′-0″**

    c.   See Figure 15-3.10.

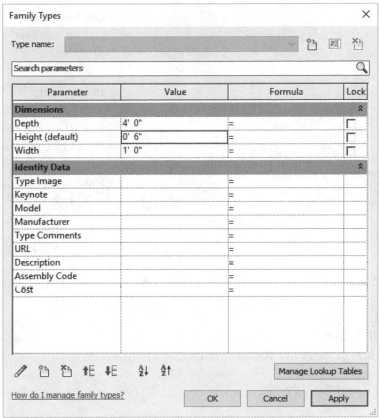

FIGURE 15-3.10 Family Types dialog - Modified

15. Move the dialog box off to one side, so the 3D view of the box can be seen, and drag on the dialog title bar; this will allow the changes to the box to be seen.

16. Click **Apply**.

TIP: Clicking Apply *within a dialog box commits the changes to the model. This allows the user to see if the modification looks correct before closing the dialog box and possibly needing to reopen it. If the changes do not need to be visually inspected first, the* OK *button can simply be selected – it is not necessary to click* Apply *first.*

The size of the box should now be modified as shown in Figure 15-3.11 below. When flexing the family, the size was changed by a large enough amount to make it unmistakable that a change occurred. Just changing the numbers by an inch or two might make it hard to notice the adjustment visually.

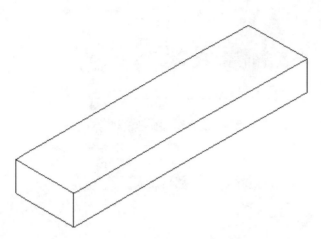

Next the box will be loaded into the Project environment; there it will be shown how the family can be adjusted via the properties.

Because the *Width* and *Depth* are **Type Parameters**, any change will affect all instances of the box. On the other hand, with the *Height* being an **Instance Parameter** – one change will only affect the selected objects.

FIGURE 15-3.11 3D Box - Modified

17. Open the *Project* from the previous exercise **Box Project**. Do <u>not</u> close the **Box** *Family*.

18. Press **Ctrl + Tab** until the **Box** *Family* is current (Ctrl + Tab switches between the currently open views).

Load into Project

19. Click the **Load into Project** button on the *Ribbon*.

The **Box Project** is a Revit *Project* that already contains a *Family* named **Box**, which was loaded at the end of the previous exercise. When Revit notices that a *Family* is being loaded with the same name as one previously loaded, it will present the user with the **Family Already Loaded** dialog, see Figure 15-3.12, which provides a few options to choose from. Revit will not assume that the file being loaded is a replacement for the previously loaded *Family*. Rather, the user must decide by selecting one of two overwrite options or clicking *Cancel* to not load the *Family* at all. It could be that a totally different box was created or downloaded and should not replace the currently loaded box; in this case you would click *Cancel* and rename the *Family* so it does not conflict with the previously loaded one.

Most of the time the first overwrite option should be selected, as the second may overwrite project specific changes to the *Family* such as the **Cost** parameter.

20. Click **Overwrite the existing version**.

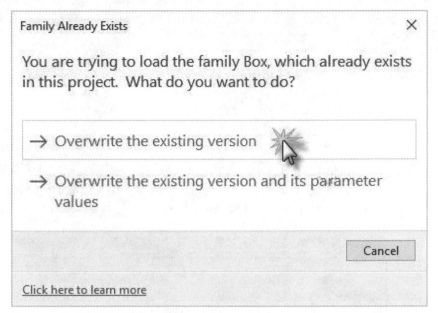

FIGURE 15-3.12 Family Already Exists prompt

TIP: Revit will only display the "Family Already Exists" prompt if the Family *file being loaded has been modified as compared to the version already in the* Project *file.*

If the user has more than one Revit project open, a dialog displays in which the user selects which project(s) to load the *Family* into.

The **Box Project** file becomes the current view on the computer screen and the **Box** *Family* changes size to match the values of the newly added *Parameters.*

21. In any view (i.e., Level 1, Elevations, 3D), **select** the previously placed **Box** within the *Drawing Window*.

Notice the selected element temporarily turns blue and the *Modify Generic Models* contextual tab is displayed on the Ribbon.

22. With the Box selected, notice the *Properties Palette*.

Of the three parameters created in this exercise, only one shows up under the *Instance Parameters* (see Figure 15-3.13). The other two parameters, <u>Width</u> and <u>Depth</u>, will be visible in the *Type Parameters* dialog which will be explored next.

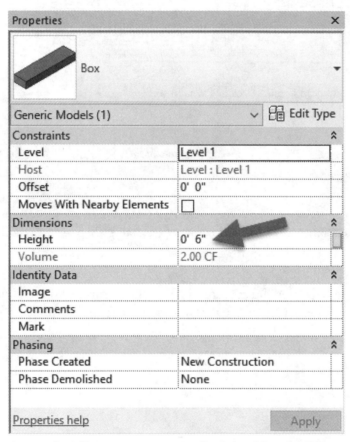

FIGURE 15-3.13 Instance Properties for updated Box family

23. In the *Properties Palette* dialog box, click **Edit Type**.

The two *Type Parameters* show up here in the *Type Properties* dialog; see Figure 15-3.14. An example will be shown next on exactly what the difference is between an *Instance Parameter* and a *Type Parameter*. However, it may be helpful to go back, at this time, and review the **Instance and Type Parameters** discussion earlier in this chapter (see page 15-8) before proceeding.

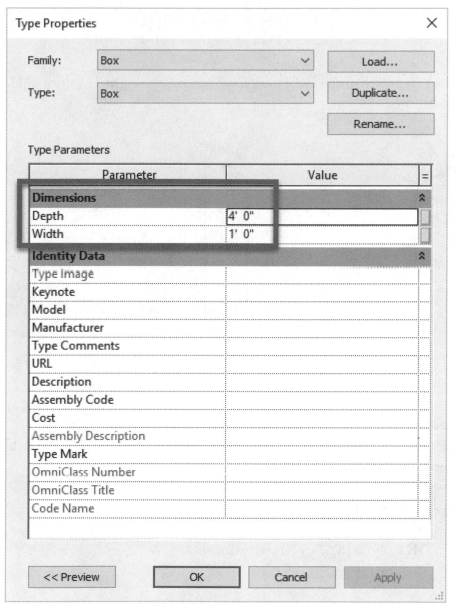

FIGURE 15-3.14 Type Properties for updated Box family

24. Close the dialog box by selecting **Cancel**.

Next, two additional boxes will be placed to show the difference between *Instance and Type Parameters*. This can be accomplished by clicking the *Place a Component* button on the *Architecture* tab, as was done to place the first box, or by simply copying the existing box.

25. Select the box, if not already selected, and then click **Copy** on the *Ribbon*.

> *WARNING: Do not click the* Copy *command in the* Clipboard *panel; this is for copies between views or projects! Rather, use the* Copy *command on the* Modify *panel.*

26. Make two copies of the *Box Family* **6'-0"** towards the **right** as shown in Figure 15-3.15 below.

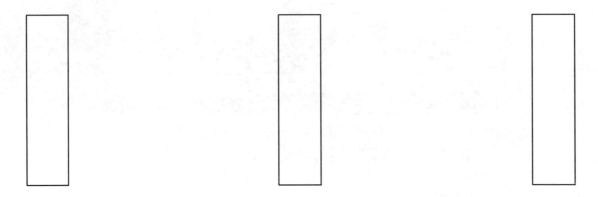

FIGURE 15-3.15 Box family copied for *Type* versus *Instant* study

27. Select the *Box* instance on the far left, and then view its *Type Parameters*; click **Edit Type** on the *Properties Palette*.

28. Change the **Width** *Parameter* from 1'-0" to **4'-0"**.

29. Click **OK** to close the *Type Parameter* dialog.

Because the <u>Width</u> *Parameter* is a *Type Parameter*, all instances of the type <u>Box</u> are updated instantly; see Figure 15-3.16 below. The <u>Depth</u> and <u>Height</u> for each remain unchanged. It makes no difference which box is selected anywhere in the *Project*. If this box existed on 20 different floors and in multiple phases, they would all have been updated. This is an important concept to understand because, just as it is easy to make several corrections or revisions with this technique, it is just as easy to make several errors in the design.

If additional widths are needed, either an additional *Type* would need to be added to the *Family* or a totally new *Family* would need to be created. ***FYI:*** *If the geometry is the same, and just the dimensions need to change, one would typically make a new* Type *within an existing* Family *rather than a new* Family.

FIGURE 15-3.16 Type parameter <u>Width</u> changed

Next, the *Instant Parameter* <u>Height</u> will be studied. To see this change a vertical (i.e., elevation) or 3D view needs to be opened.

30. Switch to the **South** exterior elevation via the *Project Browser*.

FIGURE 15-3.17 South exterior elevation

Notice the Boxes are all **6″** tall (Figure 15-3.17) which was set previously in the *Family Editor*, prior to being loaded into the current *Project*.

31. Select the box in the middle and view its *Instance Properties* via the *Properties Palette*.

32. Change the **Height** from 6″ to **2′-3½″**.

Notice that just the selected instance was changed.

FIGURE 15-3.18 South exterior elevation middle box modified

TIP: It is possible to change multiple instances at once if they are all selected prior to editing the Properties Palette *dialog. Additionally, if EVERY instance in the entire Project needs to be modified, simply select one and then right-click and pick the "Select all instances" option. Caution should be used when employing the second tip as things change in ALL views and on ALL levels, and in ALL phases of the current Project when the "entire Project" option is selected.*

33. **Save** both the *Project* and the *Family* as they will be further developed in the next exercise.

Revit Model Content Style Guide

A number of years ago Autodesk released a document intended to help standardize how Revit content is made. This will help when sharing projects between offices and when downloading content from manufacturer's websites. Specifically, adding tags and scheduling will be greatly streamlined. This file is not readily available from Autodesk anymore but can be found by searching "Revit Model Content Style Guide" in your web browser.

Exercise 15-4:
The Box: Formulas and Materials

The ability to add formulas and materials will be covered in this exercise.

The ability to add **formulas** means that the <u>Width</u> *Parameter* could be "programmed" such that it is always half the size of the depth *Parameter*; this is the example that will be explored in this exercise. Additionally, one could control the spacing of shelving brackets based on the length of the shelf, or the size of a lintel based on the width of the window below it; the lintel would need to be in the window family in this case. As should be obvious, the possibilities are many.

Defining **materials** in the *Family Editor* means that the component is ready to be rendered the moment it is placed in the project. It may not always be possible to anticipate what the material should be, but it is often more convenient to have something selected rather than nothing, which renders a flat gray color.

Cabinets with different materials applied (same family)

Adding a formula:

The following steps show how to add a formula to a parameter.

1. Open the **Box** *Family* created in the previous exercise.

    a. It is recommended that a copy of the **Box** *Family* be saved for each exercise in case problems arise and an older file is needed to revert back to. Maybe copy the **Box** *Family* file and rename it to **Box 15-3** to save a copy of the *Family* from the previous exercise.

2. Click the **Types** button on the *Ribbon* (see image to right).

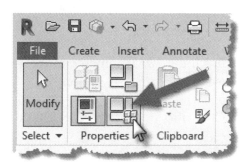

Notice in the *Family Types* dialog below that a column named *"Formula"* exists. *Whenever* something is entered in this column the *Value* is grayed out for that *Parameter* (i.e., row); the result of the formula becomes the *Value* for that *Parameter*. A preliminary example is shown below.

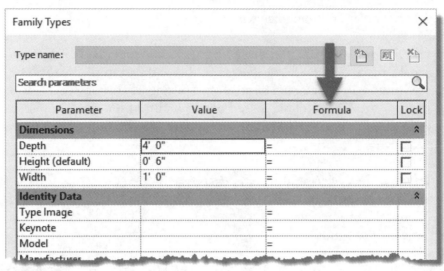

FIGURE 15-4.1 Family Types dialog

It is often helpful to open and explore the various *Families* that come with Revit; this can provide much insight on how to do certain things. The image below shows several formulas used to define a pipe elbow *Family* that comes with Revit. The values next to the formulas are the result of the adjacent formula; it is not possible to manually enter a value in this case, when a formula is present. Notice how other parameters can be used in formulas; this is case sensitive.

Parameter	Value	Formula
Dimensions		
Tick Size (default)	115/256"	= Fitting Outside Diameter * 0.4 * tan(Angle / 2)
Nominal Radius (default)	1/2"	=
Nominal Diameter (default)	1"	= Nominal Radius * 2
Insulation Radius (default)	9/16"	= Fitting Outside Radius + Insulation Thickness
Fitting Outside Diameter (default)	1 1/8"	= text_file_lookup(Lookup Table Name, "FOD", Nominal Dia
Center to End (default)	13/16"	= Center Radius * tan(Angle / 2)
Angle (default)	90.000°	=

FIGURE 15-4.2 Sample Family showing complex formulas

3. For the *Width* parameter, in the *Formula* column, enter the following: **Depth/2** (see Figure 15-4.3).

4. Click into a different cell within the *Family Types* dialog to see the *Value* update.

Notice the <u>Width</u> *Value* is now half of the <u>Depth</u>, i.e., 2'-0" (see Figure 15-4.3).

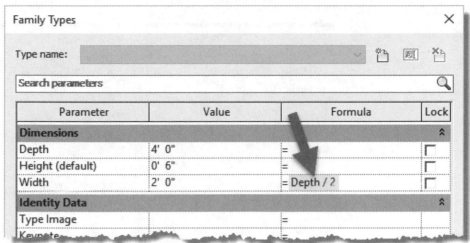

FIGURE 15-4.3 Adding formula to control the width

TIP: *When using parameters in a formula, keep the following in mind:*

- *Parameters in formulas must match the case (i.e., uppercase, lowercase, mixed) of the Parameters being referenced. Typing "depth" rather than "Depth" in the example above would not work – Revit would show an error message.*
- *A Parameter used in a formula must exist in the Family before it can be used.*
- *If a Parameter is deleted, any formulas that use that Parameter will also be deleted. Revit will present a warning first.*
- *If a Parameter is renamed, Revit will automatically rename the Parameter in all formulas in which it is used.*

Below are a few of the basic symbols Revit can use in formulas. See Revit's *Help System* for a full list:

- *Addition*: **+**
- *Subtraction*: **-**
- *Multiplication*: *****
- *Division*: **/**
- *Tangent*: **Tan**
- *Cosine*: **Cos**
- *Sine*: **Sin**

> Conditional Formatting
>
> It is also possible to use conditional statements in a formula such as IF, AND, OR, <, >, =
>
> Here are a few examples:
>
> - if(Depth > 3', 2', 1')
> - if(and(Depth > 3', Height = 1'), 2', 1')
> - See Revit's *Help* for more info.

Next the *Family* will be loaded into the *Project Environment* to see the formula in action.

5. Open the project file **Box Project**, and then switch to the **3D view**.

6. Press **Ctrl + Tab** until the **Box Family** file is current.

7. Click the **Load into Project** button on the *Ribbon*.

Notice the size of the box changes. When the <u>Depth</u> is now changed, the <u>Width</u> will automatically be updated. This will be tested next.

8. Select one of the boxes and, via its *Type Properties*, change the **Depth** to **6'-0"** and then

    a. Click in an adjacent cell.

    b. Notice the <u>Width</u> instantly updates.

9. Click **OK** to close the dialog and accept the changes.

All three boxes change size: 6'-0" depth and 3'-0" width. The height is still an *Instance Parameter* so it remained unchanged.

10. Switch back to the box family.

Next, a brief look at a few variations.

The following revisions will not be saved; the Cancel button will be selected to discard the next two steps.

11. Change the formula for *Width* to simply read **Depth** (see Figure 15-4.4).

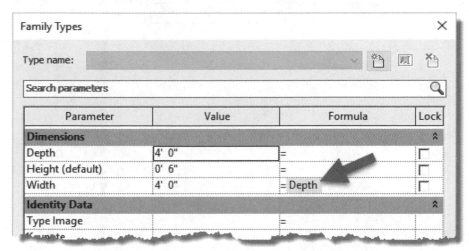

FIGURE 15-4.4 Modifying formula to control the width

Thus, it is possible to have one *Parameter* simply equal another one.

12. Click **Cancel** to discard the formula change.

Another way to achieve the results shown in the previous image and steps is to have one *Parameter* control two dimensions. This would reduce the number of *Parameters* from three to two as the *Family* currently stands. The image on the next page (see Figure 15-4.5) shows an example of this. Simply select a previously drawn dimension and then select the desired *Parameter* from the *Label* drop-down list on the *Options Bar*. This is only visible when a dimension is selected.

One *Parameter* could control several dimensions. A *Parameter* has to be of type *Length* (versus *Text, Currency, Volume, Yes/No,* etc.) to work with a *Dimension*. The *Label* drop-down list will automatically filter the list.

It is best if the dimensions reference the *Reference Planes* rather than the 3D geometry.

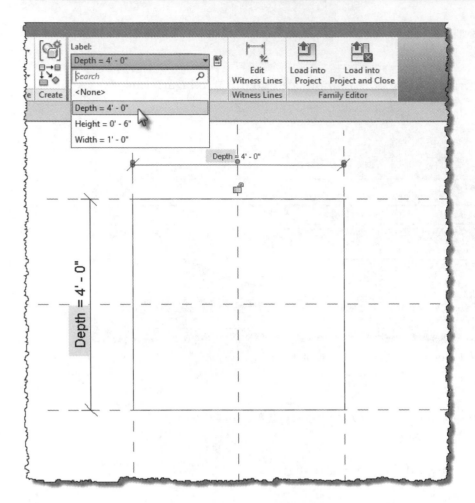

FIGURE 15-4.5 One Parameter controlling multiple dimensions

This concludes the brief introduction to adding formulas to a *Family*. The remainder of this exercise will shift gears a bit and look at adding materials to the box family.

Adding a Material

Adding a *Material* in the *Family* can save time for the design teams using the content when it comes to creating renderings.

13. Select the 3D extrusion that represents the box in the *Family Editor*.

14. View the *Properties Palette* dialog box.

Notice, in Figure 15-4.6, that the *Material* is set to *By Category*. This will be changed to something specific to this *Family*.

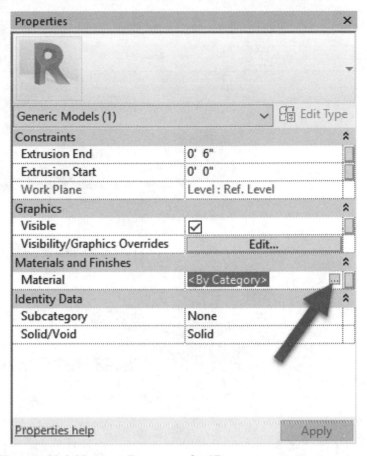

FIGURE 15-4.6 Instance Parameters for 3D geometry

15. Click in the cell with *<By Category>* to reveal the small link button to the right; see Figure 15-4.6.

16. Click the **link-button** to open the *Materials* dialog.

Notice the *Family* has a limited number of *Materials* loaded; see Figure 15-4.7. The following sequence of steps will show how to create a new *Material* and define it.

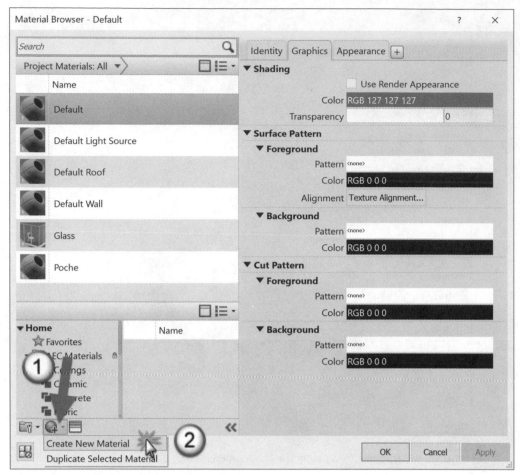

FIGURE 15-4.7 Materials for the Box Family, not the Box Project

17. Click the **Create New Material** icon in the lower left corner; see arrow in Figure 15-4.7.

18. Right-click the new *Material* and **Rename** it to **Box Material**; see Figure 15-4.8.

19. Click the **Foreground Surface Pattern** preview area (currently set to "none"); see Figure 15-4.9.

> ***TIP:*** *Make sure the* Graphics *tab is selected to see the* Surface Pattern *area.*

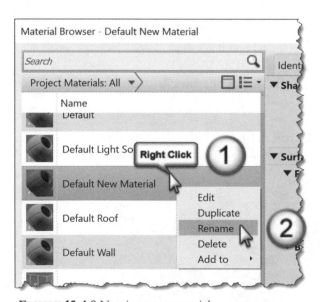

FIGURE 15-4.8 Naming new material

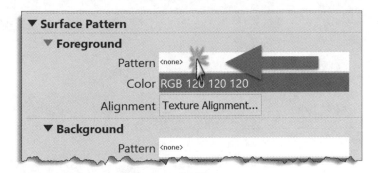

FIGURE 15-4.9 Setting surface pattern

A new "Model" *Surface Pattern* will be created. A "Model" pattern will not change scale with the *View Scale* in the *Project Environment*; this is meant for real-world items like siding, tile, CMU, etc. as seen in elevation. Conversely, the "Drafting" patterns do change scale with the *View Scale* and are meant for representative patterns typically used in sections to imply a certain material.

20. Select **Model** for *Pattern Type* and then follow the steps shown in Figure 15-4.10 to create a new "Model" based *Fill Pattern* named **3″ Tile**.

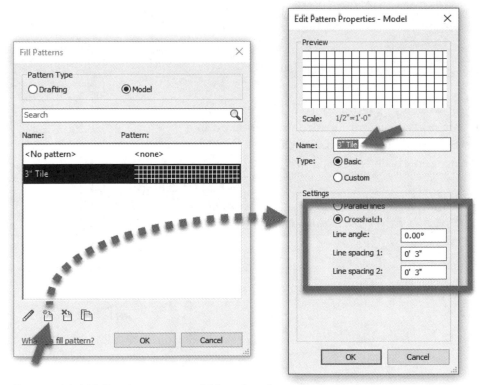

FIGURE 15-4.10 Creating a new model based surface pattern

21. Click **OK** to close the *Fill Patterns* dialog but not the *Materials* dialogs!

Next you will add an *Appearance* asset so you can control how the material will appear in a rendering.

22. Click to add an **Appearance** asset (Figure 15-4.11).

23. Set the *Render Appearance* to **3in Square - Terra Cotta** (from the *Appearance Library*).

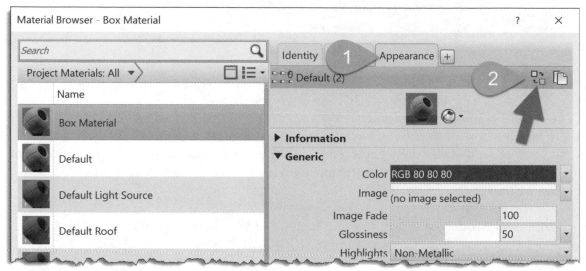

FIGURE 15-4.11 Selecting rendering appearance

24. Click **OK** to close the *Materials* dialog box.

The fill pattern is now visible, Figure 15-4.12. However, the render appearance is not visible until loaded and rendered in a project. Per steps previously covered in the textbook, load the family into the project and do a rendering in a 3D view.

TIP: When loading a Family, if the project already has a material named "box material," per the above example, the material definition in the project will take precedence if the two are not identical.

25. **Save** the Box *Family*.

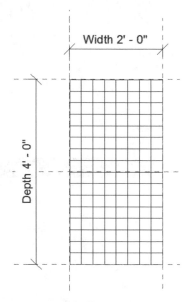

FIGURE 15-4.12 Family Editor

26. Load the updated *Family* into the **Box Project** (*Project* file).

When the newly updated *Family* is loaded into the project, the **Fill Pattern** shows up right away (Figure 15-4.13). The fill pattern can be selected on each face of each box and be rotated and repositioned as desired in the *Project Environment*.

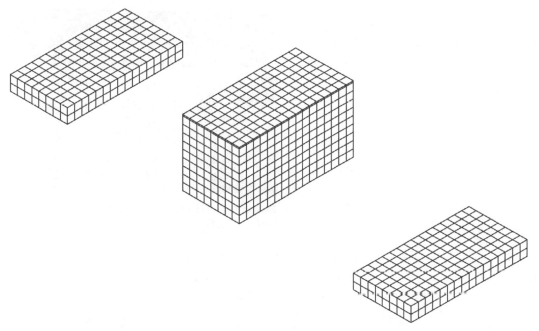

FIGURE 15-4.13 Box family in project environment

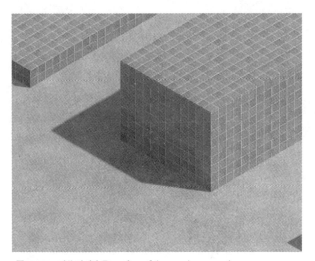

27. Switch to a 3D or Camera view and create a **rendering** to see the *Render Appearance* material.

   a. Adding a floor below the boxes will allow the shadows to show up as in the image to the right.

28. **Save** the *Project*.

FIGURE 15-4.14 Rendered in project environment

Changing the Material in the Project Environment

So far, the *Material* for the Box has been "hard wired" within the *Family* and cannot be changed once in the *Project Environment*. In this last set of steps, the techniques required to achieve this goal will be covered.

Just to make things more clear: it is possible to create several 2D shapes in a single *Family*. For example, a door has a solid for the frame, the door panel and a vision panel; see the image to the right (Figure 15-4.15).

Therefore, it would not make sense for Revit to provide one built-in parameter that controls the *Material* for everything in the *Family*.

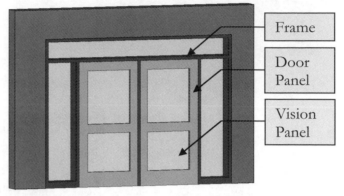

FIGURE 15-4.15 Door family with multiple materials defined

In the door example, a parameter is created for each *Material* needed in the *Family*. Next, each solid in the *Family* is mapped to one of the three *Parameters*. It is possible for one *Parameter* to control the *Material* of several 3D elements, e.g., a "glass material" *Parameter* controls the *Material* for all seven pieces of glass in Figure 15-4.15.

29. In the Box *Family* click the **Types** button.

30. Click the **Add…** button to make a new *Parameter*.

31. Modify the *Parameter Properties* dialog as shown in Figure 15-4.16.

    a. *Name*:
       Finish Material
    b. *Group parameter under*:
       Materials and Finishes
    c. *Type*:
       Material

32. Click **OK**.

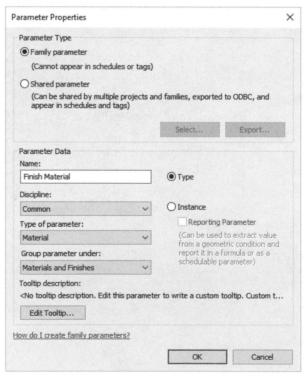

FIGURE 15-4.16 Creating a new parameter

33. In the *Family Types* dialog, change the "value" for **Finish Material** to **Box Material**. (See Figure 15-4.17.)

Parameter	Value	Formula	Lock
Materials and Finishes			⌃
Finish Material	Box Material ... =		
Dimensions			⌃
Depth	4' 0"	=	⌐
Height (default)	0' 6"	=	⌐
Width	2' 0"	= Depth / 2	⌐
Identity Data			⌃
Type Image		=	
Keynote		=	

Family Types dialog with Type name and Search parameters fields.

FIGURE 15-4.17 Setting material for new parameter

At this point, a new *Parameter* has been created but does not control anything yet. In the next few steps you will map the *Material* for the 3D box to the *Material* parameter just created in the previous steps. After this, the *Family Types* parameter will be the only way to change the *Material* for the box.

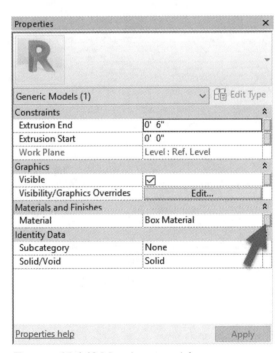

34. Click **OK** to close the *Family Types* dialog box.

35. Select the box and then look at the **Properties Palette**.

36. Click the small mapping icon to the far right of the **Material** *Parameter* (see Figure 15-4.18).

FIGURE 15-4.18 Mapping material to parameter

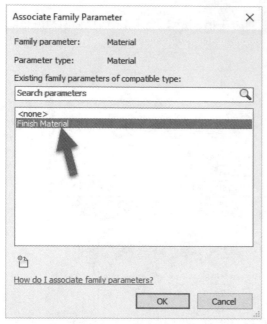

FIGURE 15-4.19 Mapping material to parameter

The *Associate Family Parameter* dialog is now displayed. Revit presents the user with a list of *Parameters* that are of the *Type* "material" (versus text, currency, etc.). At this point, the only option is the *Parameter* just created.

37. Select **Finish Material** from the list and then click **OK** (see Figure 15-4.19).

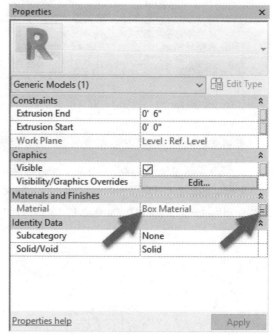

FIGURE 15-4.20 Mapped material

Now that the *Material* is mapped to a *Parameter*, notice that the mapping icon now has an **equal** sign in it and the rest of the row is grayed out; see Figure 15-4.20. The grayed out *Material* means that the *Family Type Parameter* is now controlling this *Instance Parameter*.

38. **Save** the **Box** *Family* again.

39. **Load** the *Family* into the box Project.

40. Select one of the boxes and view its **Type Parameters**.

41. Notice **Finish Material** is now an option.

42. **Save** the **Box Project** file.

Exercise 15-5:
The Box: Family Types and Categories

This exercise will study *Family Types*. A *Type* is simply the ability to save various parameter settings in a *Family* so they do not need to be entered manually within each Project.

Thus far, in this chapter, a *Family* with just a single *Type* has been created. Any new *Family* automatically has one *Type* if none are specifically created; the *Type* name is the same as the *Family* name once loaded into the *Project Environment*. This exercise will look at how to create several predefined *Types* within the *Family Editor* and how to create additional *Types* on the fly within the *Project Environment*.

Why use Types? Let us say a "Box" manufacturer offers several standard sizes. When creating the *Family* for the "Box," it would be most expedient to create a *Type* for each standard option. This would save the end-user(s) time in more ways than one. If a design firm had trusted custom Revit content, an end-user could load the "Box" *Family* and pick a size from the predefined *Types* knowing that they are real options. Maybe some rarely used or overpriced options are intentionally omitted. However, just not having to enter the data manually can be a great benefit.

The reader may wish to turn back to the first exercise in this chapter and review the information initially presented on this topic; see page 15-5.

Finally, the use of **Categories** will be studied. Revit uses *Categories* to control visibility and to determine which command will be used to place a *Family* (e.g., Door tool, Window tool, Mechanical Equipment tool, Component tool).

1. **Open** the *Family* named **Box**.

2. Click the **Types** button on the *Ribbon*.

Notice, in Figure 15-5.1, that the *Name* drop-down at the top is blank. Clicking the *New* button on the right allows one or more named *Types* to be created.

Once *Types* exist, all the *Parameter* **Values** below relate specifically to the selected named *Type*.

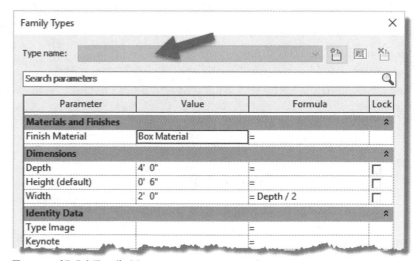

FIGURE **15-5.1** Family Types – no types created yet

Next, four *Types* will be created in the <u>Box</u> *Family*. Each *Type* will have a different <u>Depth</u> assigned to it.

3. In the *Family Types* dialog, click the **New…** button.

4. Enter **2'-0" x 4'-0"** for the name of the first *Type* (Figure 15-5.2).

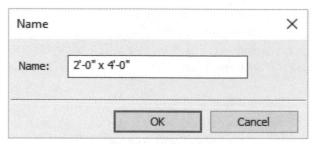

FIGURE 15-5.2 Naming the family type

Type names should be descriptive and not duplicate any part of the *Family* name (i.e., the name of the file on the hard drive). In this example the exact size will be used as the *Type* name. In other cases it might be more useful to list the model number or some other descriptive wording.

5. Click **OK**.

Now, one *Type* exists in the *Name* drop-down list. Note that the *Type* name matches the values below; this was intentional. However, it should be stated that the name and values could be inconsistent. The name has no direct control over any of the *Parameter* values, i.e., the name 2'-0" x 4'-0" does

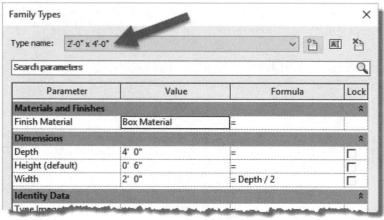

FIGURE 15-5.3 Named family type

not automatically make the <u>Depth</u> and <u>Width</u> *Values* match. Thus, the person creating the *Family* needs to do some QA (quality assurance) checking. This will be made clear in the next few steps.

6. Click the **New…** button again.

7. Enter the name **3'-0" x 6'-0"**.

   a. Remember, because of the formula added in the previous exercise, the <u>Width</u> is always half the <u>Depth</u>.

The **Box** *Family* now has two *Types* created; each can be easily selected within the *Project Environment.*

Revit is now managing two complete lists of *Parameters*, one for each *Type*.

As will be proven momentarily, the *Values* (i.e., <u>Width</u> and <u>Depth</u>) can vary from one list to the other.

As seen in Figure 15-5.4, the new *Type* did not change the *Parameter Values* below. Next this will be changed.

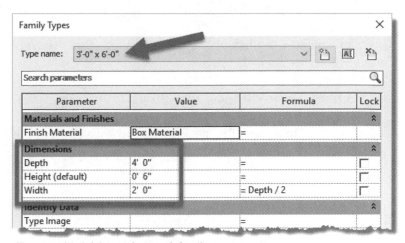

FIGURE 15-5.4 Second named family type

8. With the *Type* 3'-0" x 6'-0" current, change the <u>Depth</u> to **6'-0"**.

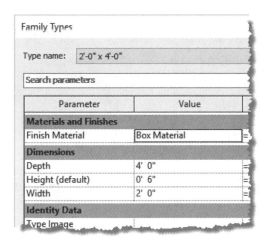

FIGURE 15-5.5A First type settings

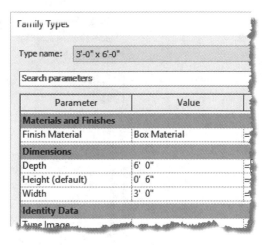

FIGURE 15-5.5B Second type settings

9. Using the *Name* drop-down list, switch back and forth between the two options; notice the *Parameter Values* change (Figures 15-5.5A and B).

10. Create two more *Types* per the information below:
    a. *Name*: **4'-0" x 8'-0"**   **5'-6" x 11'-0"**
    b. *Depth*: **8'-0"**   **11'-0"**

Now that several *Types* have been defined in the **Box** *Family* it will be loaded into the *Project Environment* to see how they work.

11. **Open** the **Box Project** file, if not already open.

12. In the <u>Box</u> *Family* file, click the **Load into Project** button.

13. Select **Overwrite the existing version**.

14. Select the **Component** tool from the *Architecture* tab on the *Ribbon*.

15. Select the **Type Selector** (Figure 15-5.6).

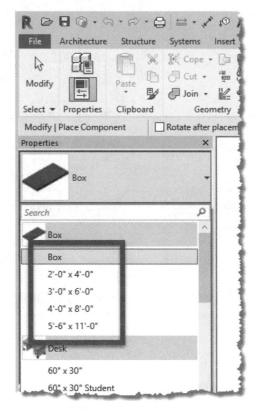

FIGURE 15-5.6 *Type Selector* in the project environment

Looking at Figure 15-5.6, notice all the *Types* created in the *Family Editor* are now available under the *Family* name **Box**. A *Type* named **Box** is also listed because the *Family* was previously loaded without any specific *Types* created, and as mentioned previously, when no *Types* exist Revit will create one with the same name as the *Family*. This can be deleted or renamed in the *Project Browser* within the *Families* section.

Selecting a Category

Selecting a *Category* is actually one of the first things typically done when starting a new *Family*. However, this change was left until now to make it perfectly clear what this setting does.

NOTE: Some Family *templates will already have the correct* Category *selected (e.g., door, window, casework templates). However, the* **Box** *Family was started with the* **Generic Model** *template, so the* Category *needs to be set manually as this template can be used for many things.*

16. **Open** the **Box** *Family* (if not already open).

17. On the *Create* tab, click **Category and Parameters**.

This list represents all the *Categories* used by Revit to control visibility and determine which tool inserts any given *Family*. *Categories* are also used by *Filters* and for scheduling. This list is "hard wired" and cannot be modified in any way.

This list is filtered for Architecture categories. However, clicking the drop-down list at the top allows each discipline to be toggled on individually.

At this point we will pretend that our **Box** *Family* is *Furniture*. Once set to be "furniture," the **Box** will show up in the *Furniture* schedule(s) and be visible (or not visible) based on the visibility and filter settings for any given view.

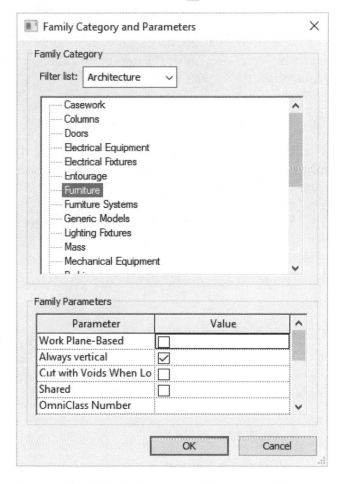

18. Click **Furniture** and then click **OK**.

FIGURE 15-5.7 Family Category and Parameters dialog

TIP:

The **Furniture** *category is generally for freestanding items that are not fastened to the building. Examples might be chairs, tables, desks, beds, etc.*

The **Furniture Systems** *category is for modular desks, often called cubicles. These can be fastened to the wall and/or have power and data hardwired to them. Thus they are different enough from regular furniture to warrant a separate section.*

19. **Save** the **Box** *Family.*

Next the modified *Family* will be loaded into the project and tested.

20. Load the *Family* into the **Box Project** file per steps previously covered.

21. In the **Box Project**, switch to the default **3D view**.

The next few steps will show how the visibility of the furniture category now affects the <u>Box</u> family.

22. Type **VV** to control visibility settings for the current view.

23. Uncheck *Furniture* on the *Model Categories* tab (Figure 15-5.8).

Notice there is a category for "Generic Models" as well. This would have allowed control of the **Box** *Family* previously. However, this category should be used sparingly due to its ambiguity.

24. Click **OK**.

The three boxes should have been hidden from the 3D view. They should still be visible in other views as long as the view's *Visibility Graphics Override* dialog did not have the *Furniture Category* turned off.

25. Per the previous steps, **turn the Furniture Category back on** in the *3D view.*

26. **Save** the **Box Project** file.

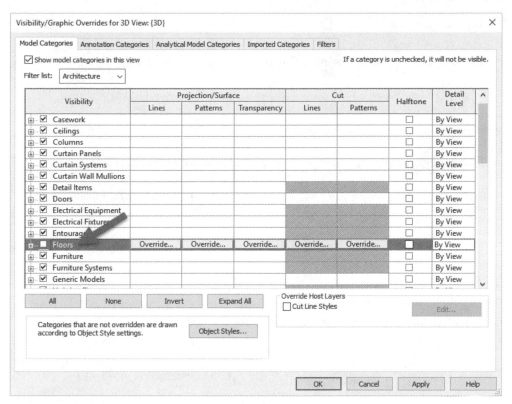

FIGURE 15-5.8 VISIBILITY control in the project

Revit does not have a specific *Furniture* tool on the *Ribbon* so new instances of the **Box Family** would still be placed using the *Component* tool.

Care should be taken to select the correct *Category* for Revit *Families*. If someone is in a hurry they may hastily select a wrong *Category* which could have a huge negative impact on a project in terms of budget and completion time.

For example, a "smart board" for a classroom might be placed on the *Furniture Category* when it should have been placed on *Electrical Equipment*. Near the end of the project the design team decides to turn off all the *Furniture* for the construction documents set, as it was laid out for design only and is not part of the bid set. Now, the final set is missing all of the "smart boards" in the floor plans, ceiling plans and interior elevations.

Much more could be said and taught about *Family* creation but is beyond the introductory scope of this text.

Smart board family with materials

Self-Exam:

The following questions can be used as a way to check your knowledge of this lesson. The answers can be found at the bottom of this page.

1. *Loadable Families* can be imported into a project as needed. (T/F)

2. How a *Family* is named is not important. (T/F)

3. The <u>sill height</u> of a window is usually an *Instance* parameter. (T/F)

4. Use the _____ tool to create the framework for a new *Family*.

5. The solid tool, used in this chapter, to create the 3D box: _____.

Review Questions:

The following questions may be assigned by your instructor as a way to assess your knowledge of this section. Your instructor has the answers to the review questions.

1. The 3d box, drawn in this chapter, was aligned and then locked to the reference planes as part of the process of creating parametric content. (T/F)

2. *Loadable Families* are typically preferred over in-place *Families*. (T/F)

3. A <u>family name</u> and a *Family's* <u>type name(s)</u> should not have redundant information. (T/F)

4. It is possible to specify if a parameter is an <u>instance</u> or <u>type</u> while in the process of creating it. (T/F)

5. When the geometry changes, a new *Family* is typically required, rather than being able to use named *Types*. (T/F)

6. The tool used to get a completed family into a project (while in the family editor environment): _____.

7. Of the two types of *Fill Patterns*, only the _____ type patterns do not change size when the view scale changes.

8. A *Family's* _____ setting controls/determines its visibility and which tool is used to place the item in the project environment.

9. When a *Parameter* is associated to one or more dimensions, the *Parameter's* value controls the length of the dimension (and object being dimensioned). (T/F)

10. It is not possible to make one parameter to always be one half the size of another. (T/F)

Lesson 16
Introduction to Phasing and Worksharing:

This chapter will start a small new exercise designed to explore two important features in Revit: Phasing and Worksharing. Unless a project is brand new, it will be using Revit's phasing feature. Additionally, if a project needs to be worked on by more than one person at a time, it will be utilizing Revit's Worksharing feature.

Note: This tutorial can be done at any time separate from the rest of the book.

Exercise 16-1:
Introduction to Phasing

Any project that involves remodeling or an addition should use the **Phases** feature in Revit. Revit is often referred to as three-dimensional modeling software. However, Revit's ability to manage elements over time is considered a fourth-dimension. This section will provide an introduction to Revit's phasing feature.

The **Phasing** dialog (Figure 16-1.1) is accessible from the Manage tab.

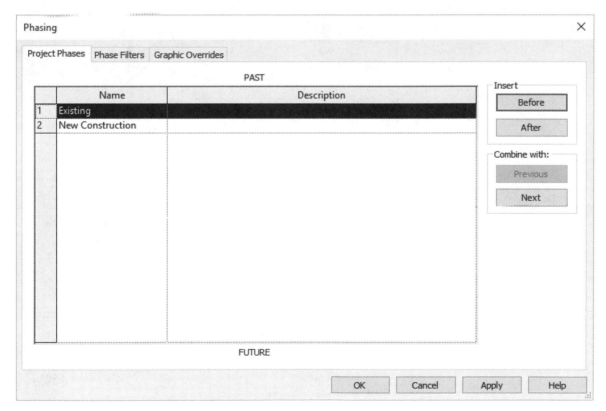

FIGURE 16-1.1 Project Phases tab; Phasing dialog

Most existing/remodel projects have two phases (Figure 16-1.1):
- Existing
- New Construction

Additional phases can be created, such as Phase 2 and Phase 3 (in this case, New Construction might be renamed to Phase 1). There should never be a phase called "Demolition," as Revit handles this automatically (more on this later).

Revit manages Phases with two simple sets of parameters:
- **Elements:**
  - Phase Created
  - Phase Demolished
- **Views**
  - Phase
  - Phase Filter

Among these four parameters, Revit is able to manage elements over time and control when to display them.

Element Phase Properties

As just mentioned, every model element in Revit has two phase-related parameters as shown for a selected wall in Figure 16-1.2.

The **Phase Created** parameter is automatically assigned when an element is created—the setting matches the phase setting of the view the element is created in. Put another way, in a floor plan view with the Phase set to *New Construction*, all model elements created in that view will have their Phase Created parameter set to *New Construction*.

The **Phase Demolished** parameter is never set automatically. Revit has no way of knowing if something should be demolished.

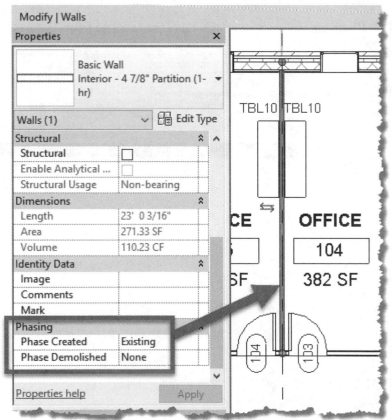

FIGURE 16-1.2 Element phase properties

View Phase Properties

Every view has two phase-related parameters, as shown for the floor plan view in Figure 16-1.3. Any elements added in this view would be designated as existing.

The **Phase** parameter can be set to any Phase that exists in the project. This setting can be changed at any time; however, it is best to have one floor plan view for each phase—thus, the Phase setting is not typically changed.

This setting represents the point in time the model is being viewed.

TIP: Create existing and demolition views in a company/personal template.

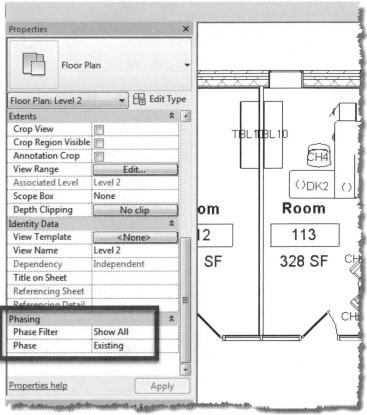

FIGURE 16-1.3 View phase properties – nothing selected

Phase Filters:

The **Phase Filter** setting controls which model elements appear in the view based on their phase settings.

To understand how the Phase Filter setting affects the view, look at **Manage → Phasing → Phase Filters** (See Figure 16-1.4).

Elements will appear in one of three ways:

- **By Category**   - Displayed as normal, no changes
- **Overridden**   - Modified based on Graphic Overrides tab settings
- **Not Displayed**   - These elements are hidden

Be careful not to confuse the *Phase Filter* column headings, New, Existing, Demolished and Temporary, as literal phases—they are not. Rather, they are a 'current condition' (aka **Phase Status**) based on a view's Phase setting and the element's Phase Created & Phase Demolished settings.

For example, the 'current condition' of an *Existing* wall in an *Existing* view is considered <u>New</u> because the phases are the same. Think of it as if you are standing in the year 1980, looking at a wall built in 1980—it is a new wall. Similarly, a *New Construction* wall in a *New Construction* view is also considered <u>New</u> in terms of how the Phase Filters work (now all *Existing* walls are considered <u>Existing</u>). You are now standing in the year 2019 looking at a wall built in 1980.

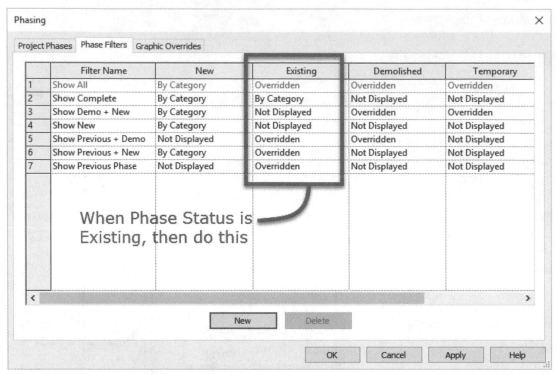

FIGURE 16-1.4 Phase Filters; Phasing dialog

The three images below are of the same model as seen in three different views—each with a different combination of Phase and Phase Filter settings. Each condition will be explained in depth on the next pages.

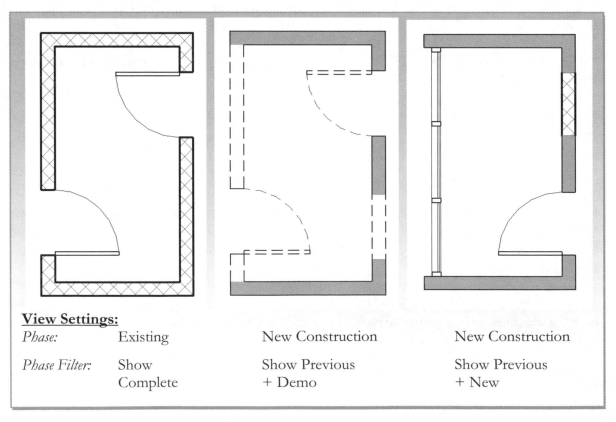

Existing Conditions:
When modeling an existing building, create a
Level 1 – Existing plan view; do this for each
level in the building. This view will have the
following phase-related settings:

- Phase: **Existing**
- Phase Filter: **Show Complete**

Any model element created in this view will
automatically have Phase Created set to Existing.

Everything in the drawing to the right has the
Phase Created set to Existing. FYI: One wall and
both doors have their Phase Demolished set to
New Construction—however, we cannot visually
see that here.

**The Phase Status is actually considered New
in this view.** The phase of the element matches
the phase of the view. Thus, because the Show
Complete Phase Filter has New set to By
Category, there are no overrides applied to this
view. We see the normal lineweights and fill
patterns.

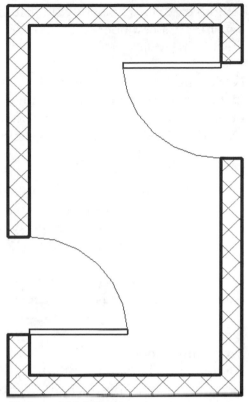

FIGURE 16-1.5 Existing view

Demolition Conditions:
When demolition is required, create a **Level 1 –
Demo** plan view; do this for each level in the
building. This view will have the following phase-
related settings:

- Phase: **New Construction**
- Phase Filter: **Show Previous + Demo**

Often, new Revit users think it strange that the
demo view needs to be set to New
Construction—given no new elements appear in
this view. However, looking at the existing view
just covered—all existing elements (even ones set
to be demolished) are considered "new."
Therefore, the "time slider," if you will, needs to
be moved past Existing to invoke the Phase
Demolished setting. If this is still confusing, it
should make more sense in the tutorial.

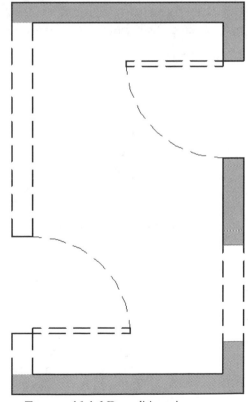

FIGURE 16-1.6 Demolition view

In Figure 16-1.6, the wall and two doors which appear dashed have their Phase Demolished set to New Construction. Recall that the view's phase is set to New Construction. In the lower right, a new door (i.e. Phase Created = New Construction) automatically demolishes the existing wall.

Notice the Phase Filter **Show Previous + Demo** has the Demolished *Phase Status* set to **Overridden**. So if any element with the Phase Created setting is set to a phase which occurs prior to the Phase setting of the current view, **and** the same element has a Phase Demolished setting that matches the current view's Phase setting, then it will show demolished.

FYI: If this project had a Phase named Phase 2, no existing elements set to be demolished in Phase 1 (i.e., New Construction in our example) could be shown in a view with a phase set to Phase 2. Those elements simply do not exist anymore, and there is no reason to ever show them in this future context.

One final note on this demolition view: because Revit inherently understands the need to demolish elements, **it is never necessary to have a Demolition phase.**

New Conditions:

All Revit templates come with the default plan views set to New Construction. For projects with existing and demolition conditions, it might be helpful to rename these views to Something like **Level 1 – New** (similar for each level). This view will have the following phase-related settings:

- Phase: **New Construction**
- Phase Filter: **Show Previous + New**

The door and curtain wall shown in the Figure 16-1.7 have their Phase Created set to New Construction. The existing walls are shaded due to the Phase Filter having an override applied to existing elements (more on this later).

Demolished openings present a unique situation in Revit. When an opening, e.g., door or window, is demolished, Revit automatically infills the opening with a wall. By default, this wall is the same type as the host wall (this can be selected and changed to another type). This special wall does not have any phase settings and it cannot be deleted. If an opening in the wall is required, then the wall needs to be hidden or an opening family added. If the demolished door is deleted this special wall will also be deleted.

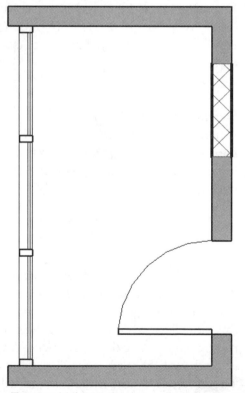

FIGURE 16-1.7 New construction view

The Phase Filter used here has the Demolished items set to Not Displayed. Also, the Existing items are set to Overridden, which is why the existing walls are filled with a gray fill pattern. The overrides will be covered next.

TIP: Smaller projects might show the demo in the new construction plans, in which case the <u>Show All</u> Phase Filter might work better.

Phase Related Graphic Overrides

When the Phase Filter has a **Phase Status** set to **Overridden**, we need to look at the Graphic Overrides tab in the Phasing dialog to see what that means (Figure 16-1.8).

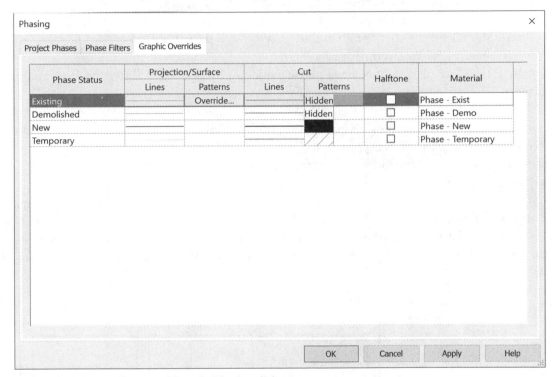

FIGURE 16-1.8 Graphic Overrides tab; Phasing dialog

In the context of a given view, if elements are considered Existing, and the selected Phase Filter has existing set to Overridden—then the settings for the Existing row shown above are applied.

All the walls being cut by the View Range in plan will have **black** lines with a lineweight of **3** (Figure 16-1.9). Simply click in the box below Cut/Lines to see these settings.

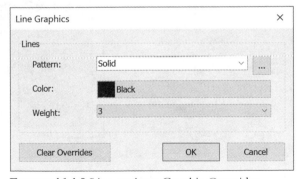

FIGURE 16-1.9 Line settings; Graphic Overrides

Existing walls, in this example, will also have a solid fill [background] pattern set to gray (Figure 16-1.10). This convention is helpful to clearly delineate existing walls from new walls in printed bidding and construction documents.

FYI: Phase state overrides will override nearly all other graphic settings in the view (view Filter being an exception.)

In Figure 16-1.8, Graphics Overrides tab, the **Hidden** setting for the Demolished row (Phase Status) means that the Visibility option has been unchecked (compare Figure 16-1.10). Any blank boxes have no overrides and Revit will still use the By Category equivalent.

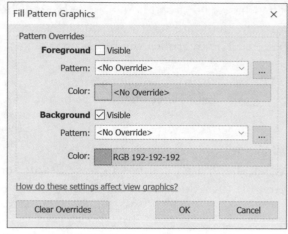

FIGURE 16-1.10 Fill Pattern settings; Graphic Overrides

Also in Figure 16-1.8, Graphics Overrides tab, notice there are several overrides applied to the New phase status. However, notice in Figure 16-1.4, Phase Filters tab, nothing in the New column is set to Overridden. Thus, the overrides for New do not apply to anything in the entire project currently.

Phasing and Rooms

The Room element has only one phase parameter: Phase. This parameter is set based on the Phase setting of the view it is placed in. However, this parameter is read-only and cannot be changed.

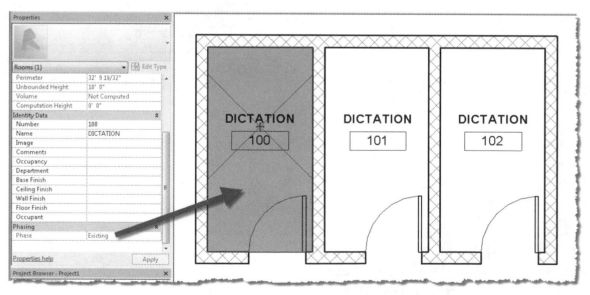

FIGURE 16-1.11 Phase setting for Room elements in existing view

In the image above, the view phase is set to Existing. If the view's phase is changed to anything other than existing, the Rooms and Room Tags will be hidden. Rooms are not really model elements that are built in the real world, so they are handled a little differently. As we will see in a moment, when walls are set to be demolished, Revit would not be able to maintain to existing Room elements in the new, larger area in the building. Thus, Rooms only exist per phase. The unfortunate side effect to this is that existing Rooms which have no changes need to have another Room element added for each Phase in the project. The room name and number are manually entered each time and are not connected in any way between phases.

> **TIP:** It is possible to Copy/Paste existing Rooms with no changes into a new view. This will save time retyping room names and numbers. However, there is still no connection between the two phases.

In the image below, we see the same model with a wall and door demolished. The Rooms and Room Tags shown are completely separate elements from those shown in the existing view.

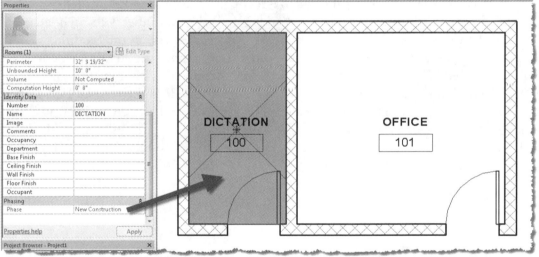

FIGURE 16-1.12 Phase setting for Room elements in new construction view

Phasing and Annotation

Most annotation elements are not affected by phasing; they are view specific and can tag an element whether it is New Construction or Existing. However, beware of a few exceptions.

View tags such as Sections, Elevations and Callouts all have Phase and Phase Filter settings. Actually, when these elements are selected, the properties presented in the Properties Palette are the same as the view properties seen when that item's view is active. The tag and view are connected—that is why the view is deleted when the section tag is deleted.

If a View's Phase setting is changed from New Construction to Existing, all the view tags will be hidden in that view (not deleted).
One more exception is that tags will disappear if the element they are tagging disappears due to phase-related view setting adjustments.

Exercise 16-2:
Introduction to Worksharing

When a Revit project requires more than one person to work on it, the Worksharing feature must be used. The feature is fairly simple, but everyone with access to the project must know a few basic rules to ensure things go smoothly.

Projects started from a template do not have Worksharing enabled. These projects can only have one person working in them at a time—just like most other files, e.g., a Word, Excel, AutoCAD, etc. file.

Worksharing Concept
The basic concept of Worksharing can be described with the help of the graphic below. When a regular Revit project file has Worksharing enabled, it becomes a **Central File**. The Central File is stored on the server where all staff have access (normal Read/Write access). **Once a Central File is created, it is typically never opened directly again.** Individual users work in what is called a **Local File**, which is a copy of the Central File, usually saved on the local C drive of the computer they are working on. When modifications are made to the Local File, the Central File is NOT automatically updated. However, the modified elements ARE checked out in that user's name—other users cannot modify those elements until they are checked back in. When a Local File is **Synchronized with Central**, the delta changes are saved; that is, only changes the user made are 'pushed' to the Central File, and then only changes found in the Central File, since the last Sync, and 'pulled' down (thus, the two way arrow). All elements checked out are typically relinquished during a Sync with Central. There is no technical limit to the number of users, though when there are more than 10 users the project can become sluggish.

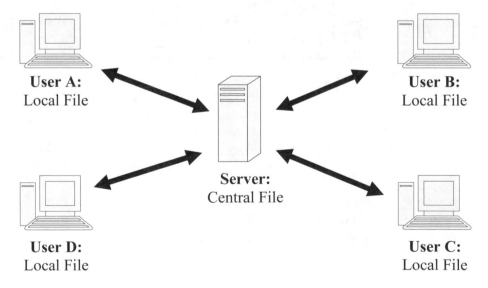

FIGURE 16-2.1 Revit project Worksharing concept

Overall, this system is very stable and has very few issues.

> **FYI:** Revit Worksharing does not work on cloud storage services such as Dropbox. There is too much latency in the network and several problems will be created including possible file corruption. There have been recent developments in using Revit Worksharing over the internet, but they involve Revit-specific software and/or hardware.

Enabling Worksharing

In just a few clicks of the mouse, Revit's Worksharing feature can be enabled; this section will describe how.

The main Worksharing tools are conveniently located on the Status Bar (Figure 16-2.2). These same tools, plus a few others, are located on the Collaboration tab in the Ribbon. The average user can get by with just the Status Bar tools, but we have to start on the Ribbon.

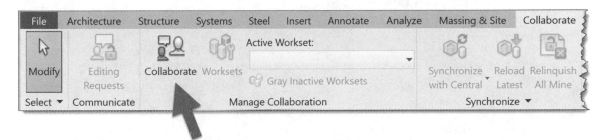

FIGURE 16-2.2 Worksharing tools on the Ribbon

To enable Worksharing do this:
- Ensure project file is in a shared location on a server
- Open the Revit project
- Click **Collaborate** on the Ribbon shown in Figure 16-2.2
- With **Within your network** selected, click **OK** (Figure 16-2.3)
- **Save** the project
- Click **Yes** (Figure 16-2.4)

The project file now has Worksharing enabled. **At this point the project file should be closed and all users will create a Local File** (covered in the next section). There is typically no reason to ever open this file again, even if only one person will be working on the project. It is not necessarily bad for someone to open the Central File, even with others working in Local Files, but following this rule will help minimize issues.

FIGURE 16-2.3 Enabling Worksharing (Collaboration)

The option to collaborate using A360, shown in Figure 16-2.3, is a paid cloud service offered by Autodesk to allow design firms from different offices to work on the same Revit model.

Once Worksharing is enabled, two folders are created next to the Central File—one contains the same name as the project file (Figure 16-2.5). The "temp" folder only exists while the Central File is open.

Important: **The Central File cannot be moved and the folders defining the path to the file cannot be changed.**

There is a special set of steps required if the Central File needs to be moved. This involves opening the Central File, or a Local File, **Detached From Central**—which is discussed more later on.

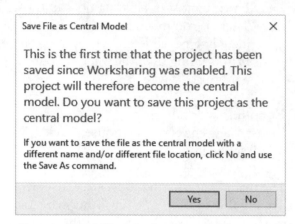

Save File as Central Model   ×

This is the first time that the project has been saved since Worksharing was enabled. This project will therefore become the central model. Do you want to save this project as the central model?

If you want to save the file as the central model with a different name and/or different file location, click No and use the Save As command.

Yes   No

FIGURE 16-2.4 Worksharing related notice

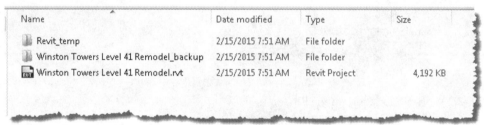

Name	Date modified	Type	Size
Revit_temp	2/15/2015 7:51 AM	File folder	
Winston Towers Level 41 Remodel_backup	2/15/2015 7:51 AM	File folder	
Winston Towers Level 41 Remodel.rvt	2/15/2015 7:51 AM	Revit Project	4,192 KB

FIGURE 16-2.5 Folder created for Worksharing

Creating a Local File

Once a Central File has been created there is really no reason to ever open it directly again. Rather, each user creates a Local File to work in. This section describes how to do that.

The steps to create a Local File are simple:
- Open Revit
  - Use the desktop icon or start menu
  - <u>Do not</u> double-click on the Central File from Windows Explorer
- Click **Open** within Revit
- Browse to, and select, the Central File (caution, single click on it)
- Ensure **Create New Local** is <u>checked</u> (Figure 16-2.6)
  - This should be checked by default when a Central File is selected
- Click the **Open** button

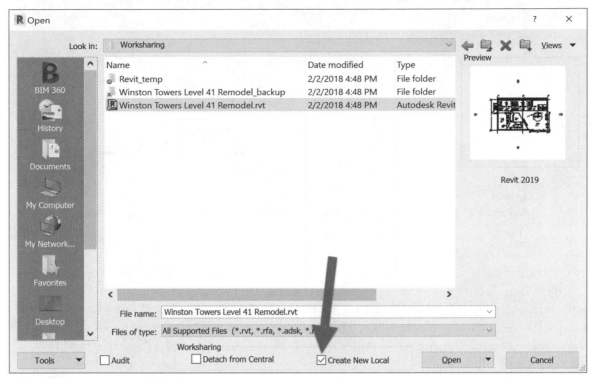

FIGURE 16-2.6 Create New Local check box; Revit Open dialog

The **Create New Local** option copies the Central File to the user's local hard drive (Figure 16-2.7). The file is located based on the "Default path for user files" setting in the **Options → File Locations** dialog. The local file has a slightly different name: **Central Filename + Username**. Once the open process is done, this user is in the newly copied Local File.

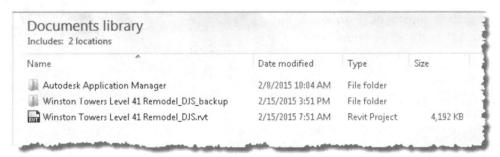

FIGURE 16-2.7 Local file created on user's hard drive

Each local file has one **Username** associated with it; the local filename includes the username on the end to help keep this straight. The Username is specified in the Options dialog (Figure 16-2.8). This is initially set to match the username of the Windows login/username. The username can be changed—in fact, Revit will change it to match your **Autodesk A360/BIM 360** username when you log into the Autodesk Cloud Services (the username will not change if you are in a Local File).

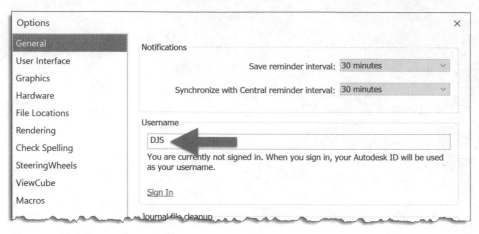

FIGURE 16-2.8 Username setting in Options dialog

If the Username saved within the Local File (which does not really have anything to do with the filename) does not match the current username specified in the Options dialog, Revit will warn you that this situation is not acceptable. In this case, the Local File will not be able to synchronize with the Central File. This problem can arise if/when Revit changes the username to match the Autodesk A360 account as just mentioned. In this case, the file should be closed, and a new Local File created per the steps previously covered.

The Revit Username is the designator used to track who has what checked out. No two users may have the same username. Duplicate usernames, either at the same time or at different times, would cause corruption in the Revit database.

Saving: Local versus Central
Once Worksharing is enabled there are two ways in which each user working in their Local Files may save their work: **Local Save** and **Synchronize with Central** (Fig. 16-2.9). These two options will be covered here.

FIGURE 16-2.9 Save icons on QAT

While in a Local File, the user has two save options on the Quick Access Toolbar.

- **Save Local**
  The first one is the normal **Save** icon (looks like a floppy disk). Every time this is selected, the Revit project is saved to the local hard drive. This ensures data will not be lost if Revit crashes or if the power goes out. A local save is fast as it does not involve the network and should be done often while working actively in the Local File.

- **Synchronize with Central**
  The second option is only available when Worksharing is active in the current project file. This command saves all the delta changes between the user's Local File and the Central File. This save takes more time as data is being saved across the network. This type is performed less frequently: once an hour, at the end of the day and when someone needs access to elements you have checked out.

The **Synchronize with Central** (SWC) command opens the dialog shown below (Figure 16-2.10). <u>Avoid using the **Synchronize Now**</u> option listed in the drop down as this option may not relinquish rights to items you have checked out.

While in the SWC dialog, **it is important to check all of the 'relinquish' boxes** which are not grayed out (except Compact Central). If this is not done, elements will remain checked out in your name even when you are not working on the project. This can be a big problem for teammates working in the evenings or while you are on vacation/holiday!

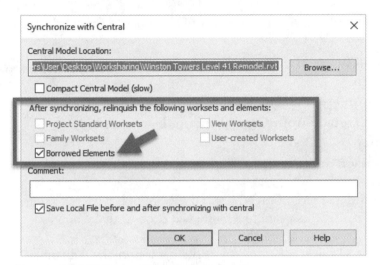

FIGURE 16-2.10 Syncronize with Central dialog

<u>Relinquish</u>
In the image above, only one category has elements checked out for the current user: User-created Worksets. Checking this box will ensure all the checkout elements are available to others on the project team.

<u>Save Before and After SWC</u>
The Save local before and after SWC is checked by default. This is good practice, as Revit can sometimes crash during SWC (especially on large projects on 32bit computers). Saving local before SWC helps prevent data loss. Saving local after SWC ensures the new local file is up to date.

<u>Compact Central</u>
This option is often used once a week on large, complex projects. It only needs to be performed by one person. Doing this database management helps to keep the file optimized for performance and file size. When selected, the SWC takes significantly longer.

<u>Comment</u>
This allows the user to add a brief note about what changes they made. This is helpful if the central file needs to be rolled back to a previous version; the comments are listed in the related dialog.

> **TIP:** This author recommends users create a new Local File each day. The existing local file does not benefit from the Compact Central command, so creating a new

Local File allows the user to work in the more optimized file. This also ensures the user is working on the latest version of the model, e.g., there are changes in the Central File which have not been downloaded yet—maybe another user worked all night on the project.

Detach from Central

Whenever a new Central File is needed or a user wants to look at a file without making changes, then **Detach from Central** should be used.

If a Central File becomes corrupt, any Local File can be used to replace the Central File. To do this:

- Open Revit (do not double-click on the project file)
- Click **Open**
- Browse to and select the Local File
- Check the **Detach from Central** option (Figure 16-2.11)
- Click **Open**
- Select **Detach and Preserve Worksets**

You are now in an unnamed, unsaved Central File. Simply save the file to the server and it will be a Central File. At this point, everyone will need to make new Local Files.

If someone wants to look at a Revit Project file without making changes, they need to open the file Detached from Central. When this is done, there is no way to save any changes back to the original Central File as Revit does not know about it. It is important to understand that even copying a Central File to your desktop will automatically turn it into a Local File when opened. Some make this mistake, thinking they will copy a file to see how others have created a Revit model—not realizing they are checking things out with their username! The consequence of this mistake is that staff working on the project can no longer edit parts of the model until the elements are checked back in.

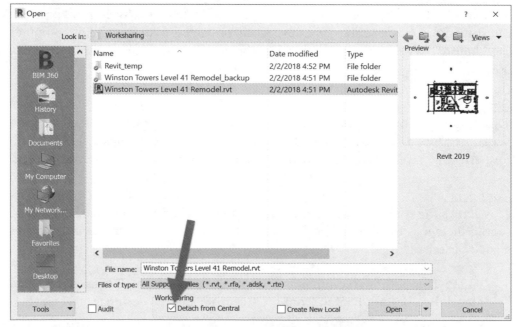

FIGURE 16-2.11 Detach from Central checked; open dialog

Know the Red Flags

While in the Open dialog, if you select what you think is a Central File and the Create New Local is not checked, that is a red flag! There are two reasons why this would not be checked:

- The file is actually a Local File
  - o You may have accidently selected the Local File rather than the Central File
  - o Someone may have accidentally saved a local file over the Central File
    - In this case, they need to do a Save As; in Options check 'Make New Central' and save the file to the same location. Everyone needs to make a new Local File.
- The file is a different version of Revit
  - o A Revit project should not be upgraded unless everyone on the team is ready to do so
  - o If the file is saved in a newer version of Revit, that version must be used

Worksets

Once Worksharing has been enabled in a Revit project, every element, view and family is associated with a Workset. Here are a few things Worksets are used for:

- The underlying mechanism allowing multiple people to work in same file/model
- Opening specific Worksets limits geometry loaded, conserving computer resources
- Controlling visibility project-wide or per view

Looking back at Figure 16-2.3, notice that modeled elements, if any exist, are divided up between two default Worksets when Worksharing is activated:

- **Shared Levels and Grids**
- **Workset1**

All the Levels and Grids in the model are associated with the same named Workset. FYI: when creating new Levels or Grids, the user needs to make sure they go on the correct workset.

All other elements in the model are placed on Workset1. Additionally, each view and family is associated with its own Workset—that is, one unique Workset per view and per Family.

This default setup is all that is ever needed on many projects. On large, complex projects additional Worksets are created. For Example:

- Core and Shell
- Level 1
- Level 1 Furniture
- Level 2
- Level 2 Furniture
- Etc.

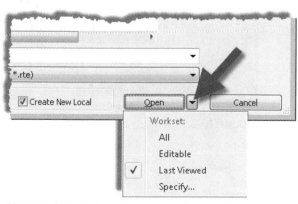

FIGURE 16-2.12 Specify worksets; Open dialog

Segregating model elements on separate Worksets is mainly done to conserve computer resources. For example, in the Open dialog, a user can choose to open only the <u>Level 1</u> and <u>Level 1 Furniture</u> worksets (Figure 16-2.12). For a 30-story building this can save a significant amount of RAM/Memory.

The Worksets in the current Revit project are listed in the Worksets drop-down list on the status bar (Figure 16-2.13). The selected Workset is what all new elements will be placed on. In this example, all model elements would go only on Workset1.

If elements are created on the wrong Workset, it is simple to change. Once Worksharing is enabled all elements have a Workset parameter (Figure 16-2.14). Simply select a different workset from the list in the Properties palette.

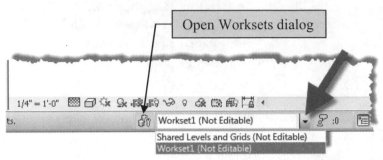

FIGURE 16-2.13 Worksets drop-down on status bar

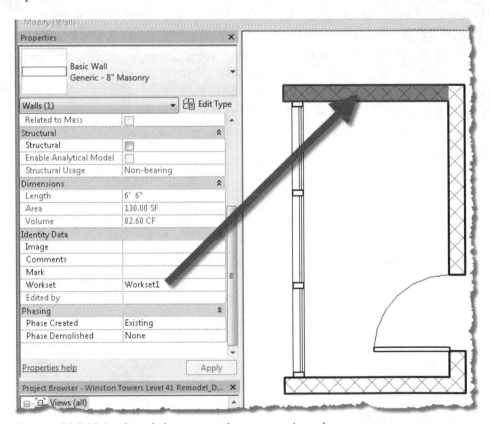

FIGURE 16-2.14 A selected elements workset; properties palette

Worksets Dialog

The Worksets dialog (Figure 16-2.15) can be accessed from the Worksets icon on the status bar (Figure 16-2.13).

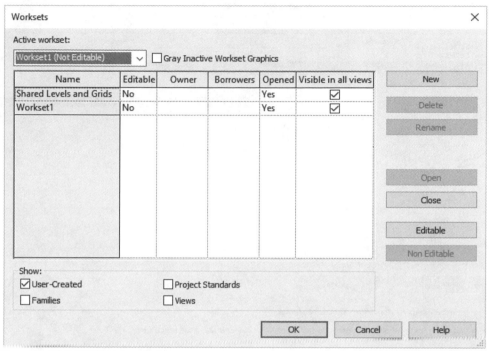

FIGURE 16-2.15 Worksets dialog

Clicking the **New** button allows additional Worksets to be created. Worksets can also be opened and closed when selected using the buttons on the right. The two default Worksets cannot be deleted. When deleting Worksets, which contain elements, Revit prompts to select which Workset to move the elements to – deleting a Workset will not delete elements.

The **Editable** column should typically be set to **No**. This allows individual elements to be checked out by multiple people. When the Editable column is set to Yes, then only that person listed may edit anything on that Workset. This is sometimes done to limit who can change certain elements in the model. For example, if the structural model is in the same model as architectural and interior design, the structural engineer/designer might leave the structural Worksets checked so no one on the design team can accidentally move a column.

Unchecking the **Visible in all views** option will hide those elements in the entire project.

Workset visibility by View

Once Worksharing is enabled, each view's visibility/graphics override dialog has a Worksets tab (Figure 16-2.16). This allows elements to be hidden based on which Workset they are on. The default setting for visibility is to rely on the Worksets dialog (i.e., Use Global Setting). Two other options, Show and Hide, allow for a view specific setting independent of the global setting.

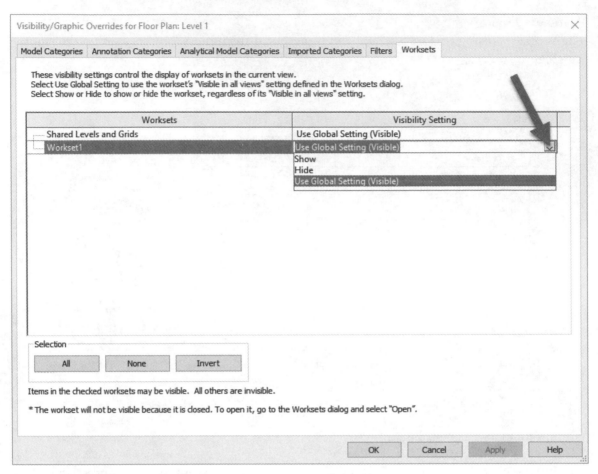

FIGURE 16-2.16 Workset options per view

Worksharing Display

One of the toggles on the View Control Bar provides a way to highlight a few Worksharing items (Figure 16-2.17). Revit is able to highlight, with different colors, which Workset elements are on, or who has what checked out at any given time. Clicking the **Worksharing Display Off** turns the highlights off.

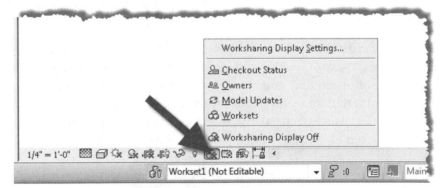

FIGURE 16-2.17 Worksharing display visual toggles

Making Worksets Editable

It was already mentioned that an entire Workset should not be made editable, as no one else can edit any element on that Workset.

Most of the time, Revit will automatically check out elements when you begin to edit/modify them. However, some edits are more indirect and require the user to manually make the elements editable before changes may be made. When an element, or elements, is selected and a right-click is performed, there are two related options (Figure 16-2.18):

- Make Worksets Editable
- Make Elements Editable

Typically, only the second option is used— which checks out just the selected elements rather than the entire workset.

Worksets and linked models
If a linked model and the host model have the same named Workset, then Revit automatically connects the two. For example, if the structural grids in the structural model are on a Workset called <u>Struct Grids</u>, and the architectural model also has a Workset called <u>Struct Grids</u>, then they will be connected. Turning off the global visibility of <u>Struct Grids</u> in the architectural model will turn off the structural grids, leaving the architectural grids visible. This would be done when the architects have used Copy/Monitor to create

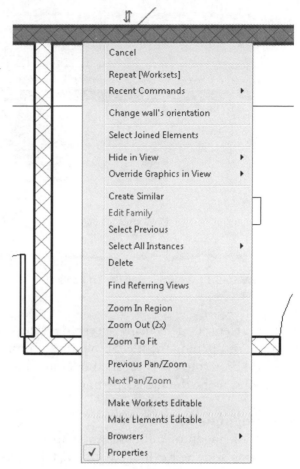

FIGURE 16-2.18 Making worksets editable

their own semi-independent copy of grids they can adjust lengthwise. Structural engineers might also want to do something similar for the architects' levels.

Each linked Revit model will also have a separate list of Worksets contained with the linked model itself. This provides granular control over Workset visibility per linked Revit model. For example, the structural model sometimes contains duplicate walls, compared to the architectural model, to highlight just the structural/bearing walls in their model. If those duplicate walls are placed on a separate Workset, then the other disciplines may turn them off.

To see the Worksets in linked models:
- In any view, type **VV**
- Click the **Revit Links** tab (Figure 16-2.19)
- Click the *Display Settings* button for the desired link
- Click the **Workset** tab

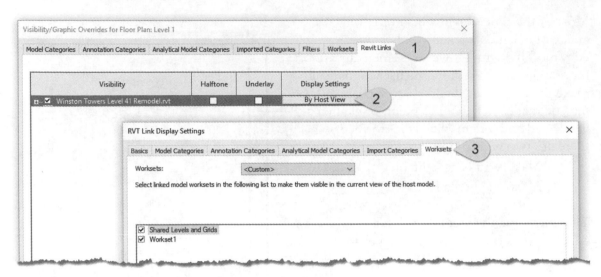

FIGURE 16-2.19 Worksets in linked models

While a Central file is being accessed others cannot access it – this includes Syncing with Central or opening the file. Sometimes prompts will appear during opening that need to be closed before the file is finished opening. Ideally each user should not walk away from the computer while the file is opening.

Logging into A360

When logging into Autodesk A360, to access cloud services, Revit requires the A360 username and the Revit username match. If the Revit username is changed, via Options, then the user needs to create a new Local File.

Conclusion

This concludes the introduction to Worksharing. It is important that everyone working on a Worksharing project understands this information to have a successful project without significant issues and/or downtime.

Exercise 16-3:
Phasing Exercise

This section will provide a hands-on tutorial of Revit's **Phasing** functionality. The exercise involves remodeling an existing building.

1. Open the Revit project file, from the online content (see inside front cover of book), called **Existing Widget Engineering Building.rvt** – this file is in the **Phasing and Worksharing** folder. There is another file with a similar name, so be sure you have the correct file.

2. Ensure the **Level 1** floor plan view is open (Figure 16-3.1).

Notice the Phase setting for this view is set to Existing. Thus, any elements added in this view will be existing. In this tutorial, the boxed rooms will be converted to a conference room. Level 2 will also have some modifications, which will be discussed next.

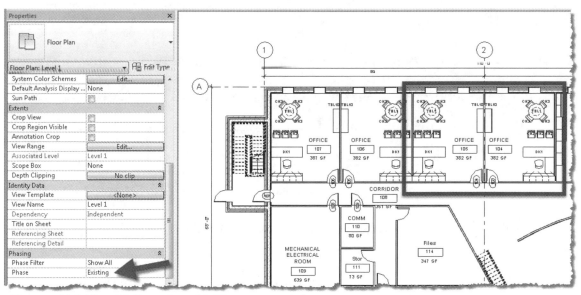

FIGURE 16-3.1 Level 1 floor plan with Phase set to existing

3. Switch to the **Level 2** floor plan view (Figure 16-3.2).

On this level, the goal is to add offices, with walls and doors, in the boxed area.

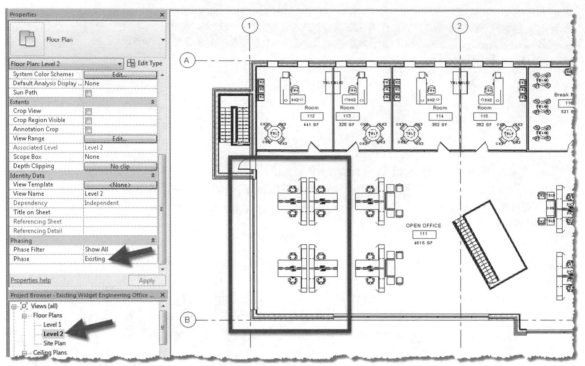

FIGURE 16-3.2 Level 2 floor plan with Phase set to existing

The next step will be to set up separate views for the Existing and New Construction phases.

4. In the Project Browser, right-click on **Level 1** and select **Rename**.

    a. Change the view name to **Level 1 – Existing**.

    b. Click **No** to rename the corresponding level and views.

5. Repeat the previous step for Level 2, creating a new plan named **Level 2 – Existing**.

Next you will duplicate the two plan views and adjust their Phase setting for New Construction.

6. Right-click on *Level 1 – Existing* and select **Duplicate View → Duplicate**.

7. Rename the newly created view to **Level 1 – New**; click **No** to rename the levels.

8. With nothing selected in the *Level 1 – New* plan view, change the following in the Properties Palette (16-3.3):

    a. Phase Filter: **Show Previous + New**
    b. Phase: **New Construction**

9. Repeat the previous three steps to create **Level 2 – New**.

Due to the selected Phase Filter, the existing walls appear filled in the new views. Next, we will adjust the solid fill to be a shade of gray, rather than solid black.

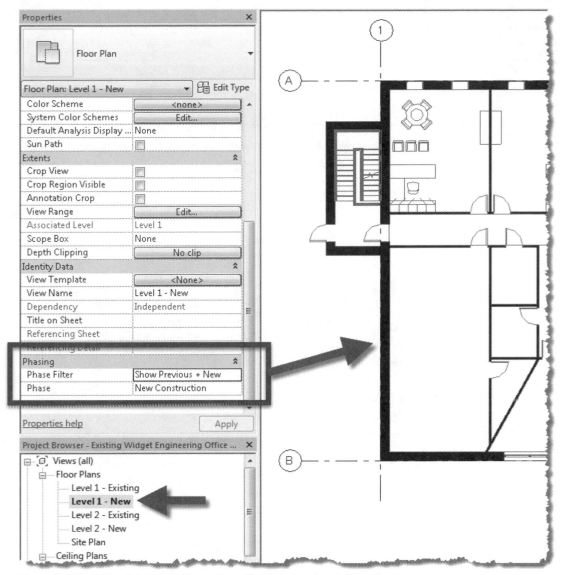

FIGURE 16-3.3 Level 1 – New floor plan with Phase settings adjusted

10. Select **Manage → Phasing**.

11. Select the **Phase Filter** tab (Figure 16-3.4).

Notice the Phase Filter we set for <u>Level 1 – New</u>, which is **Show Previous + New**. The existing items will be **Overridden**. This is what is causing the walls to be filled in black. Next, we will adjust the overrides on the next tab.

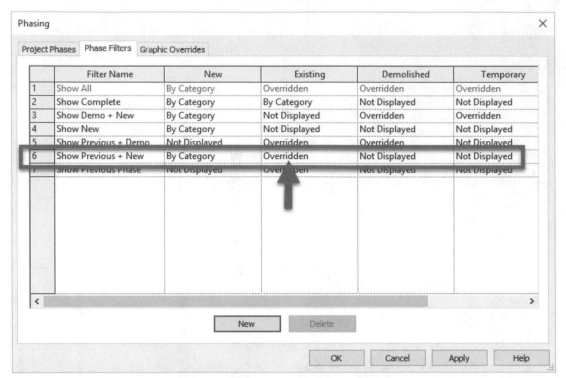

FIGURE 16-3.4 Phase Filters tab; Phasing dialog

12. Select the **Graphic Override** tab.

13. Follow the steps shown in Figure 16-3.5 to change the fill color to gray.

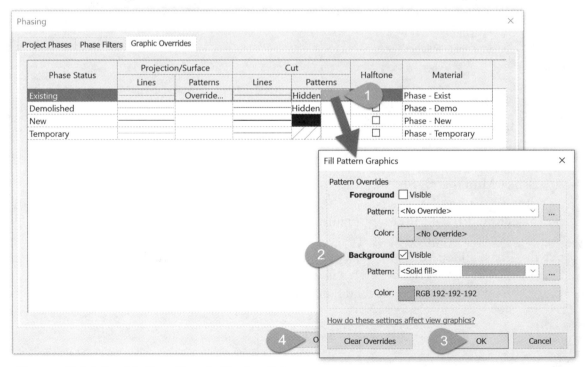

FIGURE 16-3.5 Graphic Overrides tab; Phasing dialog

You should notice, in ALL of the 'new' views, that the walls are now filled in gray rather than black. However, the wall lines are also gray and would be better black. You will also make that change.

14. Make the following change in the Phases dialog to adjust the lines to black for existing elements (Figure 16-3.6).

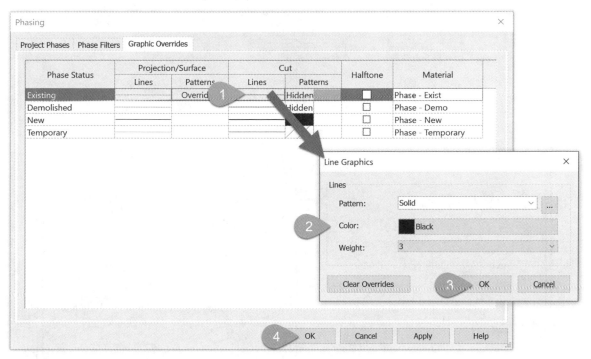

FIGURE 16-3.6 Graphic Overrides tab; Phasing dialog

Notice the existing walls are now filled with a solid gray pattern but the lines are solid black (Figure 16-3.7). This will make the existing elements contrast well next to new elements.

Now that the views are set up the original existing conditions can always be viewed in the 'existing' views and the new work in the 'new' views.

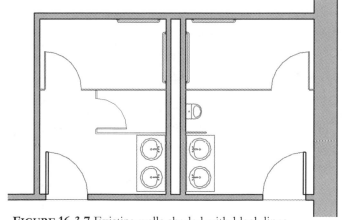

FIGURE 16-3.7 Existing walls shaded with black lines

Demolition

Next, some elements will be set to be demolished, and a third set of views will be created to see the demolition work.

15. Switch to the **Level 1 – Existing** floor plan view.

16. Zoom into offices 104 and 105 as shown in Figure 16-3.8.

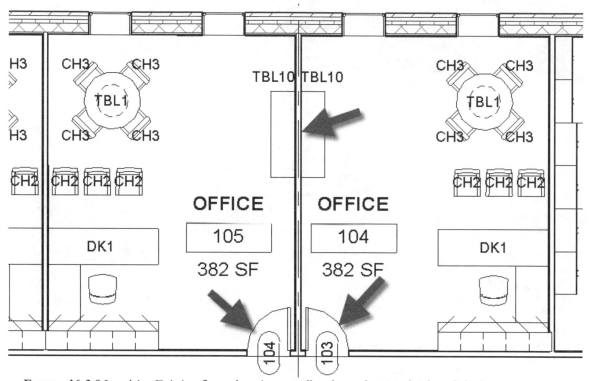

FIGURE 16-3.8 Level 1 – Existing floor plan view – wall and two doors to be demolished

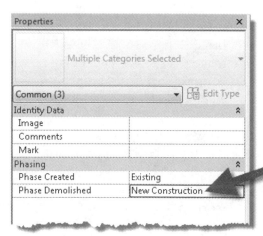

FIGURE 16-3.9
Phased Demolished set to New Construction

17. Holding the Ctrl key, select the wall and two doors pointed out in Figure 16-3.8.

18. With the three elements selected, set the Phase Demolished to **New Construction** (Figure 16-3.9).

Notice, with the wall and two doors selected, that the Type Selector says 'Multiple Categories Selected' and the drop-down list below (aka Properties Filter) indicated three elements are currently selected. When multiple items are selected, only the Common properties between those elements are displayed. In this case there are only five common parameters, two of which are related to phasing.

19. Move your cursor back into the drawing area and press **Esc**, or the Modify button, to unselect the elements.

Because you are in the existing view, nothing has changed graphically in the view.

20. Switch to **Level 1 – New** floor plan view (Figure 16-3.10).

Notice, due to the phase settings of the New view, the wall and two doors are hidden in this view. The Phase Filter **Show Previous + New** is set to hide demolished elements. One interesting thing that happened is Revit automatically infills demolished openings, both doors and windows, with a wall that matches the host wall type. This wall can be selected and adjusted to another wall type, but it cannot be deleted. If a "plain" opening is needed, with no door or frame, the infill wall needs to be manually hidden via right-click → hide element or using a view filter.

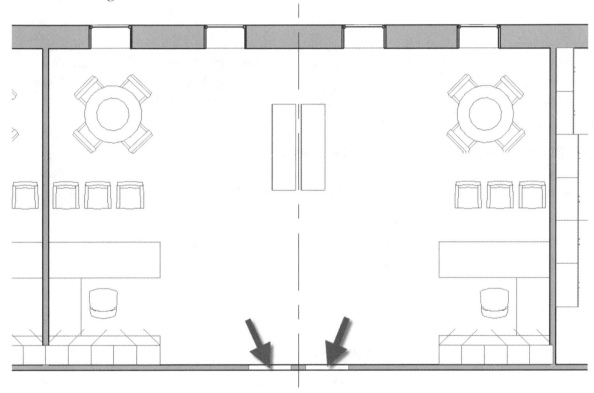

FIGURE 16-3.10 Level 1 – New floor plan view – wall and two doors hidden

21. Select one of the infill walls pointed out in Figure 16-3.10.

22. Notice the following in the Properties Palette:

   a. The default wall Type matches the demolished door's host

   b. The Type Selector allows different types to be selected

   c. There are no phase parameters

   d. This wall may not be deleted without deleting the door

Next you will demolish the furniture. First, we will see that existing items may be set to be demolished in the New Construction views. Second, a problem related to demolishing elements within Groups will be reviewed.

23. While still in the **Level 1 – New** floor plan view, select one of the chairs at the desk in Office 105.

24. Set the Phase Demolished to **New Construction**.

The chair instantly disappears in the 'new' floor plan view.

25. Switch back to the Level 1 – Existing view to see the chair is still there.

26. In either Level 1 floor plan view, select one of the round tables with chairs in Office 105.

Notice the selected element is a Group and no phase properties are listed. The group can be edited and elements within the group changed to be demolished, but that would affect all instances of the Groups—which is not what we want. Thus, there are two options: create another Group for demolition or just un-group the elements to be demolished. For this smaller project we will do the latter.

27. With the Group still selected, click **Ungroup** on the Ribbon.

28. With the table and four chairs selected, set the Phase Demolished parameter to **New Construction**.

29. Repeat these steps to demolish all furniture in Offices 104 and 105. Do not delete the furniture tags in the existing views.

> **FYI:** Sometimes it is easier to demolish elements in a 'new' view as things will disappear as they are set to be demolished. This way you know you have everything once all elements are hidden in the view. This is especially helpful with things like system furniture which has smaller elements hidden under larger elements.

Adding New Elements
The next step in converting the two offices to a larger conference room is to add a door.

30. In the Level 1 – New floor plan view, add a **Door** as shown in Figure 16-3.11.

31. Press **Esc** to end the door command, and then **select** the door.

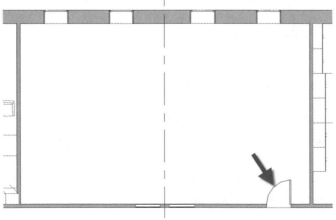

FIGURE 16-3.11 Door added in Level 1 – New floor plan view

Notice, in properties for the selected door, that its Phase Created is New Construction, which matches the Phase setting of the current view. Also, the Phase Demolished is set to none.

32. Looking at the **Level 1 – Existing** view, nothing has changed graphically at the new door location.

Next a view will be created to document the demolition work. This is a view, placed on a sheet, which indicates the elements to be demolished. Contractors need to know this for bidding and during construction.

33. Right-click on the **Level 1 – New** view and select **Duplicate View → Duplicate**.

34. **Rename** the newly created view to **Level 1 – Demolition**.

35. Change the Phase Filter to **Show Previous + Demo** for Level 1 – Demolition; the Phase setting should already be set to New Construction.

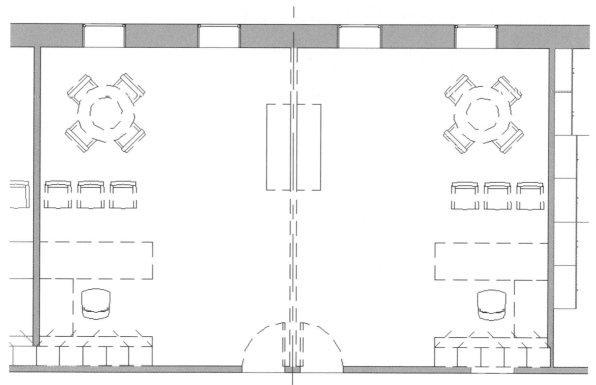

FIGURE 16-3.12 Level 1 – Demolition floor plan view

This new view shows all the elements to be demolished with dashed lines—this is an industry standard convention (Figure 16-3.12). Often, movable items such as furniture are not shown. Notice the opening for the new door is also identified. This tells the contractor that a portion of the existing wall needs to be removed; dimensions should be added to the demolition view, so the new door is accurately positioned.

In the demolition view, notice the Phase is set to New Construction. The view phase must be past Existing in order for the demolish setting to kick in. Also, due to the Phase Filter (Show Previous + Demo), only Existing and Demo elements appear. All new elements are hidden, even though the view's phase is set to New Construction.

36. Using the information just covered, create a **Level 2 – Demolition** plan view.

37. In the **Level 2 – New** floor plan view, demolish the eight desks shown dashed in Figure 16-3.13.

Next, new walls and doors will be added.

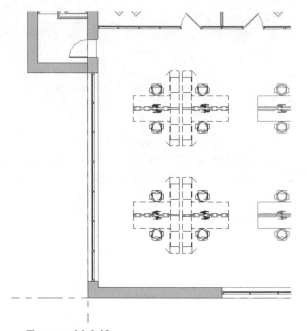

FIGURE 16-3.13
Level 2 – Demolition floor plan view

38. In the **Level 2 – New** floor plan view, add the walls and doors shown in Figure 16-3.14; center the two walls on the exterior curtain wall mullions.

All the walls and doors just created in the 'new' view are set to New Construction for Phase Created. These elements do not appear in the existing view or the demolition views.

Room Elements
The final thing to consider for this introduction to phasing is the Room elements.

39. In the **Level 1 – Existing** view, select the **Room** element for Office 105.

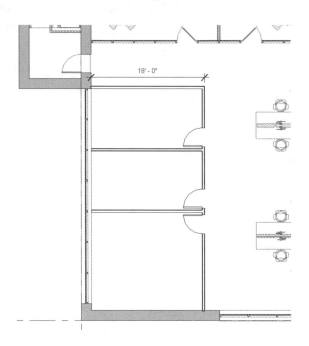

FIGURE 16-3.14
Level 2 – New floor plan view

In the Properties Palette, notice the Phase is set to Existing and that it cannot be changed. Existing Rooms only appear in existing views and new Rooms only appear in new views. The phase is set based on the view the Room was placed in.

40. Switch to **Level 1 – New** and zoom in on the new conference room.

Notice there are no Room elements in this entire view. Given our specific remodel project, this logic starts to make sense. When two rooms become one larger room, Revit cannot automatically make the two existing rooms become one – especially without affecting the fidelity of the existing view.

Thus, Room elements must be added separately in the 'new' floor plan view.

41. In the **Level 1 – New** view, add a **Room** to the new conference room.

42. Edit the name to **CONFERENCE ROOM** and the number to **105**.

43. In the **Level 1 – New** view, add **Rooms** as shown in Figure 16-3.15; name and number as shown.

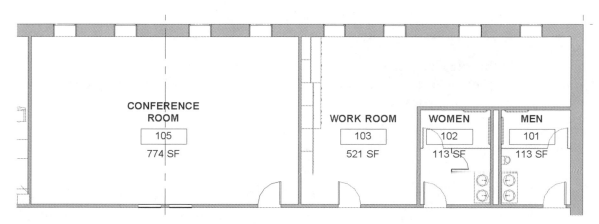

FIGURE 16-3.15 Rooms added; Level 1 – New floor plan view

The four Rooms just added have their Phase set to New Construction. These Rooms do not appear in the 'existing' views. Given these steps, it should be noted that the existing rooms have no connection to the new rooms in the same location. Thus, any changes to existing room names and/or numbers must be carefully coordinated.

> **TIP:** For existing room elements that don't change, i.e., the room name and number are the same in both 'existing' and 'new' views, they can be copied and then pasted from one view/phase to another. This is a quick way to transfer the name and number data.

44. Add three **Rooms** to the **Level 2 – New** floor plan as shown in Figure 16-3.16.

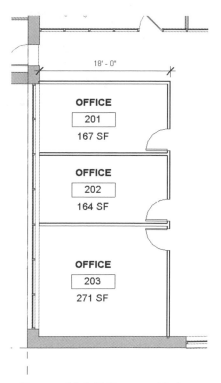

FIGURE 16-3.16 Rooms added, Level 2 – New floor plan view

Room Tags in Demolition Views

The demolition views will only be able to show the 'new' Rooms because the view's phase is set to New Construction. This is a challenge as we want to show the existing rooms to make sure the contractor enters the correct area and removes the right stuff!

One trick, or workaround, is to create another 'existing' view and hide everything in the view except the Room and Room Tag categories. This new special view can be placed on top of the demolition view on the sheet. In fact, Revit will allow you to snap the two models together when placing one on top of another (you will see an alignment snap for both the X and the Y axis).

No Demolition Phase

It is important to reiterate that Revit handles the demolition phase automatically. There should never be an actual demolition phase in the Phases dialog.

Conclusion

This concludes the introduction to Phasing. There is more that can be covered on this topic, but the material presented should be enough to get started with Phasing in any project.

Exercise 16-4:
Worksharing Exercise

This section will provide a hands-on application to worksharing, more than one person working in the project at one time. The tricky thing about this tutorial is that it is best followed with two people, at the same time, with access to the same network.

FYI: Worksharing does not work over the internet, using file sharing tools such as Dropbox, A360, Google Drive or Microsoft One Drive.

If this tutorial cannot be performed by two people, <u>at the same time</u>, on the same network, there is a way to simulate this on one computer, by one person. As mentioned at the beginning of this chapter, Revit tracks each user based on their Username set in Options. Thus, opening two sessions of Revit on one computer and changing the Username for each allows Revit to 'see' two different users accessing the Central file.

Steps to simulate two users:

<u>User A:</u>
- Open Revit
- Application Menu → Options
  - Change Username to **User A**
- Open project starting at step #1 below
- Continue following steps as User A

<u>User B</u>
Important: don't follow these steps until second user instructions begin – starting on page 16-40:
- Open a second session of Revit
  - This does not mean open another Revit project file
  - Double-click the desktop icon
  - This starts a second session of Revit on your computer
- Application Menu → Options
  - Change Username to **User B**
- Open the project file as instructed for User B
- Follow steps for User B using this session of Revit
  - Notice the Local File name, seen on the Title Bar, contains 'User B' as a suffix

Before Closing the Revit Sessions
- Application Menu → Options
  - Restore your original Username

FYI: Do not log into A360 as that will change your username.

The steps just listed, allowing Revit to 'see' two users on the same computer, must be performed each time Revit is used. So, if this entire tutorial cannot be completed in one sitting, these steps must be followed each time. The reason has to do with how Revit stores the username. When the username is changed, a text file (Revit.ini) is adjusted. Each time Revit is opened, it reads this file to determine what the username is. Obviously, there can only be one username, and it is based on the last session of Revit to change it.

Warning: Be Patient and Take Turns

The steps MUST be followed in the order listed by BOTH users, or this tutorial will not work. When User A has a step, User B should be idle—not doing anything. There are no steps where both users are working at the same time. Once the basics are understood, students are encouraged to "go at it" with multiple users working at the same time. However, to highlight the way Worksharing actually works, we need to have one user make a change and then, after that, have another user do something so the sequence of events can be clearly seen.

Enabling Worksharing

The first step is to enable Worksharing. **This is only done by only one person.** If a second person is involved in this tutorial, that person is encouraged to watch User A perform these steps.

> **TIP:** It might be best to work through this tutorial twice—switching roles the second time.

1. <u>**USER A**</u>: Open the completed phasing project **Existing Widget Engineering Building.rvt** from the previous exercise.

   a. Make sure the file is on the network first, if two people will be following this tutorial.

   b. While in the **Open** dialog, notice the Worksharing options at the bottom—they are grayed out because the selected file does not have Worksharing enabled (Figure 16-4.1).

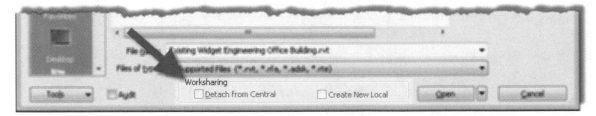

FIGURE 16-4.1 Worksharing options in the Open dialog

The process to enable Worksharing starts by simply opening the Worksets dialog.

2. <u>**USER A**</u>: Click the **Collaborate** tool on the Ribbon (Figure 16-4.2).

FIGURE 16-4.2 Collaborate tool on Ribbon

This process can take a minute or two, depending on the speed of your computer (and the size of the project).

The Collaborate dialog is now open (Figure 16-4.3).

3. **<u>USER A</u>**: With **Within your network** selected, click **OK** to close the dialog.

At this point the Worksharing file has not been saved yet, so no one else can work in it.

4. **<u>USER A</u>**: Click the **Save** icon on the Quick Access Toolbar (Fig. 16-4.4).

   a. Notice the icon just to the right of the Save icon... this is currently grayed out.

5. **<u>USER A</u>**: Click **Yes** to the final worksharing related prompt (Figure 16-4.5).

The file, in the location you opened it from, is now a Central File. Be careful not to edit, move or rename the folders created next to the central file.

On the Quick Access Toolbar, notice the regular Save icon is now grayed out and the icon to the right, **Synchronize with Central**, is active (Figure 16-4.6). In the future, keep in mind that this is a clue that you are in the Central File—which you should not be in. **In fact, once this file is closed, you typically never need to open it directly again.**

6. Click **Synchronize with Central** on the QAT (Figure 16-4.6).

FIGURE 16-4 3
Enabling collaboration options dialog

FIGURE 16-4.4 Save icon

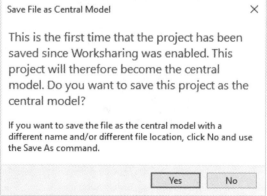

FIGURE 16-4.5 Final prompt for worksharing

7. Check all checkable boxes, except the *Compact Central Model*; ignore the grayed-out boxes and then click **OK** (Figure 16-4.7).

8. **USER A**: Close the project <u>and</u> close Revit.

FIGURE 16-4.6 Sync with Central

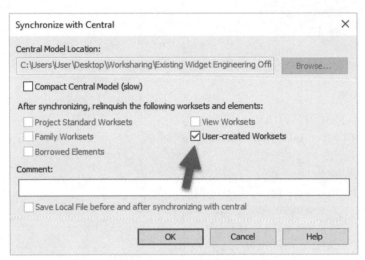

FIGURE 16-4.7 Synchronize with Central dialog

The Revit project has officially been converted into a Worksharing file. User A did not really have to completely close Revit to continue with the next step. However, to clearly delineate the Central File setup steps from the multiple workers in the Central File, the application was closed.

9. **USER A**: Open Revit – do not open the project file yet; **do not** double-click on the file via *Windows Explorer.*

10. **USER A**: Select the **Open** icon on the Quick Access Toolbar.

11. **USER A**: Browse to and select the Central File, which is the same file and location from which this section was started (See Figure 16-4.8). **Do not double-click or open the file yet.**

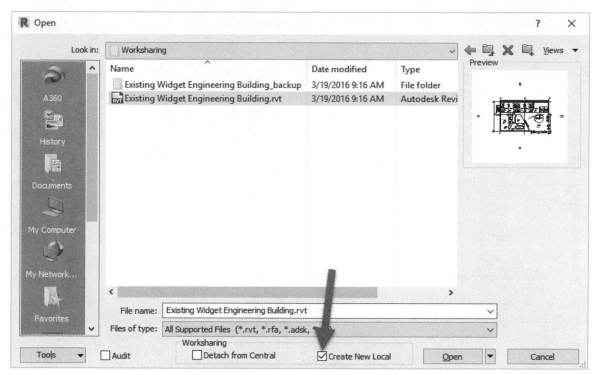

FIGURE 16-4.8 Central file selected; Open dialog

In the Open dialog (Figure 16-4.8) Revit can detect that the selected file is a Central File. When a Central File is selected, and the file version matches the version of Revit you are using, the **Create New Local** option is automatically checked. This is good—each user will always work in a Local File.

12. **USER A**: With the Central File selected and **Create New Local** checked, click **Open**.

User A is now in a Local File, which has a special connection back to the Central File. In the background, the Central File was copied to the local Documents folder and renamed to include the username as a suffix (Figure 16-4.8).

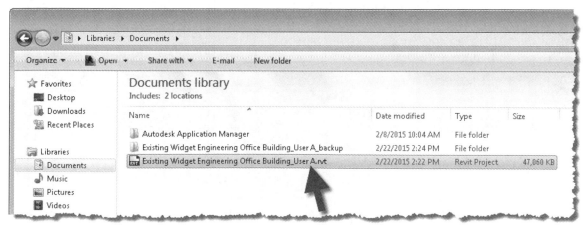

FIGURE 16-4.9 Windows Explorer; Local File

Warning: If the Central File is copied manually using Windows Explorer and then opened, it will also become a Local File with a connection to the Central File.

Notice while in a Local File, that both save icons are active on the Quick Access Toolbar (Figure 16-4.10). Click on the one on the left for a quick save to the Local File on your hard drive (Save once every 15 minutes). Click the Sync with Central (maybe once an hour) when someone needs access to elements you have checked out and for sure before closing Revit.

FIGURE 16-4.10
Save Local and Sync w/ Central

Now User B will also open a Local File.

13. <u>**USER B**</u>: Open Revit – do not open the project file yet.

    a. <u>Important</u>: If the same person is simulating a second user, refer to the steps at the beginning of this section on opening a second session of Revit and adjusting the username in Options.

    b. If a second user will be following these steps on a second computer there is nothing extra to do.

14. <u>**USER B**</u>: Select the **Open** icon on the Quick Access Toolbar.

15. <u>**USER B**</u>: Browse to the Central File and select it; **Do not double-click or open the file yet.**

16. <u>**USER B**</u>: With the Central File selected and **Create New Local** checked, click **Open.**

User B is now in a Local File. Similar to User A, the Local File has a connection to the Central File on the server. Changes made to the Local File do not automatically change the Central File—however, modified elements are marked as checked out in the Central File. This prevents more than one person from making changes at the same time. We will see an example of this next.

17. <u>**USER B**</u>: Switch to the **Level 2 – New** floor plan view.

18. <u>**USER B**</u>: Zoom into the new offices in the Southwest corner of the building.

19. Adjust the wall shown in Figure 16-4.11 from 18' – 0" to **14' – 0".**

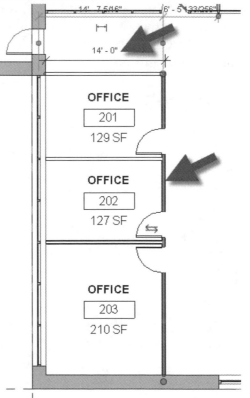

FIGURE 16-4.11 Wall adjusted by user B

The adjusted wall has been checked out in the Central File, but the wall has not changed in the Central File yet. In fact, the wall has not changed in the Local File either. The modification only exists in your computer's memory (i.e., RAM). Next User B will do a Save Local so their work is committed to the hard drive.

20. **USER B**: Click the **Save** icon on the Quick Access Toolbar; see image to right.

The change is now saved to User B's hard drive. However, the Central File is still out of date and User A does not have access to the modification to the model.

21. **USER B**: Click the **Synchronize and Modify Settings** icon to the right of the Save icon.

> **TIP:** Never click the sub-option 'Synchronize Now' as elements may remain checked out.

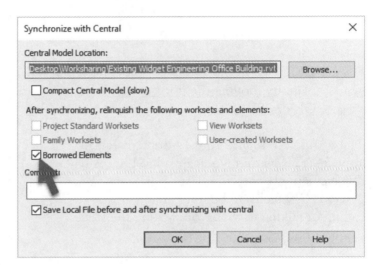

FIGURE 16-4.12 Synchronize with Central dialog

22. **USER B**: Make sure all checkable boxes **are checked** except 'Compact Central Model' (Figure 16-4.12).

23. **USER B**: Click **OK**.

The Central File now has the updated wall; however, User A still cannot see the change as their Local File is still based on the original Central File and any changes they have made. (User A should not have made any changes yet in this tutorial.)

24. **USER A**: Click the **Synchronize and Modify Settings** icon to the right of the Save icon.

25. **USER A**: Make sure all checkable boxes **are checked** except 'Compact Central Model'.

User A can now see the changes User B made. Every time a Sync with Central (SWC) is done, the delta changes are saved between the Local and Central models. Designers are often working on different parts of the building, so it is not necessary to see each other's changes in real-time (plus the computers are not really that powerful, yet!).

Next, we will look at what happens when users "step on each other's toes" and try to edit the same elements. There are really only two warnings anyone will ever see.

Two Worksharing-related warnings (paraphrased):
- Another user has this element checked out
- The element you are trying to change is outdated in your Local File

The first warning requires BOTH users to SWC—the user who changed the element(s) must SWC first. The second only requires the person who received the warning to SWC.

26. <u>**USER A**</u>: In the **Level 2 – New** floor plan, <u>change the door</u> swing in Office 203 as shown in Figure 16-4.13.

    a. **Do not save**

27. <u>**USER B**</u>: In the **Level 2 – New** floor plan, <u>try moving the door to the center</u> in Office 203.

The warning in Figure 16-4.14 immediately appears. This warning indicates that User A has this element checked out. User B must ask User A to SWC, which can be done manually (by calling or talking to them) or clicking the Place Request button.

The next steps will walk through the resolution to this issue.

28. <u>**USER B**</u>: Click **Cancel** and ask USER A to SWC.

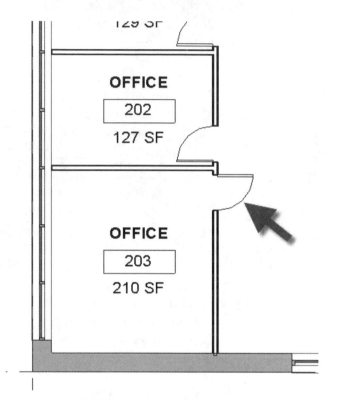

FIGURE 16-4.13 User A changed door swing

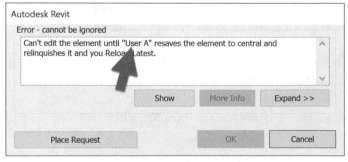

FIGURE 16-4.14 Worksharing warning

29. <u>**USER A**</u>: Perform a **Synchronize with Central**.

30. **USER B**: DO NOT do an SWC yet!!!

The Central File is up to date; however, User B's Local File still has the door shown with the wrong swing. Next, you will see the second and last Worksharing-related warning.

31. **USER B**: Try moving the door in Office 203 to the center of the room again.

The warning, shown in Figure 16-4.15, indicates User B now has the element checked out. However, Revit will not let the element(s) be changed until User B is seeing the elements in the correct location; the correct location being the most recent change to the element among all users.

FIGURE 16-4.15 Worksharing warning

32. **USER B**: Click **Cancel** and then perform an **SWC**.

User B's Local File has now been updated and they can now freely edit the door.

33. **USER B**: Try moving the door to the center of the room (Figure 16-4.16).

That is all there is to know about Worksharing! These same steps work for any number of users; however, project performance can suffer when more than about 8-10 users are in the model at one time.

The only thing left to discuss is what to do at the end of the day and what to do the next day.

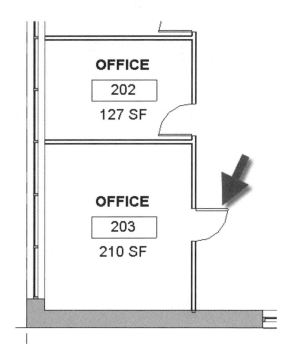

FIGURE 16-4.16 User B moved door

We will assume the end of day has come and the design team is ready to go home.

34. **USER B**: Perform an SWC and then close Revit.

   a.   For a single user simulating two users, change the Username, in Options, to User A before closing Revit.

35. **USER A**: Perform an SWC and then close Revit.

Now we will assume it is the next day and User A is ready to start working on the project. This author recommends creating a new Local File each day.

36. **USER A**: Open Revit.

   a.   Ensure the Username is set to User A via Options.

37. **USER A**: Click the **Open** icon.

38. **USER A**: Browse to the **Central File** (not the Local File).

39. **USER A**: Select the file, and with **Create New Local** checked, click **Open**.

User A is prompted to either overwrite yesterday's file or copy the old file and add a 'timestamp' to the file name—and then create a new Local File. If you are sure you synced your changes at the end of the day yesterday, then use the first option. If you want to be safe, you can use the second option. However, the second option can fill your hard drive up pretty quickly.

40. **USER A**: Click **Append timestamp to existing filename** (Figure 16-4.17).

User A is now in a 'fresh' Local File which will have any changes which may have been made by others working last night. This new Local File will also benefit from **Compact Central** having been run by another user.

The old Local File has now been renamed and is accessible if needed (Figure 16-4.18).

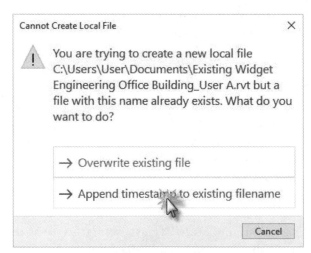

FIGURE 16-4.17 Cannot Create Local File dialog

As mentioned previously in this chapter, any Local File can replace the Central File if needed. Also, if you want to just 'look' at a Revit model, be sure to open it with Detach from Central checked in the open dialog. This will open an unnamed file with no connection to the original central file – keep in mind no changes can ever be saved from the detached file to the original central.

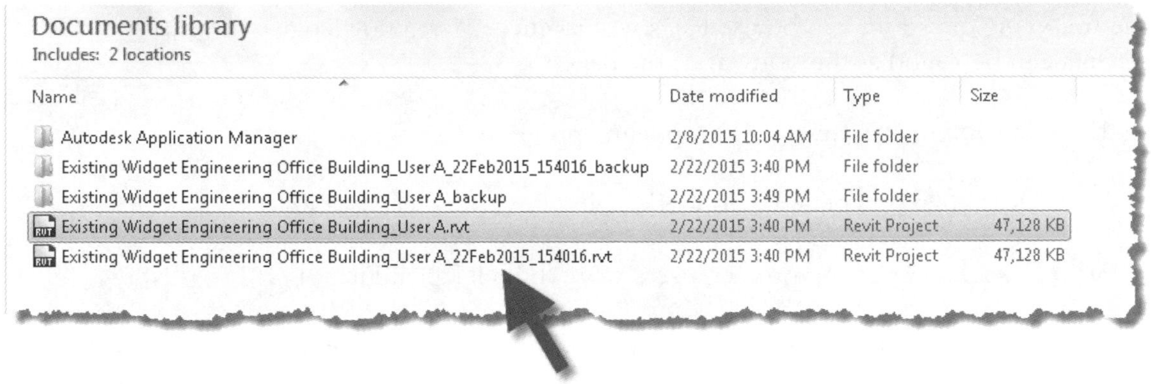

FIGURE 16-4.18 Time stamped local file in Documents folder

41. **USER A**: SWC and then close Revit.

This concludes the brief introduction to Worksharing in Revit. Students are encouraged to continue experimenting in the sample project, making several changes by all users at the same time.

Autodesk BIM 360 Design

A cutting-edge development in Revit Worksharing is Autodesk **BIM 360 Design**. This technology allows multiple users, or companies, to work in the same Revit model over the internet. The Revit model is storage in the cloud. The designer needs a good **internet** connection, **Revit** installed on their computer and a BIM 360 Design **entitlement** to access and work in the cloud-based model. This "entitlement" is not available for students yet; however, it may be in the future so be sure to visit the Autodesk **student portal** for more information.

See this location for commercial access: https://www.autodesk.com/bim-360/platform/design-collaboration-software/

Autodesk User Certification Exam

Be sure to check out Appendix A for an overview of the official Autodesk User Certification Exam and the available study book and software from SDC Publications and this author.

Successfully passing this exam is a great addition to your resume and could help set you apart and help you land a great a job. People who pass the official exam receive a certificate signed by the C.E.O. of Autodesk, the makers of Revit.

Self-Exam:
The following questions can be used as a way to check your knowledge of this lesson. The answers can be found at the bottom of the page.

1.   Phasing makes Revit a 4D modeling program. (T/F)

2.   More than one person can work in any Revit project file. (T/F)

3.   For a Worksharing project, users work in their Local File. (T/F)

4.   A Demolition phase should be created for remodel projects. (T/F)

5.   The author recommends creating a new Local File each day. (T/F)

Review Questions:
The following questions may be assigned by your instructor as a way to assess your knowledge of this section. Your instructor has the answers to the review questions.

1.   Elements to be demolished can just be deleted. (T/F)

2.   Worksets can be used to optimize computer resources. (T/F)

3.   Additional Worksets can be created. (T/F)

4.   A Revit project can only have two phases: Existing and New Construction. (T/F)

5.   To casually look at a Revit project set up for Worksharing, simply copy it to the Desktop using Windows Explorer. (T/F)

6.   The Phase setting for a demolition view is New Construction. (T/F)

7.   For a Worksharing project, everyone will see all changes in real-time. (T/F)

8.   Compact Central should be run once a day. (T/F)

9.   Every element in Revit has a Phase Filter parameter. (T/F)

10.   Revit requires the A360 username and the Revit username match. (T/F)

Appendix A
Autodesk® Revit® Certification Exam Introduction

Next Steps
The author of this book has prepared a separate textbook and practice software to help Revit users' study for, and improve the likelihood of passing, the official Autodesk User Certification Exam for Revit Architecture.

In this appendix you will learn more about the certification and the study guide and practice software published by SDC Publications (www.SDCPublications.com).

Study guide:	Autodesk Revit for Architecture Certified User Exam Preparation
Practice software:	Included with study guide, downloaded from SDC Publications
Autodesk exam:	Autodesk User Certification Exam for Revit Architecture

Overview
In the competitive world in which we live it is important to stand out to potential employers and prove your capabilities. One way to do this is by passing one of the Autodesk Certification Exams. A candidate who passes an exam has credentials from the makers of the software that you know how to use their software. This can help employers narrow down the list of potential interviewees when looking for candidates.

Benefits
There are a variety of reasons and benefits to getting certified. They range from a school/employer requirement to professional development and resume building. Whatever the reason, there is really no downside to this effort if you are, or hope to be, working in the architecture or construction industry.

Here are the benefits listed on Autodesk's website:
- Earn an industry-recognized credential that helps prove your skill level and can get you hired.
- Develop your skills with sample projects and exercises that emphasize real-world applications.
- Accelerate your professional development and help enhance your credibility and career success.
- Validate your skills and join an elite team of Autodesk Certified professionals.
- Display your Autodesk Certified certificate, use the Autodesk Certified logo, highlight your achievement and get noticed by listing your name in the Autodesk Certified Professionals database.

Certificate

When the ACU exam is successfully passed, a certificate signed by Autodesk's CEO is issued with your name on it. This can be framed and displayed at your desk, copied and included with a resume (if appropriate) or brought to an interview (not the framed version, just a copy!).

A high-quality **print** of your official certificate may be purchased via certiport.com in the *My Transcript* section (login using the same username and password for the exam).

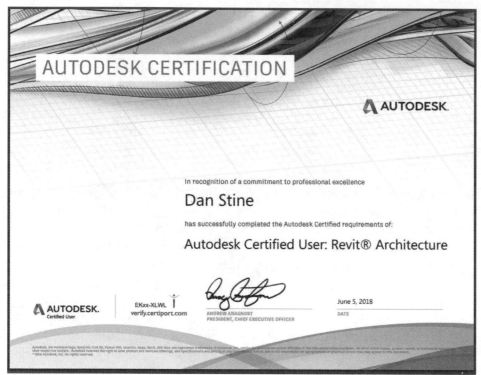

Certificate Example

Badging

In addition to a certificate, a badge is issued. Badging is a digital web-enabled version of your credential by **Acclaim**, which can be helpful to potential employers. This is a quick proof that you know how to use the Revit features covered by the ACU exam.

Practice Exam

A **practice exam** is included with the book, *Autodesk Revit for Architecture Certified User Exam Preparation*; which can be downloaded from the publisher's website using the **access code** and instructions found on the inside-front cover of that book. This is a good way to check your skills prior to taking the official exam, as the intent is to offer similar types of questions in roughly the same format as the formal ACU exam. This practice exam is taken at home, work or school, on your own computer. You must have Revit installed to successfully answer the in-application questions.

The first image below highlights the practice exam software, while the second is an example of the results which are optionally emailed to you for your records.

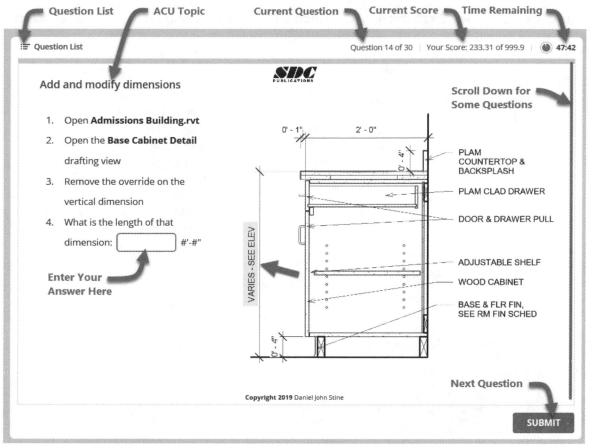

Practice exam software by SDC Publications - highlights

Practice exam results, which can also be optionally emailed to you

The ACU study guide is broken down into chapters and section based on the published exam topics and elements covered, as indicated in the Table of Contents listed below.

Sample Pages from the Study Guide

Below are a few sample pages highlighting the fact that **Autodesk Revit for Architecture Certified User Exam Preparation** is a study guide and not a tutorial. Notice the chapter tabs in the margin and that there are several images showing *before* and *after* examples and how to interpret the results. For more information, visit www.SDCPublications.com.

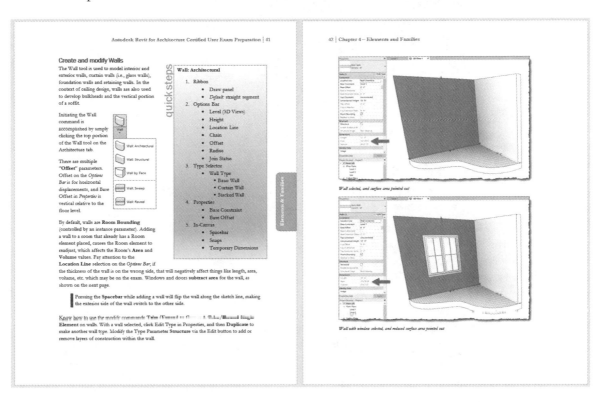

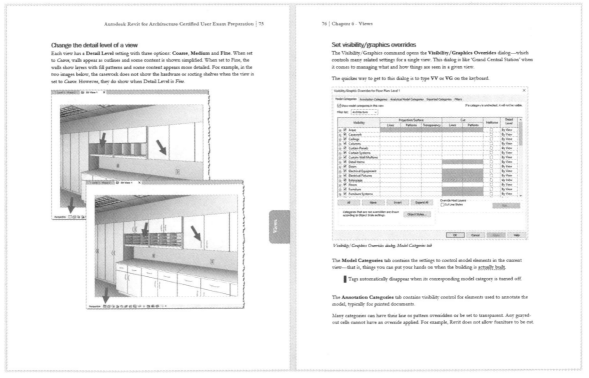

Notes:

Index